UCLA Symposia on Molecular and Cellular Biology, New Series

Series Editor
C. Fred Fox

UCLA Symposia Published Previously

(Numbers refer to the publishers listed below.)

1972
Membrane Research (2)

1973
Membranes (1)
Virus Research (2)

1974
Molecular Mechanisms for the Repair of
 DNA (4)
Membranes (1)
Assembly Mechanisms (1)
The Immune System: Genes, Receptors,
 Signals (2)
Mechanisms of Virus Disease (3)

1975
Energy Transducing Mechanisms (1)
Cell Surface Receptors (1)
Developmental Biology (3)
DNA Synthesis and Its Regulation (3)

1976
Cellular Neurobiology (1)
Cell Shape and Surface Architecture (1)
Animal Virology (2)
Molecular Mechanisms in the Control of Gene
 Expression (2)

1977
Cell Surface Carbohydrates and Biological
 Recognition (1)
Molecular Approaches to Eucaryotic Genetic
 Systems (2)
Molecular Human Cytogenetics (2)
Molecular Aspects of Membrane Transport (1)
Immune System: Genetics and Regulation (2)

1978
DNA Repair Mechanisms (2)
Transmembrane Signaling (1)
Hematopoietic Cell Differentiation (2)
Normal and Abnormal Red Cell
 Membranes (1)

Persistent Viruses (2)
Cell Reproduction: Daniel Mazia Dedicatory
 Volume (2)

1979
Covalent and Non-Covalent Modulation of
 Protein Function *(2)*
Eucaryotic Gene Regulation (2)
Biological Recognition and Assembly (1)
Extrachromosomal DNA (2)
Tumor Cell Surfaces and Malignancy (1)
T and B Lymphocytes: Recognition and
 Function (2)

1980
Biology of Bone Marrow Transplantation (2)
Membrane Transport and Neuroreceptors (1)
Control of Cellular Division and
 Development (1)
Animal Virus Genetics (2)
Mechanistic Studies of DNA Replication and
 Genetic Recombination (2)

1981
Immunoglobulin Idiotypes (2)
Initiation of DNA Replication (2)
Genetic Variation Among Influenza Viruses (2)
Developmental Biology Using Purified
 Genes (2)
Differentiation and Function of Hematopoietic
 Cell Surfaces (1)
Mechanisms of Chemical Carcinogenesis (1)
Cellular Recognition (1)

1982
B and T Cell Tumors (2)
Interferon (2)
Rational Basis for Chemotherapy (1)
Gene Regulation (2)
Tumor Viruses and Differentiation (1)
Evolution of Hormone-Receptor Systems (1)

Publishers

(1) Alan R. Liss, Inc.
 41 East 11th Street
 New York, NY 10003

(2) Academic Press, Inc.
 111 Fifth Avenue
 New York, NY 10003

(3) W.A. Benjamin, Inc.
 2725 Sand Hill Road
 Menlo Park, CA 94025

(4) Plenum Publishing Corp.
 227 W. 17th Street
 New York, NY 10011

Symposia Board

C. Fred Fox, Ph.D., Director
Professor of Microbiology
Molecular Biology Institute
UCLA

Members

Ronald Cape, Ph.D., M.B.A.
Chairman
Cetus Corporation

Pedro Cuatrecasas, M.D.
Vice President for Research
Burroughs Wellcome Company

Luis Glaser, Ph.D.
Professor and Chairman
of Biochemistry
Washington University School
of Medicine

Donald Steiner, M.D.
Professor of Biochemistry
University of Chicago

Ernest Jaworski, Ph.D.
Director of Molecular Biology
Monsanto

Paul Marks, M.D.
President
Sloan-Kettering Institute

William Rutter, Ph.D.
Professor of Biochemistry and Director of
the Hormone Research Institute
University of California, San Francisco,
Medical Center

Sidney Udenfriend, Ph.D.
Member
Roche Institute of Molecular Biology

The members of the board advise the director in identification of topics for future symposia.

CELLULAR AND MOLECULAR BIOLOGY OF PLANT STRESS

CELLULAR AND MOLECULAR BIOLOGY OF PLANT STRESS

Proceedings of an ARCO Plant Cell Research Institute—UCLA Symposium, Held at Keystone, Colorado, April 15–21, 1984

Editors

JOE L. KEY
Department of Botany
University of Georgia
Athens, Georgia

TSUNE KOSUGE
Department of Plant Pathology
University of California
Davis, California

Alan R. Liss, Inc. • New York

Address all Inquiries to the Publisher
Alan R. Liss, Inc., 41 East 11th Street, New York, NY 10003

Copyright © 1985 Alan R. Liss, Inc.

Printed in the United States of America.

Library of Congress Cataloging in Publication Data
Main entry under title:

Cellular and molecular biology of plant stress.
  (UCLA symposia on molecular and cellular biology;
new ser., v. 22)
    Includes index.
    1. Plant cells and tissues—Congresses.   2. Plants,
Effect of stress on—Congresses.   3. Plant genetics—
Congresses.   4. Molecular biology—Congresses.
I. Key, Joe L.   II. Kosuge, Tsune.   III. ARCO
Plant Cell Research Institute.   IV. University of
California, Los Angeles.
QK725.C394   1984   581.1   84-28849
ISBN 0-8451-2621-0

Contents

Contributors

James D. Anderson, Plant Hormone Laboratory, Plant Physiology Institute, BARC, ARS, USDA, Beltsville, MD 20705 **[263]**

Charles J. Arntzen, MSU-DOE Plant Research Laboratory, Michigan State University, East Lansing, MI 48824 **[51]**

Lorette Aspart, Laboratoire de Physiologie Végétale, E.R.A. 226 du C.N.R.S., Université de Perpignan, 66025 Perpignan Cedex, France **[227]**

Arthur R. Ayers, Department of Cellular and Developmental Biology, Harvard University, Cambridge, MA 02138 **[447]**

Paul D. Bishop, Institute of Biological Chemistry, Washington State University, Pullman, WA 99164 **[319]**

Thomas Boller, Botanisches Institut, Universität Basel, CH-4056 Basel, Switzerland **[247]**

Helga Börner, Department of Biochemistry, Institut für Biologie II, Universität Freiburg, D-7800 Freiburg, Federal Republic of Germany **[275]**

John S. Boyer, Department of Plant Biology and Agronomy, University of Illinois, Urbana, IL 61801; present address: Department of Soil and Crop Sciences, Texas A & M University, College Station, TX 77843 **[41]**

Willis E. Brown, Institute of Biological Chemistry, Washington State University, Pullman, WA 99164 **[319]**

George Bruening, Department of Biochemistry and Biophysics, University of California, Davis, CA 95616 **[401]**

Edo Chalutz, Department of Horticulture, University of Maryland, College Park, MD 20742; present address: Division of Fruit and Vegetable Storage, ARO, The Volcani Center, Bet Dagan, Israel **[263]**

Steven Clouse, Department of Plant Pathology, University of California, Davis, CA 95616 **[367]**

Eric E. Conn, Department of Biochemistry and Biophysics, University of California, Davis, CA 95616 **[351]**

Laszlo N. Csonka, Department of Biological Sciences, Purdue University, West Lafayette, IN 47907 **[115]**

Eva Czarnecka, Department of Botany, University of Georgia, Athens, GA 30602 **[161]**

Paul DeAngelis, Department of Cellular and Developmental Biology, Harvard University, Cambridge, MA 02138 **[447]**

Michel Delseny, Laboratoire de Physiologie Végétale, E.R.A. 226 du C.N.R.S., Université de Perpignan, 66025 Perpignan Cedex, France **[227]**

E. S. Dennis, CSIRO, Division of Plant Industry, Canberra, ACT 2601, Australia **[217]**

The number in brackets is the opening page number of the contributor's article.

Virginia Joan Dunlap, Department of Biological Sciences, Purdue University, West Lafayette, IN 47907 **[115]**

Leonard Edelman, Department of Botany, University of Georgia, Athens, GA 30602 **[161]**

J. Ellis, CSIRO, Division of Plant Industry, Canberra, ACT 2601, Australia **[217]**

M. T. Esquerré-Tugayé, Centre de Physiologie Végétale, UA 241 CNRS, Université Paul Sabatier, 31062 Toulouse, France **[459]**

E. J. Finnegan, CSIRO, Division of Plant Industry, Canberra, ACT 2601, Australia **[217]**

Hector E. Flores, Biochemistry Group, ARCO Plant Cell Research Institute, Dublin, CA 94568 **[93]**

Robert T. Fraley, Division of Biological Sciences, Corporate Research and Development, Monsanto Chemical Company, St. Louis, MO 63167 **[181]**

Arthur W. Galston, Department of Biology, Yale University, New Haven, CT 06511 **[93]**

W. L. Gerlach, CSIRO, Division of Plant Industry, Canberra, ACT 2601, Australia **[217]**

David Gilchrist, Department of Plant Pathology, University of California, Davis, CA 95616 **[367]**

Jody J. Goodell, Department of Cellular and Developmental Biology, Harvard University, Cambridge, MA 02138 **[447]**

John S. Graham, Institute of Biological Chemistry, Washington State University, Pullman, WA 99164 **[319]**

Hans Grisebach, Department of Biochemistry, Institut für Biologie II, Universität Freiburg, D-7800 Freiburg, Federal Republic of Germany **[275]**

Rebecca Grumet, MSU-DOE Plant Research Laboratory, Michigan State University, East Lansing, MI 48824 **[71]**

William B. Gurley, Department of Microbiology and Cell Science, University of Florida, Gainesville, FL 32611 **[161]**

Marie-Luise Hagmann, Department of Biochemistry, Institut für Biologie II, Universität Freiburg, D-7800 Freiburg, Federal Republic of Germany **[275]**

Michael G. Hahn, Department of Biochemistry, Institut für Biologie II, Universität Freiburg, D-7800 Freiburg, Federal Republic of Germany; present address: Department of Plant Biology, The Salk Institute, San Diego, CA 92138-9216 **[275]**

Raymond Hammerschmidt, Department of Botany and Plant Pathology, Plant Biology Laboratory, Michigan State University, East Lansing, MI 48824 **[291]**

Andrew D. Hanson, MSU-DOE Plant Research Laboratory, Michigan State University, East Lansing, MI 48824 **[71]**

Cathy M. Hironaka, Division of Biological Sciences, Corporate Research and Development, Monsanto Chemical Company, St. Louis, MO 63167 **[181]**

Paul J. Jackson, Genetics Group, Life Sciences Division, Los Alamos National Laboratory, Los Alamos, NM 87545 **[145]**

Donald F. Jin, Department of Chemistry and Biochemistry, University of California, Los Angeles, CA 90024 **[335]**

Joe L. Key, Department of Botany, University of Georgia, Athens, GA 30602 **[xvii,161]**

Janice A. Kimpel, Department of
Botany, University of Georgia, Athens,
GA 30602 [161]

P. E. Kolattukudy, Institute of
Biological Chemistry, Washington State
University, Pullman, WA 99164-6340
[381]

Tsune Kosuge, Department of Plant
Pathology, University of California,
Davis, CA 95616 [xvii]

Gwen G. Krivi, Division of Biological
Sciences, Corporate Research and
Development, Monsanto Chemical
Company, St. Louis, MO 63167 [181]

Joseph Kuć, Department of Plant
Pathology, University of Kentucky,
Lexington, KY 40546 [303]

David J. Kyle, MSU-DOE Plant
Research Laboratory, Michigan State
University, East Lansing, MI 48824 [51]

Joachim Leube, Department of
Biochemistry, Institut für Biologie II,
Universität Freiburg, D-7800 Freiburg,
Federal Republic of Germany [275]

Paul H. Li, Department of Horticultural
Science and Landscape Architecture,
Laboratory of Plant Hardiness,
University of Minnesota, St. Paul, MN
55108 [201]

Chu Yung Lin, Department of Botany,
University of Georgia, Athens, GA
30602 [161]

D. Llewellyn, CSIRO, Division of Plant
Industry, Canberra, ACT 2601,
Australia [217]

G. Loebenstein, Virus Laboratory,
Agricultural Research Organization,
50250 Bet Dagan, Israel [413]

Augusto F. Lois, Department of
Chemistry and Biochemistry, University
of California, Los Angeles, CA 90024
[335]

Michael A. Mansfield, Department of
Botany, University of Georgia, Athens,
GA 30602 [161]

Ann Martensen, Department of Plant
Pathology, University of California,
Davis, CA 95616 [367]

Autar K. Mattoo, Department of
Botany, University of Maryland,
College Park, MD 20742 [263]

D. Mazau, Centre de Physiologie
Végétale, UA 241 CNRS, Université
Paul Sabatier, 30162 Toulouse, France
[459]

Peter R. McClure, Environmental
Studies Group, Life Sciences Division,
Los Alamos National Laboratory, Los
Alamos, NM 87545; present address:
Allied Corporation, Syracuse Research
Laboratory, Solvay, NY 13209 [145]

Beverly McFarland, Department of
Plant Pathology, University of
California, Davis, CA 95616; present
address: Chevron Biotechnology Group,
Chevron Chemical Company,
Richmond, CA 94804 [367]

Bernard Mocquot, Station de
Physiologie Végétale, I.N.R.A., Centre
de Recherches de Bordeaux, 33140 Pont
de la Maye, France [227]

Peter Moesta, Department of
Biochemistry, Institut für Biologie II,
Universität Freiburg, D-7800 Freiburg,
Federal Republic of Germany; present
address: Department of Chemistry and
Biochemistry, University of California,
Los Angeles, CA 90024 [275,335]

Christiane Morisset, Laboratoire de
Cytologie Végétale Expérimentale,
Université Pierre et Marie Curie, 75230
Paris Cedex 05, France [227]

Thomas J. Mozer, Division of Biological Sciences, Corporate Research and Development, Monsanto Chemical Company, St. Louis, MO 63167 **[181]**

Ronald T. Nagao, Department of Botany, University of Georgia, Athens, GA 30602 **[161]**

Cleo M. Naranjo, Genetics Group, Life Sciences Division, Los Alamos National Laboratory, Los Alamos, NM 87545 **[145]**

Oliver E. Nelson, Department of Genetics, University of Wisconsin-Madison, Madison, WI 53706 **[1]**

Itzhak Ohad, Department of Biological Chemistry, Hebrew University, Jerusalem, Israel **[51]**

W. J. Peacock, CSIRO, Division of Plant Industry, Canberra, ACT 2601, Australia **[217]**

B. Pélissier, Centre de Physiologie Végétale, UA 241 CNRS, Université Paul Sabatier, 31062 Toulouse, France **[459]**

Alain Pradet, Station de Physiologie Végétale, I.N.R.A., Centre de Recherches de Bordeaux, 33140 Pont de la Maye, France **[227]**

D. William Rains, Department of Agronomy and Range Science, University of California, Davis, CA 95616 **[129]**

Philippe Raymond, Station de Physiologie Végétale, I.N.R.A., Centre de Recherches de Bordeaux, 33140 Pont de la Maye, France **[227]**

James K. Roberts, Department of Botany, University of Georgia, Athens, GA 30602 **[161]**

D. Roby, Centre de Physiologie Végétale, UA 241 CNRS, Université Paul Sabatier, 31062 Toulouse, France **[459]**

Dean E. Rochester, Division of Biological Sciences, Corporate Research and Development, Monsanto Chemical Company, St. Louis, MO 63167 **[181]**

D'Ann M. Rochon, Department of Biological Sciences, Wayne State University, Detroit, MI 48202 **[435]**

E. Jill Roth, Department of Biology, University of Utah, Salt Lake City, UT 84112 **[145]**

D. Rumeau, Centre de Physiologie Végétale, UA 241 CNRS, Université Paul Sabatier, 31062 Toulouse, France **[459]**

Mary L. Russell, Department of Biochemistry and Biophysics, University of California, Davis, CA 95616 **[401]**

Clarence A. Ryan, Institute of Biological Chemistry, Washington State University, Pullman, WA 99164 **[319]**

M.M. Sachs, CSIRO, Division of Plant Industry, Canberra, ACT 2601, Australia; present address: Department of Biology, Washington University, St. Louis, MO 63130 **[217]**

Janet L. Sanderson, Department of Biochemistry and Biophysics, University of California, Davis, CA 95616 **[401]**

Dilip M. Shah, Division of Biological Sciences, Corporate Research and Development, Monsanto Chemical Company, St. Louis, MO 63167 **[181]**

Robert E. Sharp, Department of Plant Biology, University of Illinois, Urbana, IL 61801; present address: Department of Land, Air, and Water Resources, University of California, Davis, CA 95616 **[41]**

Albert Siegel, Department of Biological Sciences, Wayne State University, Detroit, MI 48202 **[435]**

Deborah Siler, Department of Plant Pathology, University of California, Davis, CA 95616; present address: Department of Biochemistry and Biophysics, University of California, Davis, CA 95616 **[367]**

Charles L. Soliday, Institute of Biological Chemistry, Washington State University, Pullman, WA 99164–6340 **[381]**

Suzan J. Stavarek, Department of Agronomy and Range Science, University of California, Davis, CA 95616 **[129]**

Adina Stein, Virus Laboratory, Agricultural Research Organization, 50250 Bet Dagan, Israel **[413]**

Bruce A. Stermer, Department of Botany and Plant Pathology, Plant Biology Laboratory, Michigan State University, East Lansing, MI 48824 **[291]**

David C. Tiemeier, Division of Biological Sciences, Corporate Research and Development, Monsanto Chemical Company, St. Louis, MO 63167 **[181]**

A. Toppan, Centre de Physiologie Végétale, UA 241 CNRS, Université Paul Sabatier, 31062 Toulouse, France **[459]**

Yoash Vaadia, Department of Genetics and Tissue Culture, ARCO Plant Cell Research Institute, Dublin, CA 94568 **[13]**

Elizabeth Vierling, Department of Botany, University of Georgia, Athens, GA 30602 **[161]**

Mary Walker-Simmons, Institute of Biological Chemistry, Washington State University, Pullman, WA 99164 **[319]**

Charles A. West, Department of Chemistry and Biochemistry, University of California, Los Angeles, CA 90024 **[335]**

Karen A. Wickham, Department of Chemistry and Biochemistry, University of California, Los Angeles, CA 90024 **[335]**

Nevin D. Young, Department of Plant Pathology, Cornell University, Ithaca, NY 14853 **[93]**

Preface

A variety of environmental and biological stresses interfere with normal plant development and have negative impacts on production and quality of the food, fiber, and ornamental plants grown in this country and abroad. Besides causing economic losses amounting to millions of dollars annually, these stresses reduce efficiency of crop production to a level where it is uneconomical to raise certain crops in many geographical regions of the world. In the face of increasing needs for more efficient production of food and fiber crops, development of plants more resistant to diseases and environmental stresses remains high priority.

The advances being made in molecular and cellular biology offer new approaches to the study of how plants interact and react to stresses; these promising new approaches offer the hope of not only the development of exciting new knowledge in plant biology, but also the development of new plant varieties with enhanced tolerance to stresses.

The aim of this ARCO Plant Cell Research Institute-UCLA Symposium, held at Keystone, Colorado, April 15–21, 1984, was to bring together researchers to discuss the state-of-the-art of research on cellular and molecular responses of plants to stresses imposed by biological agents such as microorganisms, viruses, and environmental factors such as temperature, salinity, and water stresses. Key papers presented at the conference are included in this volume; readers will be given a view of the advances being made in the several basic areas of plant science research and the potential for exciting breakthroughs in the future.

We wish to thank ARCO Plant Cell Research Institute for its generous sponsorship of this meeting. We also gratefully acknowledge gifts from: Agrigenetics Research Corporation, Monsanto Corporate Research Laboratories, Rohm and Haas Company, Zoecon Corporation, Ciba-Geigy Corporation, Phytogen, Syracuse Research Laboratory and Molecular Applied Genetics Laboratory of Allied Corporation, Biogen Research Corporation, Calgene, Inc., E.I. du Pont de Nemours & Company, Pioneer Hi-Bred International, Inc., Celanese Research Company, Stauffer Chemical Company, Chevron Chemical Company-Biotechnology group, and a grant (CRCR-1-1361) from the United States Department of Agriculture.

Joe L. Key
Tsune Kosuge

ABERRANT RATIO REVISITED[1]

Oliver E. Nelson

Department of Genetics, University of Wisconsin-Madison
Madison, WI 53706

ABSTRACT   The Aberrant Ratio (AR) phenomenon, which
appeared in advanced generations of maize plants
systemically infected with barley stripe mosaic
virus and resulted in deviations from expected
proportions (1:1) of seeds with nonmutant or
mutant phenotypes for several loci in matings
between X/x and x/x sibs, is summarized. The
results from two laboratories, showing that four
AR stocks involving the a locus and giving an excess
of colorless kernels are all segregating for another
complementary color gene, are discussed. A line,
descended from a virus-infected plant and which
gives excess numbers of colored kernels in crosses
between A/a and a/a sibs, has also been investigated.
The inheritance pattern can be explained by the hypothesis that a is linked to a recessive zygotic lethal
(zl) with ca. 10 percent recombination. A better fit
to the data is obtained if one invokes the involvement of a second, unlinked, recessive factor with
the zygote lethality being expressed only in zygotes
homozygous for the recessive, second factor.

These investigations provide no support for the
hypothesis that Aberrant Ratios found in the descend-
ants of plants systemically infected with barley
stripe mosaic virus result from anomalous regulation
of gene activity for the marker loci.

[1]Paper No. 2733 from the Department of Genetics, University of Wisconsin-Madison. This research was supported
by the College of Agriculture and Life Sciences. I am
grateful to George Sprague for supplying seed of the Aberrant
Ratio stocks and to Russell Huseth for many counts of
segregating progenies.

INTRODUCTION

This report examines a second aspect of the presumed consequences of infecting maize plants with barley stripe mosaic virus: i.e., the Aberrant Ratio (AR) phenomenon in which significant deviations from the expected 1:1 ratio were observed for several loci when plants of the presumed genotypes X/x and x/x from the same family were inter-crossed. Although the initial intent of Sprague et al. (1) was to test the possible mutagenic effect of the virus infection and their initial results suggested a positive effect of infection, their attention turned toward the deviations from expected ratios found in subsequent gener-ations of plants tracing back to the original infected plants.

Since most of Aberrant Ratio stocks involved loci con-cerned with seed pigmentation, a brief discussion of the genetic control of seed anthocyanin production may be use-ful. Anthocyanin pigments are synthesized only in seeds where one or more functional alleles at each of the follow-ing loci ($a1$, $a2$, $bz1$, $bz2$, $c1$, $c2$, and r) are present. If the seed is homozygous recessive (e.g., $a1/a1/a1$) at any one of these loci ($a1$, $a2$, $c1$, $c2$, and r), then the seed is colorless. The homozygous recessive condition at either $bz1$ or $bz2$ produces tan or bronze-colored seeds, and we are not concerned with this phenotype here.

In the initial experiments, Sprague et al. (1) in-fected plants from a line with colored seeds and homozygous for the functional alleles $A1/A1$, $A2/A2$, $C1/C1$, $C2/C2$, and R/R, then pollinated plants that were multiply reces-sive ($a1/a1$, pr/pr and su/su) but were homozygous dominant ($A2/A2$, $C1/C1$, $C2/C2$, R/R) at the other loci required for pigmentation. The pr allele conditions the production of red seeds (pelargonin) vs the purple seeds (cyanin) pro-duced when Pr is present. The $su1$ (sugary-1) mutant is the usual sweet corn of commerce, and homozygous sugary seeds have a distinctive, wrinkled phenotype. In these experiments, the investigators noticed a higher frequency of mutation at the target loci in infected plants as com-pared to control (uninfected) plants. Then the F2 pro-genies from about 1% of the plants displayed marked devia-tions from expected percentages of seeds displaying the recessive phenotype. When this occurred for an F2 progeny, the investigators adopted the experimental regimen of making sib pollinations in subsequent generations between plants from seeds of the dominant genotype (presumably X/x) and plants from seeds of the recessive genotype (presumably x/x). Since we're concerned here with seed anthocyanin pigmentation as a phenotype, we can be specific about the crosses in which we're interested - reciprocal crosses between presumed A/a and a/a plants which should produce colored and colorless seeds in a 1:1 ratio if the genotypes of the plants with regard to the other genes affecting anthocyanin pigmentation is $A2/A2$, $C1/C1$, $C2/C2$, R/R.

In two subsequent papers, Sprague and McKinney (2, 3) showed that in families where significant deviations from 1:1 ratios occurred: (1) the deviations were observed whether plants from colored seeds were the male or female parents; (2) when the plant from a colored seed was outcrossed to an $a1$ tester ($a1/a1$, $A2/A2$, $C1/C1$, $C2/C2$, R/R), a 1:1 ratio of colored to colorless seeds was always obtained; (3) a and A gametes were produced in equal numbers by A/a plants; (4) genes closely linked to a gene displaying aberrant ratios segregated normally; (5) when a plant from a colorless seed from an Aberrant Ratio (AR) stock was outcrossed to an unrelated A/A stock and backcross segregations observed, some progenies gave an excess of the colorless phenotype, and this was interpreted as transmission of AR by an allele (a) never directly exposed to the virus; (6) within an AR stock in which crosses of presumed A/a plants by sibling a/a plant gave usually an excess of colorless kernels, there were observed some crosses in which there were expected percentages (50) of colorless kernels. Considering these various observations, Sprague and McKinney (3) suggested that the loci being followed were giving evidence of abnormal regulation of their activity as a result of the parental material being infected with a virus.

This interesting possibility led several laboratories (Myron Brakke's at Nebraska and mine) to request from George Sprague the stocks, in which crosses between presumed A/a and a/a sibs gave rise to excess numbers of colorless kernels, and to initiate investigations of these stocks. These investigations led both laboratories to the same conclusions -- that the AR stocks we were investigating were also segregating for a second complementary color factor and that the segregation of this second factor was capable of accounting for all of Sprague and McKinney's observations. It was unnecessary to postulate any anomalous regulation of gene activity. In Brakke's investigations, one AR stock was segregating for $\underline{C1}$ and $\underline{c1}$ in addition to $\underline{A}$ and $\underline{a}$. A second AR stock was segregating for $\underline{R}$ and $\underline{r}$ (4). In the two AR stocks which I investigated, one was also segregating for $\underline{C1}$ and $\underline{c1}$ and the second for $\underline{C2}$ and $\underline{c2}$ (5). Thus, for these stocks, the Aberrant Ratio phenomenon has a prosaic explanation although it leaves open the question of whether the mutant alleles at $\underline{C1}$, $\underline{C2}$, and $\underline{R}$ were consequences of systemic viral infection of parental material. Brakke (6) has a review of Aberrant Ratio investigations currently in press.

Over the past several years, I have been investigating the genetic basis of another stock which concerns the $\underline{a}$ locus and which I received from George Sprague. In this stock, crosses between plants from $\underline{A/a}$ seeds and their $\underline{a/a}$ siblings often produce a deficiency of colorless ($\underline{a/a}$) kernels rather than an excess as in the stocks previously discussed. Although Sprague and McKinney (3) reported AR stocks that produced more colored kernels than expected in $\underline{A/a}$ times $\underline{a/a}$ sib crosses, this stock was apparently not one that had been reported in that paper. It did, however, descend from a virus-infected parent. Sprague and McKinney referred to AR stocks that had an excess of colored kernels in $\underline{a/a}$ X $\underline{A/a}$ sib crosses as $\underline{A^*/a}$ stocks, and I will use that convention here.

The investigation of this stock has been slowed by the fact that it is very weak and has considerable female sterility so that good seed sets have never been obtained. At the time that the results with the other stocks were reported (5), I knew only that the plants from colored seeds of this stock were $\underline{A1/a1}$, $\underline{A2/A2}$, $\underline{C1/C1}$, $\underline{C2/C2}$, $\underline{R/R}$. This finding eliminated one possible explanation of the high colored ratio. Paradoxical as it may appear, heterozygosity for two color factor loci can result in an excess of colored kernels in some sib crosses. Since it is not the basis of the deviation in this family, I won't deal with its basis here. Two factor heterozygosity as an explanation for excess numbers of colored kernels has been covered previously (5).

Possible explanations for the deficiency of colorless kernels in these progenies could be deleterious gametophyte factors such as a pollen-lethal factor or a lethal ovule factor linked to $\underline{a}$, or a nondeleterious gametophyte factor, which would give $\underline{Ga}$ gametes a competitive advantage over $\underline{ga}$ gametes on the silks of plants that are $\underline{Ga/Ga}$ or $\underline{Ga/ga}$ (2), linked to $\underline{A}$. As will be shown, these possibilities

can be ruled out as well as the involvement of any cytoplasmic component.

RESULTS

Efforts to elucidate the genetic basis of excess
colored kernels were then continued. Table 1 gives the results of the first three years' test with the A*/a stock.
The data were all that could be obtained in spite of intensive efforts. Some information is available, however, from
these limited data. When a cross gives a significant excess
of colored kernels, the ratio is ca. 6 colored: 4 colorless
(25492-11 X 491-4; 25493-6 X 491-5) or 3 colored: 1 colorless (25492-2 X 491-3). This latter ratio will be shown to
be due to the fact that 25492-2 was A/a although the seed
color indicated that the embryo genotype was a/a. Thus the
ratio is close to the expected ratio from mating A/a times
A/a, but subsequent crosses demonstrated that either
25492-2 or 25491-3 were carrying factors conditioning excess
numbers of colored kernels. Also, a plant used as a male
can produce significant deviations from a 1:1 ratio in a
cross with one a/a sib but may not with a second a/a sib.
A decision was made to examine the progeny of 25492-2
X 491-3 (24% colorless) in greater detail before it was
understood that the cross was an A/a times A/a mating,
since in addition to this cross which significantly deviates
from an expected 1 colored: 1 colorless ratio, 491-3 had
been outcrossed to an a/a tester (25487 X 491-3) with a
close approach to an expected 1:1 ratio.
The data obtained when plants from the colored seeds
of the cross (25492-2 X 491-3, 24% colorless) were   self-
pollinated are presented below. Of the 24 plants selfed,
9 were A/A (all seeds colored) and 1 was a/a. Since the
cross had been a plant from a colorless seed (presumably
a/a) times a plant from a colored seed (A/a) and the colored
seeds planted, one expects all progeny to be A/a. The observation that approximately 1/3 of the plants were A/A
indicates that 25492-2 was A/a rather than a/a. The most
plausible explanation for this observation is heterofertilization in the fertilization event that produced the colorless seed which gave rise to plant 25492-2. In double
fertilization of angiosperms, the egg nucleus and the polar
bodies in the central cell are usually fertilized by sperm
from the same pollen grain. This is not invariably so,

TABLE 1
RECORD OF RESULTS WITH AN A*/a STOCK.
THE SOURCE WAS SPRAGUE 1976 - 433_5 X 434_9 A*/a X a/a

| 1979 | 21117 | 433_5 X 434_9 | Colored seeds A/a |
| | 21118 | 433_5 X 434_9 | Colorless seeds a/a |

Cross	Colored	Colorless	% Colorless
117B X 118-4	42	42	50
118 X 117-3	93	59	39**(1:1)
118 X 117-5	82	78	49

| 1980 | 23261-262 | 21118 X 117-3 | Colored seeds A/a |
| | 23263-264 | 21118 X 117-3 | Colorless seeds a/a |

Cross	Colored	Colorless	% Colorless
263-3 X 262-10	62	32	34**(1:1)
			*(3:1)

1981	25490-491	23263-3 X 262-10	Colored seeds A/a
	25492-493	23263 X 262-10	Colorless seeds a/a
	25487	a tester (a1, A2, C1, C2, R)	

Cross	Colored	Colorless	% Colorless
491 on 492-9	49	55	47
on 487(a/a)	139	185	57
491 on 492-4	60	77	56
on 492-2	128	40	24**[a]
on 487	106	98	48
491-4 on 492-8	99	112	53
on 492-11	66	41	38**
491-5 on 493-6	62	41	40**
491-6 X 493-7	68	61	47
on 493-7	61	72	54
on 493-9	88	91	51

* Significant deviation from an expected ratio at the .05 level.

** Significant deviation at the .01 level.

[a] This mating was shown to be A/a X A/a.

however, and the polar bodies may be fertilized by a sperm
from a different pollen grain than that fertilizing the egg
nucleus. In this particular case, I suggest that the egg
nucleus was fertilized by an A sperm while the polar bodies
were fertilized by an a sperm. The a/a plant resulting from
a colored seed of 25492-2 X 491-3 suggests that the opposite
type of heterofertilization event also occurred in the next
generation. Although in maize, heterofertilization ordinar-
ily occurs in about one percent of fertilization events,
higher rates have been observed and are known to be under
genetic control (8). It is possible that this A*/a stock
has higher than usual rates of heterofertilization. I am
not able, however, to exclude the possibility that these
unexpected genotypes arose via mutation.

Of the 14 A/a plants, 11 had sufficient seed set to
estimate whether significant deviations from an expected 75
colored: 25 colorless percentage were appearing. Six plants
did not deviate significantly (24, 24, 28, 25, 27% colorless)
while four plants did (10, 6, 7, 5% colorless). Two other
plants had 17 and 16% colorless kernels, but the total numbers were relatively small so that deviations from an expected 25% were not significant at the 5% level.

Eight plants from colored seeds of this cross (25492-2
X 491-3) were crossed as males onto plants from the colorless seeds from the cross of the a/a tester X 25491-3 (48%
colorless); the data are given in Table 2. Those crosses
were 27010 X 27008. An average of three crosses were made
per male parent. For the six male parents in 27008 which
when selfed gave ca. 25% colorless kernels, the crosses
onto 27010 gave progenies with ca. 50% colorless kernels.
The two males, which deviated from 25% colorless when selfed
(27008-3 and -12) both gave some progenies with a deficiency
of colorless kernels.

TABLE 2
1982 DATA RELEVANT TO THE A*/a STOCK

27001-002	W22 a/a
27003-007	25487 X 491-3 Colored seeds
27008-009	25492-2 X 491-3 Colored seeds
27010-014	25487 X 491-3 Colorless seeds
27015-016	25487 X 491-3 Colored seeds
27017-018	25487 X 491-3 Colorless seeds

Pollination	Colored	Colorless	% Colorless
008-1 ⊗	38	0	0
-2 ⊗	112	36	24
-3 ⊗	71	13	16
-4 ⊗	105	0	0
-5 ⊗	109	35	24
-6 ⊗	105	40	28
-7 ⊗	All	0	0
-8 ⊗	126	41	25
-9 ⊗	9	6	-
-10 ⊗	6	2	-
-11 ⊗	70	0	0
-12 ⊗	152	17	10**
009-6 ⊗	0	58	100
27005 X 008-2			ca. 25 (3)
X 008-3			ca. 25 (3)
X 008-5			ca. 25 (3)
X 008-6			ca. 25 (3)
X 008-8			ca. 25 (3)
X 008-9			ca. 25 (2)
X 008-10			ca. 25 (3)
X 008-12	71	36	34*
	72	58	45**[a]
	202	78	28
X 009-6			ca. 50 (3)
27010 X 008-2			ca. 50 (4)
X 008-3	120	84	41**
	116	140	55
	128	130	50
X 008-5			ca. 50 (3)
X 008-6			ca. 50 (4)
X 008-8			ca. 50 (2)
X 008-9			ca. 50 (3)
X 008-10			ca. 50 (3)
X 008-12	79	67	46
	56	41	42

27001 X 27003 & Reciprocal
A set of paired reciprocal crosses all gave percentages of colorless kernels of approximately 50.

27015-16 X 27017-18 & Reciprocal
A set of paired reciprocal crosses all gave percentages of colorless kernels of approximately 50.

---

* Significant deviation at the .05 level.
** Significant deviation at the .01 level
[a] This mating was shown to be a/a X A/a.

The plants from colored seeds from one of the crosses [27010 (a/a tester X 25491-3) colorless seeds] times [27008-8 (25492-2 X 491-3) colored seeds] referred to in the preceding paragraph and which produced 51% colorless kernels, were planted in 1983 and selfed. Note that 27008-8 when selfed produced ca. 25% colorless kernels. The results are shown in Table 3. It is clear that in this progeny, the excess of colored kernels typical of some sib intercrosses in the A*/a stock has reappeared and that all ears with excess colored kernels are semisterile although the reverse may not be true. Thus, there is a rather complicated pattern of inheritance of the tendency to produce more colored than colorless kernels in a/a times A*/a matings or when A/a plants are selfed.

TABLE 3
PERCENTAGES OF COLORLESS KERNELS OBSERVED WHEN
PLANTS FROM 27010 X 27008-8 (COLORED SEEDS)
WERE SELF-POLLINATED

Percent Colorless		
30 +[a]	20 SS	19* SS[b]
28 +	20 SS	19* SS
27 +	20 SS	14** SS
24 +	20 SS	** SS
24 SS		** SS

23 SS	10** SS
23 +	7** SS
22 +	6** SS

---

a
 Ear not semisterile.

b
 Ear was semisterile.

*
 Deviation from 25% colorless significant at the .05 level.

**
 Deviation significant at the .01 level.

DISCUSSION

Before outlining a hypothesis that can explain the observations presented here, it would be well to eliminate the various alternative explanations offered earlier. An explanation involving cytoplasmic entities can be ruled out since the deviant ratios reappear in lineages (Table 3) in which A*/a stocks appear only as male parents. A pollen lethal factor linked to a can be discounted since A*/a plants which in crosses to a/a sibs give deviant ratios show expected ratios in crosses to an a/a tester. A lethal ovule factor linked to a would not explain the results since plants grown from colorless seeds of the cross, 25187 (a/a tester) X 25491-3 did not show semi-sterility. If a gametophyte factor (Ga) were linked to A, one should have observed pronounced distortion of the ratio of colored to colorless in almost all crosses of 257008 plants as males onto 27005 [a tester X 25491-3 (colored seeds)], and this was not the case.

In attempting to explain the behavior of the A*/a stock, the following observations seem most pertinent: (1) Plants which in intercrosses (25492-2 X 491-3) do not give deviations from expected ratios transmit factors which condition excess numbers of colored kernels in advanced generations; (2) In self-pollinations of A/a plants where excess numbers of colored kernels were produced, the percentages of colored kernels center about 90%; (3) In a/a X A/a sibcrosses, where excess kernels were produced, the percentages of colored kernels center about 60; (4) The observations for both self-pollinations and sibcrosses are compatible with expectations if a were linked with a zygotic lethal

with a recombination value of 10 percent. The linkage of a
with a zygotic lethal (zl) seems plausible since all plants
with deviant ratios have noticeable ear sterility; (5) When
deviant ratios reappear in a progeny after outcrossing and
backcrossing as in the progeny shown in Table 3, approximately one-half the progeny give high percentages of colored
kernels when selfed.

The observations above prompt the hypothesis that there
is a lethal zygote factor linked to a. One can then suggest
the genotypes of plants which have been involved in the
lineages being followed. Plant 25491-3 was A+/a zl while
25492-2 was A+/a+. When 25491-3 was crossed as a male onto
the a tester, the colored seeds were A+/a+ and the colorless
seeds a+/a zl. When plants from these colorless seeds
(27010) were crossed by individual plants in 27008 (27492-2
X 491-3, colored seeds), then no plants with an excess of
colored kernels would result when the male was A+/a+.
Such males would produce ca. 25% colorless kernels when
selfed, but the plants from colored seeds of the cross on
27010 would be A+/a zl (significantly less than 25% colorless seeds when selfed) and A+/a+ (ca. 25% colorless seeds
when selfed) in equal numbers. This is close to what is
observed for the progeny for which data are presented in
Table 3.

A better fit to the data from the various crosses can be
found if one invokes the involvement of a second hypothetical locus such that the zygotic lethality conditioned by
zl/zl is expressed only in gametes which are homozygous
for the recessive allele at the second locus. This creates
such a flexible hypothesis, however, that it is preferable
to wait until future investigations indicate necessity of
postulating a second locus before doing so.

It should be understood that the foregoing suggestion as
to the genetic basis of the excess of colored kernels in
a/a X A/a crosses is an ad hoc hypothesis to explain a
rather complicated set of observations and requires testing
by the proper crosses. It does seem clear the deficiency
of colorless kernels is associated with semisterility on
the ear, and it is probable that there is a lethal zygote
factor linked to a in these stocks. The important point is
that there is almost certainly a conventional explanation
for the observed excesses of colored kernels. Thus, the
four cases of deviant ratios in the progeny of virus-infected plants that I have investigated all have conventional

genetic explanations. None requires the assumption of anomalous regulation of gene activity as was suggested by Sprague and McKinney (1971). The interesting question that remains is whether the mutations at other loci, which influence the proportions in which homozygotes for the target loci appear in advanced generations of virus-infected plants, are induced by systemic infections with barley stripe mosaic virus. None of my data bear on this point but Mottinger (9) has presented evidence that such infection can produce mutations in maize. A second report by Mottinger et al. (10) has shown that mutations at the shrunken-1 locus in advanced generations from a virus-infected plant are associated with insertions within shrunken. It seems reasonable to believe that the high frequency of deviant ratios reported by Sprague and McKinney (2) and which Brakke et al. (4) and Nelson (5 and here) have shown to be due to mutations at other loci is attributable to virus-induced mutation.

REFERENCES

1. Sprague GF, McKinney HH, Greeling L (1963). Virus as a mutagenic agent in maize. Science 14:1052-1053.
2. Sprague GF, McKinney HH (1966). Aberrant ratio: an anomaly in maize associated with virus infection. Genetics 54:1287-1296.
3. Sprague GF, McKinney HH (1971). Further evidence on the genetic behavior of AR in maize. Genetics 67:533-542.
4. Brakke MK, Samson RG, Compton WA (1981). Recessive alleles found at R and C loci in maize stocks showing aberrant ratio at the A locus. Genetics 99:481-485.
5. Nelson OE (1981). A reexamination of the Aberrant Ratio phenomenon in maize. Maydica 26:119-131.
6. Brakke MK (1984). The Aberrant Ratio phenomenon and other possible mutations induced in corn by viruses. Ann Rev Plant Path 22. In press.
7. Nelson OE (1952). Nonreciprocal cross-sterility in maize. Genetics 37:101-124.
8. Sprague GF (1932). The nature and extent of hetero-fertilization in maize. Genetics 17:358-368.
9. Mottinger JP (1981). Mutations of Adh induced by barley stripe mosaic virus. Maize Genet Coop Newsletter 56: 89-90.
10. Mottinger JP, Dellaporta SL, Keller PB (1984). Stable and unstable mutations in Aberrant Ratio stocks of maize. Genetics 106:751-767.

Cellular and Molecular Biology of Plant Stress, pages 13–40
© 1985 Alan R. Liss, Inc.

THE IMPACT OF PLANT STRESSES ON CROP YIELDS

Yoash Vaadia

ARCO Plant Cell Research Institute
6560 Trinity Court, Dublin, California 94568

ABSTRACT. Environmental stresses reduce crop yields. The extent of losses to disease is shown to be about 12% on a global basis. The loss to insects can be as high or higher depending on the year, crop, and management practices. Actual crop yields are only a fraction of the maximum possible in a given climate. for example, in cotton under intensive cultivation this fraction is 35-50%. In other crops with less intensive cultivation this fraction may be much smaller. The losses represented by the difference between maximum and actual yields are ascribed to various unidentified environmental stresses.
The relationship between dry matter production and transpiration is discussed. The two processes are shown to be highly regulated in plants and the relationship between them appears to be fixed under conditions of crop production. Various stresses do not seem to modify the relationship because they reduce dry matter as well as transpiration. The importance of the relationship is discussed. Crop development is presented in terms of stage of growth and possible impact of stress on assimilate partitioning and harvestable grain yield. The complexity of the interaction between growth stages and the effects of stress is presented.
The approach of explanatory dynamic crop simulation models is introduced as a concept which may incorporate knowledge obtained on the cellular and molecular levels and provide an integrated picture of crop performance under differing environments. Such models may provide a tool to identify characteristics which may bring about improvement in plant performance. They may also be useful in crop management.

INTRODUCTION

This symposium is concerned with plant responses to environmental stresses. These stresses include restricted water supply, excess salt, mineral deficiencies, ion and heavy metal toxicity, chilling temperatures and heat shock, anoxia and attack by pathogens and pests. Plant responses to these stresses involve changes in photosynthesis, growth, respiration, hormonal regulation, accumulation of solutes, as well as various defense and adaptive mechanisms regulated at the molecular level including gene expression. All of these topics will be discussed in this symposium to the extent possible and with a great deal of detail.

Our interest in plant responses to environmental stress has increased greatly in recent years. Many monographs have been published on various aspects of the subject (16,18,19,24) and many symposia are held. Further, the support of public funding on the subject is on the rise.

Obviously, much of the interest is generated by the relatively rapid advances in biological research in general in recent years. Stresses such as heat shock or anoxia have been shown to induce the appearance of otherwise unexpressed proteins (1,29) and extreme temperatures have been shown to modify the frequency of DNA transposition (8). These observations, as well as others, have introduced molecular biology as a powerful tool to the studies of mechanisms of plant response and tolerance to environmental stresses. Stress has become another added factor that may help to elucidate regulation and development in plants, and thus contribute to our basic understanding of plant biology and growth.

Potential environmental stresses are a common occurrence and their impact on crop yields are considerable. They always decrease plant growth and crop yields. There is the hope, generally shared by scientists and farmers alike, that stress research would help in the development of improved crop management and cultivars and would result in increased crop production despite the presence of environmental stresses.

This paper deals with the impact of stresses on crop yields. It is not a comprehensive survey of the subject. It is intended to provide an overview from the point of view of whole plants and crop canopies in the field. Some empirical observations of yield losses due to stresses will be presented. The impact of stresses on dry matter accumulation by crops and the importance of crop phenology to crop yields will be discussed. Also, the approach of

current research on crop simulation in relation to
environmental stresses will be introduced.

EMPIRICAL OBSERVATIONS ON YIELD LOSSES
DUE TO ENVIRONMENTAL STRESSES

Crop yields vary with management, climate, incidence of
disease and many other variables. Often, economic constraints
dictate raising crops with limiting water, fertilizers and
pesticides which result in low average yields. Severe crop losses
may result from epidemics of disease, frost, heat or drought.
The assessment of such crop losses is of importance for several
reasons. These include supply predictions, crop management
and crop insurance adjustments. Crop loss assessment is
becoming a more and more sophisticated field in its own right
(21,32). For proper assessment of crop losses it is necessary to
know the potential yield and the actual yield. The potential
yield is defined as the yield of a crop surface optimally supplied
with water and nutrients in a disease- and weed-free
environment under the prevailing weather conditions (27).
Actual yield is the yield obtained by farmers with their
particular agronomic practices. Crop losses generally are
considered to be the differences between potential yields and
actual yields. The art of assessment of crop losses attempts to
ascribe portions of this difference to specific agents such as
water, disease or weeds.

As we shall discuss later, there are several explanatory
models that integrate growth over time and prevailing weather
conditions. Such models provide data for potential yields and
may be used to assess the specific impact of extreme changes in
weather. Models predicting yield under limiting water,
nutrients, and other environmental stresses are being
developed (27).

Until such models are available for practical use various
empirical and statistical approaches must suffice (21,22,34).
The Handbook of Pest Management in Agriculture (15)
presents a source for data on the impact of disease, insects and
weeds on crop yields. Table 1 (15) shows crop losses due to
disease on a global basis. These losses amount to about 12% and
are valued at $50 billion at the producer level. This is a huge
sum which could be reduced very profitably by research through
the development of new effective resistant cultivars and means
of eradication. It is noteworthy that the data presented show no
difference in losses between Asia and the U.S. for example. In

TABLE 1
GLOBAL DISTRIBUTION OF ESTIMATED
PLANT DISEASE LOSSES (15)

Region	Value of losses ($ billions) 1967	Value of losses ($ billions) 1976	Loss/region expressed as % of total global loss	% loss due to disease within each region
Asia	7.1	14.2	29	11.3
Europe	6.2	12.4	25	13.1
North/Central America	3.9	7.8	16	11.3
USSR/People's Republic of China	2.6	5.2	11	9.1
Africa	2.4	4.9	10	12.9
South America	2.1	4.2	9	15.2
Oceania	0.2	0.4	<1	12.6
Total	24.5	49.0	100	

Reproduced with permission from Handbook of Pest
Management, CC Press Inc. Boca Raton, Florida

TABLE 2
LOSSES FROM INSECTS AND MITES FOR SOME
REPRESENTATIVE CROPS IN CALIFORNIA IN 1978

	Crop value $1000	Crop loss $1000	%
Alfalfa	368,762	107,958	29.3
Almonds	267,029	31,833	11.9
Apples	45,488	5,125	11.3
Apricots	30,496	3,347	11.0
Avocado	44,100	3,528	8.0
Barley	118,651	3,511	2.96
Corn	136,875	15,382	11.2

California Dept. Food and Agriculture, 1981

fact, losses seem to be similar in all regions. This may be true, or may indicate that assessment methods may not be very accurate (15).

Table 2 shows similar data for insect damage in selected crops in California. Here the damage differs from crop to crop. In 1978, losses in barley are only 3%, whereas in alfalfa they are some 30%. Average losses for the other crops cited are about 11% and if damage to crops not included in the table are also considered, amount to hundreds of millions of dollars.

An alternative to the approach of crop simulation of potential yield and actual yield is to compare maximum observed yields and actual yields. The difference constitutes the impact of the various existing stresses on crop yields. Table 3 presents average yields of cotton in California between 1978 and 1983. With the exception of 1978 actual yields of cotton fiber are stable at about 1150 kg/ha. The low yield of 1978 represents Lygus attack, an insect which brings about fruit drop. The extent of loss due to Lygus in 1978 is about 400 kg/ha, since yields before 1978 were similar to those after 1978. Crop loss due to Lygus in 1978 was worth several hundred million dollars. Even so, the losses due to other stresses were even higher. The actual yield is roughly 1/3 of the maximum yield obtained. Cotton in California is a good example of intensive agriculture since the crop is irrigated, fertilized, and a careful program of crop protection is maintained. Even so, yields are very much lower than those possible. There are several reasons for these differences but no systematic evaluation. A certain proportion is ascribed to disease and in particular *Verticilium* wilt. Soil compaction due to the use of heavy machinery is a major factor restricting root growth. Toxic effects of herbicide residue could well be implicated and of course water and fertilizer management may well be improved. That such an improvement is possible is shown in Table 4.

The data in Table 4 represents cotton production in Israel, where the same variety and similar intensive practices and inputs are used. Maximum yields in Israel and California are similar at about 3000 kg/ha. Average yields in Israel, however, have been increasing rather than stationary and are higher than those of California by some 50%. The reasons for the differences are not documented, but rest largely in improved management practices, the most important of which are better timing and control of the use of water, fertilizer, herbicides and pesticides. Cotton management in Israel includes considerable data collection on plant growth at different growth stages and the use of computer simulation programs to recommend timing

TABLE 3
YIELD OF COTTON FIBER IN CALIFORNIA

Year	State Average kg/ha
1978	727
1979	1136
1980	1130
1981	1168
1982	1223
1983	1108

Maximum observed yield for the variety - 3000 kg/ha
Calif. Crop and Livestock Reporting Service, 1984

TABLE 4
YIELD OF COTTON FIBER IN ISRAEL

Year	State Average kg/ha
1977	1260
1978	1340
1979	1320
1980	1350
1981	1530
1982	1550
1983	1630

Maximum observed yield - 2960 kg/ha
Ministry of Agriculture, Israel, 1984

of management. The data collection and careful supervision of crop growth are very useful in the improvement of management.

Cotton in California and Israel are intensively grown crops and therefore may not be the best examples for the difference between maximum and actual yields. This difference is expected to be much greater in areas of extensive production, and where water and nutrients are more limiting to plant growth. Thus, stresses may reduce crop yields to levels that are 1/3 of potential production in intensive agriculture and to much lower values in more common and less intensive situations. A current estimate (5) suggests that actual yields per hectare on a global basis are only 3.2% of the maximum yield possible. From the point of view of resource utilization, these losses represent a considerable waste of used inputs of land, fertilizer and pesticides which are not fully utilized. A considerable effort in research in crop management should be directed to maximize yields to optimal utilization of the inputs available.

STRESS AND DRY MATTER ACCUMULATION IN PLANTS

Environmental stresses reduce crop yields. The typical relation between yield and a particular stress is shown schematically in Figure 1. This type of figure is commonly used to describe the relation between yield or growth and the level of stress applied. Similar curves are used to demonstrate growth of cells in cell cultures as a function of salinity, water potential or other stresses. This relationship is different for different plants and thus it is often used to compare stress tolerance and survival between varieties and species. More tolerant lines are taken to be those that are less affected by stress or that survive it better.

Studies of the type shown in Figure 1 have contributed considerably to the understanding of plant response to stress. Thus, variations in turgor pressure, osmotic adjustment, and cell wall elasticity have been invoked to explain the reduction in growth and to explain differences in responses between species (14). Selection of salt tolerant species are often based on such response surfaces (30).

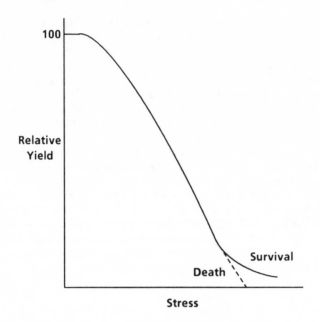

FIGURE 1.   A schematic representation of the
relationship between yield and stress (30).

Stress is not an input to growth and development. Stress
brings about a modification of reaction of rates which in turn
change plant performance in terms of its ability to process
inputs and produce outputs. Further, under normal growing
conditions varies considerably, and is rarely a constant value.
Plants experience cycles of increasing water stress between
irrigations or rainfalls. Thus the "dose" of stress in Figure 1
cannot be easily duplicated in a crop situation. Also, the
response of a crop to stress is complicated because of continuous
changes in the environment and continuous changes in the
level of stresses experienced by the crops. For these reasons the
relationship of Figure 1 for crop yields may differ from year to
year and from field to field. Its usefulness is in giving general
guidelines for tolerance and survival, and in defining the
relationship between specific physiological processes and stress
under controlled conditions.
    Plant growth and development are highly regulated
processes. This regulation occurs on all levels and allows the

plant to adapt to the continuous changes which are occurring in its environment. It is of interest to study the impact of stress on interdependent physiological processes, and evaluate its effects on the regulation of such processes. Such studies are rare, but one that may be of importance in the context of the impact of stresses on crop yields concerns the question of the relationship between dry matter accumulation and transpiration in plants. Dry matter accumulation is, in the main, a function of net photosynthesis, and therefore of CO_2 fixation. Transpiration is a function of available water, evaporative demands and stomatal conductance. The interrelationship between the two processes involve regulation of photosynthetic and respiratory rates, stomatal conductance, and the regulation of water potentials in leaf cells.

de Wit (4) was first to suggest that the relationship between dry matter production and transpiration is highly regulated in plants. He used many sources of older experimental data to plot dry matter accumulation as a function of transpiration. Figure 2 presents de Wit's plot for sorghum and wheat. The data are of various pot experiments where evaporation from the soil surface was prevented. Thus the transpiration recorded represents truly transpired water. Data were collected in different states by different workers between 1911 and 1940 in South Dakota, Colorado, Texas and India. Dry matter represents total accumulated dry weight of plant tops. Transpiration was determined by frequently weighing the pots. The data on the abscissa represent total transpiration, corrected for average pan evaporation for the growth period at the site in which the measurement was made. This correction, which normalizes transpiration to the evaporation potential of the sites, is an essential feature of the data analysis.

It is evident that there is a close correlation between dry matter and transpiration in both sorghum and wheat. Since the experiments were done in different years and locations, the data suggest that the slopes are typical to the crop species. Sorghum, a C_4 plant, is a more efficient water user in as much as its slope is steeper. Wheat, a C_3 plant, has a typically flatter slope. The experiments reported by de Wit did not purport to study the impact of environmental stresses. Yet such stresses were present. According to de Wit, the slope of the line in Figure 2 remains constant under a wide range of water stresses and is not modified by nutrient deficiency within limits. Data presented in Figure 3 suggest that the slope is not modified by different salinity stresses and water stress in guayule plants.

Other data on the stability of the slope under salinity are available for various crops (13,25).

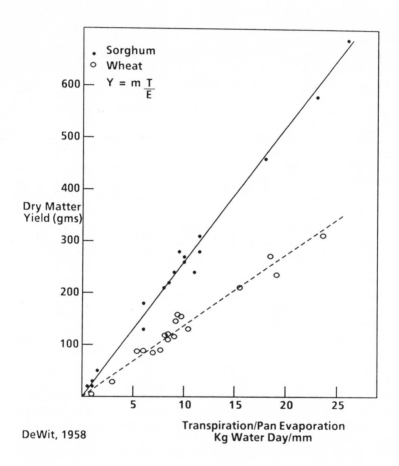

FIGURE 2. The relationship between dry matter accumulation and transpiration in sorghum and wheat. Data are from different experiments, sites and years. Transpiration is divided by the appropriate average pan evaporation for each site and year. m = slope of the line (4).

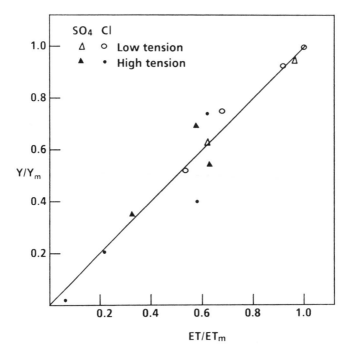

FIGURE 3. The relation between dry matter accumulation and evapotranspiration in guayule under soil water deficits and chloride and sulfate salinity. Low tension - wet, high tension - dry (35).

Fischer and Turner (11) examined de Wit's original observations and calculated the relationship between dry matter accumulation and transpiration for a different selection of experiments. They found considerable variation of the slope between different crop species when only dry matter of shoots was considered (Table 5). However, when root dry matter was included their data suggest that the slopes may possibly be similar for different species within the groups of C_3 plants and C_4 plants. Fischer and Turner discuss four selected field situations of dry matter accumulation and transpiration. Three represent C_3 plants. Of these, one, in Migda, Israel represents a mix of pasture species rather than a single crop. The slope of all these situations is similar if roots weight is included (Table 5). The slope for corn is about twice as steep as that of the C_3 plants. de Wit suggested that the slope is dependent on the plant species only, but his data did not include root dry matter.

Thus, the suggestion (11,12) that the slope is similar for C_3 plants if root weight is included in the analysis requires further confirmation.

TABLE 5
PRIMARY PRODUCTIVITY AND OTHER
PARAMETERS FOR SEVERAL PLANT COMMUNITIES (11)

	Shrubland, Utah C. Pantana 1972-1974	Migda, Israel Annual Pasture 1971-1974	Wagga Wagga, NSW Wheat cv Heron 1962	Mandan, SD Maize 1968-1970
E	0.48	0.22	0.24	0.55
T	12.3	18.6	22.9	26.8
Y_S	860	6870	1030	893
Y_T	3580	9500	11640	11170
M_S	34	81	84	183
M_T	140	112	122	229

E = Pan Evaporation, cm/day; T = Transpiration, cm; Y_S = Yield Shoots, Kg/ha; Y_T = Yield, Total, Kg/ha; M_S = Slope with Y_S, M_T = Slope with Y_T, Basic Equation $Y = mT/E$. Reproduced with permission of Annual Reviews of Plant Physiology.

Tanner and Sinclair (31) provide a detailed theoretical evaluation of the original relationship suggested by de Wit. Their treatment is based on the relationship of water and CO_2 exchange in leaves, on vapor pressure gradients, leaf area indices, light penetration through the crop canopy, existing knowledge on respiration and conversion factors of carbohydrates to dry matter, and on shoot/root partitioning. Using this theoretical treatment, they calculate the slope (k) expected for the relationship between dry matter and transpiration. They compare the theoretical values thus obtained with empirical data they collected in Table 6. The theoretical values for k are 0.118 for corn, 0.50 for alfalfa and 0.41 for soybeans. The agreement with the observed data in Table 6 is good.

TABLE 6
EXPERIMENTAL VALUES FOR YIELD OF DRY MATTER
OVER TRANSPIRATION, VAPOR PRESSURE GRADIENT
AND K FOR VARIOUS CROPS AND LOCATIONS (31).

Crop	Location	Year	Y/T	(e*-e)	k
Corn	Calif.	1974	.0046	22.0	0.100
Corn	Calif.	1975	.0054	18.2	0.098
Corn	Colo.	1974	.0055	21.7	0.120
Corn	Utah	1974	.0037	22.1	0.084
Corn	Utah	1975	.0046	18.7	0.082
Corn	Ariz.	1974	.0020	47.7	0.097
Corn	Ariz.	1975	.0026	39.3	0.104
Corn	Neb.	1912	.0047	19.0	0.089
Corn	Neb.	1912	.0029	31.3	0.091
Alfalfa	Wis.	1977	0.041	10.4	0.043
Soybean	Kansas	1970	0.0017	23.8	0.040

The basic relation is: $Y = kT/(e*-e)$, where Y = yield, T = transpiration, $(e*-e)$ = vapor pressure gradient and k = slope. Reproduced from "Limitations to Efficient Water Use in Crop Production" by permission of ASA, CSSA, and SSSA.

The data presented in Figure 3 and 4 and in Tables 5 and 6 are good evidence that dry matter accumulation is somehow closely related to the amount of water transpired. The relationship seems a fixed one for a crop (or possibly for all C_3 plants and for all C_4 plants) and does not seem to have changed by breeding (see data for corn 1912). It does not seem to be affected by water stress, salt stress, nutrient deficiency, or even nematode infection (4,6,13,25,31,35). The situation with disease is less clear because of the paucity of data on the subject. The little data available suggest that some pathogens affect yield more than transpiration whereas others affect yield and transpiration in accordance with the slope described above (31).

Obviously the relationship described between dry matter accumulation requires further study. Its importance for future development of improved crop varieties is considerable. Is it possible to develop more efficient water use? The data presented suggest that this is unlikely. Yet, the physiological nature of the relationship and the regulatory mechanisms involved require detailed study. Some attempts in this direction are being made (7).

The importance of the relationship resides in the implication that stresses that reduce transpiration also reduce dry matter production in a predictable and fixed manner. Therefore, stress tolerance may not confer an ability of the plant to accumulate more dry matter per unit water transpired under stress conditions. Tolerance, according to this interpretation, allows the crop to survive a stress, while slowing down both dry matter accumulation and transpiration. It is also possible that some plants may have the ability to transpire more than others at a given stress and thus accumulate dry matter more rapidly. However, this may be a self-defeating strategy in crop plants under conditions of limiting water, in as much as it may lead to a shortened life span of the crop.

The interaction between net photosynthesis and transpiration is very important in assessing the impact of stress on crop yields. In principle, it provides a tool to estimate dry matter production on the basis of water transpired (13). Thus, the impact of stress can be assessed without the need to identify quantitatively or qualitatively all existing stresses.

STRESS AND HARVESTABLE YIELD

The discussion heretofore concerned dry matter accumulation. Dry matter is harvestable yield mainly in pasture and feed crops. In most crops the harvestable yield is not the total dry matter produced but rather the fruit or grain or in some crops below ground parts such as in potato, sugar beets, or peanuts. Crops differ in their sensitivity to stresses at different stages of growth, from germination to maturity and this sensitivity may modify partitioning of dry matter to the harvestable yield components of the crop (12,14). Crops may be determinant and flower at a given time. Failure in reproduction at this time because of a stress cannot be corrected later as it might be in an indeterminant crop which continues flowering for an extended period (20). The interactions between growth stages may be complex and difficult to assess quantitatively.

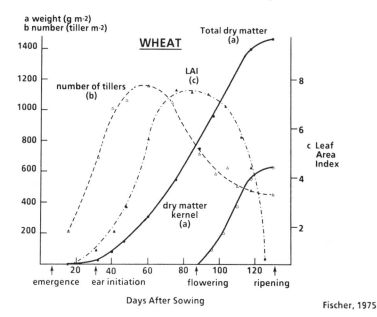

FIGURE 4. Growth parameters for a wheat crop during a growing season (10). Reproduced from Crop Science by permission of Crop Science Society of America.

An example of the development of a determinant crop such as wheat is presented in Figure 4. After emergence there is an increase of the number of shoots and, soon after, leaves begin to expand rapidly, increasing the leaf area index (ratio of leaf surface area to projected land surface). It takes some 50-60 days for the ground to be completely covered by leaves and achieve maximum possible photosynthetic rates. Environmental stresses at this time may delay leaf expansion, reduce photosynthesis, and waste water by surface evaporation rather than by transpiration. It may also influence flower initiation. As leaf area index exceeds a value of about 5 in this case, the accumulation of dry matter assumes a rapid linear rate which is not reduced at anthesis or flowering. In fact, after anthesis most of the dry matter accumulation is diverted to the grain as evidenced by the similar slopes of kernel dry matter accumulation and total dry matter accumulation. Senescence brings about a rapid reduction in leaf area index and reduced photosynthesis and dry matter accumulation. Stresses during the time of grain fill will obviously result in reduced yields. In wheat, an important parameter to indicate productivity is the harvest index. It refers to the fraction of grain in the total above ground dry matter. Since wheat diverts most of its photosynthates to the grain after anthesis, the harvest index depends to a large part on the amount of post-anthesis photosynthesis. This, in turn, is determined by the growing conditions and by life span of the leaf canopy. Delayed leaf senescence is likely to enhance harvest index.

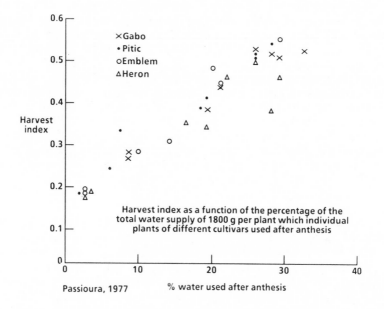

FIGURE 5. Harvest index as a function of the percentage of the total water supply transpired after anthesis (26). Reproduced with permission of the Journal of Australian Institute of Agr. Sci.

Passioura (26) clearly demonstrated the effects of stress at the post-anthesis period. A given amount of water (1800 gms) was metered out to four different varieties in such intervals that different proportions of the available water could be transpired after anthesis. Figure 5 shows that the harvest index increased with increased proportion of the total water used after anthesis. This illustrates some of the complexity of wheat response to stress. Varieties should flower early enough to have enough water to fill their grain but late enough to avoid frost damage at anthesis. Under a Mediterranean climate, varieties which tend to restrict water use early in the season and conserve moisture to post anthesis grain fill may well outperform varieties which do not conserve water early in the season. Under such situations, total dry mater may be somewhat reduced, but this loss may be well compensated by increased grain yield and harvest index.

An example of different strategies used by different cultivars is offered in Figure 6. This experiment (9) presents yields of 277 barley varieties over several seasons obtained in three different sites differing in the amount of rainfall as well as edaphic characteristics. The site mean yield, plotted on the abscissa, is compared to the yield of a given particular variety, on the ordinate in these same sites and seasons. Lower yields on the abscissa most likely represent limited water supply. The dashed line ($b = 1.0$) represents the average yield of all cultivars. The U.S. variety "Atlas" seems to yield better than average in all sites and seasons and is thus highly adaptable to favorable as well as drought-prone conditions. "Provost", an English bred variety, does very poorly under dry conditions and very well under wetter conditions. It is thus poorly adapted to drought prone environments. "Bankuti Korai", a Hungarian variety, does well under dry conditions but poorly under wetter conditions. There is no detailed physiological evaluation of the varieties to explain the different adaptability to the different sites by the different varieties. However, in view of previous discussion, the following is a possible interpretation. "Provost" is an "optimist" using its water rapidly, and, when water is limiting, little is left to fill the grain and the yield of grain is very low. "Bankuti Korai" is a "pessimist" and restricts its water use. Under dry conditions it yields well because it has enough water to fill its grain. Under wet conditions its water conservation results in water remaining in the profile at maturity. Other interpretations involving time of flowering and maturity are possible, and hard data are required to present the real picture. However, the example presented in Figure 6 should suffice to illustrate the difficulties in identifying simple traits that confer high adaptability to environmental stresses. It also provides an explanation for the tendency of breeders to develop varieties suitable for local, less variable habitats.

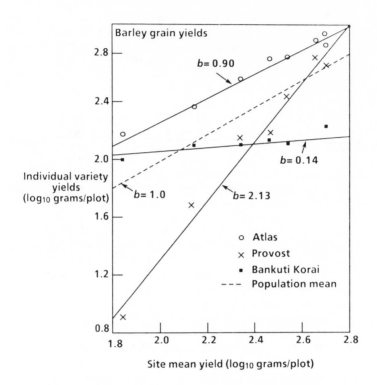

FIGURE 6. The relationship between individual grain yields of three barley varieties and the population mean of 277 barley varieties grown at different dryland sites in South Australia (9). Reproduced with permission of Aust. J. Agric. Res.

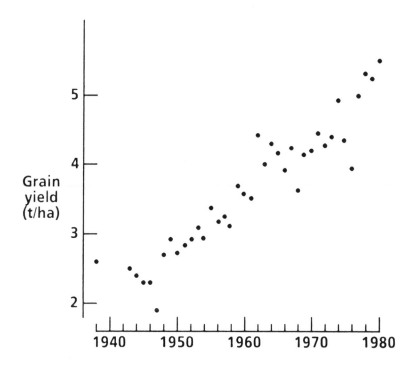

FIGURE 7. Average yields of wheat in the United Kingdom (3). Reproduced with permission of the Royal Soc. London.

Breeding has contributed greatly to the increase of crop yields in recent years. Many examples are available for the contribution of breeding to increased yields in agriculture which, it is important to reiterate, have growth conditions that commonly include environmental stress (3,9,24). One such example is given in Figure 7 for wheat in England. The data indicate doubling of the yield over the last 40 years. Much of this increase, like that in Table 4, is due to improved cultural practices. Austin (2), however, showed that new varieties are also genetically superior. The main apparent advantage is enhanced harvest index and shorter and stiffer straw, which represent some of the advantages ascribed to the new varieties of the green revolution (3,10,12) which resist lodging. This genetic improvement is one of the noteworthy achievements of agricultural science of the century. In the last century, it was apparently achieved in England by applying salt stress to the plant. Roberts (28) in 1847 suggested that: "Common salt is sometimes applied before sowing the seed at the rate of from 10 to 21 bushels per acre and is often beneficial in bringing the ears to perfection. It also causes a greater weight of grain, but seldom increases the quantity of straw". In this connection it is interesting to note that harvest indices in Roberts time were not very inferior, as shown in Table 7. Harvest indices obtained today in England in farmers fields range between 44-48% (3).

TABLE 7
YIELD OF ABOVE GROUND DRY MATTER, GRAIN AND
HARVEST INDEX FOR SEVERAL WHEAT VARIETIES (28)

Variety	Total dry matter kg/ha	Grain kg/ha	Harvest index grain/total DM %
Colne White Chaff	6,696	3,016	45.0
Bristol	6,845	2,864	41.8
Sharp's	6,754	2,862	42.4
Spaulding	7,134	2,865	40.2
Seyer	7,201	2,802	38.9
Smoothy's	6,062	2,699	44.5
Kent Red	6,434	2,667	41.5
Sewell's	6,635	2,612	39.4
Piper thickest	5,348	2,442	45.7

Yields and harvest indices of crop plants are very much influenced by the environment and by the genome. While harvest indices may be improved by breeding new varieties (2), very old varieties do have the potential to exhibit high harvest indices under the proper environment. It is clear that we are in need of better tools than we have at hand at the moment to explain crop behavior under changing environment and to identify possibilities for improvement of the genome. Such tools may be provided in the future by explanatory simulation models of plant growth and crop yields.

SIMULATION OF PLANT GROWTH
AND CROP PRODUCTION

There are several levels of organization in the biology of crop plants. We recognize plant populations, individual plants, organs, tissues, cells and molecules. When we concern ourselves with the question of the impact of stresses on crop yields we are operating at the level of organization of plant populations and individual plants. However, much of biological research is concerned with attempts to gain a basic understanding of processes at other levels of organization such as the cellular and molecular levels, the primary focus of this symposium.

One potential approach to the study of the nature of the interactions in crop plants under stress is that of explanatory models (22,27). Explanatory models attempt to explain a larger system on the basis the smaller systems and provide bridges between levels of organization. Crop behavior can be better understood by the study of individual plants, cell behavior by the study of moelecules. If the understanding of the smaller level is complete, it is very likely that the larger level is understood.

The techniques of developing explanatory models for crop behavior are still rudimentary, although considerable effort has been invested in their development. However, they offer potentially a clear way to make progress in the understanding of plant growth and crop production where it is unlikely that any level of organization will be completely understood in the near future. Such models can be tested, their results verified with those of the real world, and areas of ignorance identified as problems for study.

An example of the approach of such dynamic models is presented in Figure 8. This represents a schematic simplified relational diagram for the situation of potential crop production. Here the crop is assumed to completely cover the ground surface and not to be limited by water and nutrients and to be disease- and weed-free. The limiting factor is the weather. While temperature may restrict growth in the early part of the growing season, the main limiting factor considered is light. The basic elements involved are dry weights of shoots and roots, and leaf surface and the processes involved are photosynthesis, respiration assimilate partitioning and growth. Basic data are needed to define rates of flow (valves in Figure 8) which control the flow of materials between quantities (rectangles) and are influenced by auxiliary variables (circles) and external

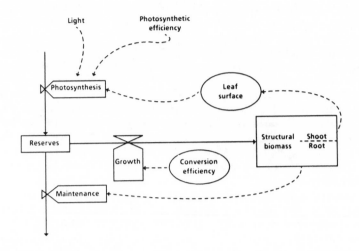

FIGURE 8. A relational diagram of the essence a system describing potential crop production where light is the main limiting factor, and nutrients, disease and weeds do not limit growth (27). Reproduced with permission of Pudoc, Wageningen, Netherlands.

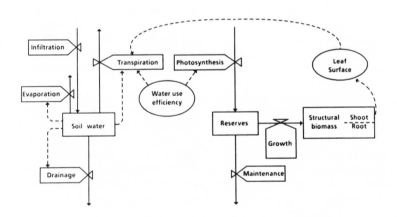

FIGURE 9. A relational diagram of the essence of a system describing crop production when water deficits are the main limiting factor to growth (27). Reproduced with permission of Pudoc, Wageningen, Netherlands.

variables. Solid lines are transfer of material and dashed lines represent transfer of information. Rates are independent of each other. Each rate depends only on the quantity associated with it and on the variables controlling it.

Figure 9 represents a relational diagram for the case where water may be limiting. A submodel is included describing the quantity of soil water and the rates controlling it. These include infiltration, drainage, evaporation and transpiration. Rates of transpiration and photosynthesis are determined by their external variables (environment) and by the auxiliary variables of leaf surface and water use efficiency.

The applicability of a detailed model of the type of Figure 9 is shown in Figure 10 and 11. The model itself, in principle, is a generally applicable model to any site. Its quantitative performance will depend on the specific variables of climate, rainfall and soil characteristics in as much as they affect rates of flow between quantities in the model. In Figure 10 and 11 it is shown that the model performs well in two different years with different rainfall where total dry matter yields were widely different.

More complex models are now being studied or are available. Such models include nutrient deficiency, partitioning of dry matter to harvestable yield in several crops, host pathogen relationships and other topics (21,22,27). A detailed book on the philosophy and methodology involved was published recently (27). These models may often be unsuccessful in explaining crop behavior which is indicative of our deficient understanding of the system. Nevertheless, they offer hope to better understand the impact of stress on crop yields. Stress modifies rate of flow of quantities and is thus expressed in such models in modified permeabilities, transport rates, modified enzyme kinetics and other variables. Basic studies on stress response on the tissue and molecular levels can thus be incorporated into such models usefully.

If and when the development of such models is successful, they may be used for the support of crop management, the estimate of crop losses and the prediction of the impact of expected weather changes on agricultural production. Simplified models of various types are in use for such purposes already (21,22,34).

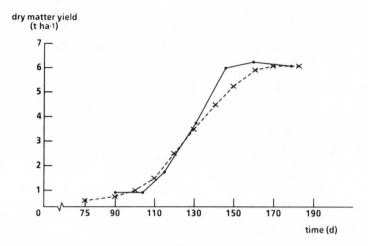

FIGURE 10. Comparison between measured (●) and simulated (x) dry matter accumulation of natural pasture in Migda, Israel in a wet year (1972/73) (33). Reproduced with permission of Pudoc, Wageningen, Netherlands.

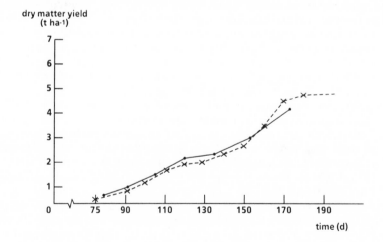

FIGURE 11. Comparison between measured (●) and simulated (x) dry matter accumulation of natural pasture in Migda, Israel in a dry year (1975/76) (33). Reproduced with permission of Pudoc, Wageningen, Netherlands.

CONCLUSIONS

Crop production normally occurs under environmental stresses of various kinds. Deficient water, excess salt, poor nutrition, low and high temperature, weeds, insects and diseases are more or less common. Some of the deleterious effects on yield can be used for advantage. For example, tomatoes grown under salt may be smaller but are tastier and more valuable (23). Stress tolerance can be selected for and crop plants can be grown under conditions where the wild-types perish. Of the many examples, a good one is the very salt-sensitive avocado which can now be grafted onto selected salt-excluding root stocks. Such grafted trees do very well on land formerly unsuitable for avocado production (17). Similar examples can be cited for other crops and stresses.

The main impact of stresses on crop yields are on existing agricultural lands, and these yields can be improved (Table 4). Despite the great progress made in agricultural science, pests take a regular toll of crop yields and actual yields are only a fraction of the maximum yields possible under similar practices. One possible solution suggested in this paper is to promote our understanding of the regulation and behavior of whole plants. This may be achieved by further understanding of physiological and biochemical processes on the tissue and molecular levels, and the incorporation of such knowledge into dynamic models describing crop response to the environment.

REFERENCES

1.   Altschuler M and Mascarenhas (1982). The synthesis of heat-shock and normal proteins in high temperatures in plants and their possible roles in survival under heat stress. In Schlesinger MJ, Ashburner M, Tissieres A (eds): "Heat Shock from Bacteria to Man", Cold Spring Harbor Labortory, p 321.

2.   Austin RB (1980). Physiological limitations to cereal yields and ways of reducing them by breeding. In Hurd RG, Biscoe PV and Dennis C (eds): "Opportunities for Increasing Crop Yields," London: Pitman, p 3.

3.   Bingham J (1981). The achievements of conventional plant breeding. Phil Trans R Soc Lond B 292:441.

4.   de Wit CT (1958). "Transpiration and crop yields", Wageningen: Pudoc.

5. de Wit CT, van Laar HH, van Keulen H (1979). Physiological potential of crop production. In Sneep J, Hendriksen AJT (eds): "Plant Breeding Perspectives", Wageningen: Pudoc, p 47.
6. Elkins, CBR, Haaland RL, Rodriguez-Kabana R, Hoveland CS (1979). Plant-parasitic nematode effects on water use and nutrient uptake of a small and a large rooted tall fescue genotype. Agro J 71:497
7. Farquhar GD, Sharkey TD (1982). Stomatal conductance and photosynthesis. Ann Rev Plant Physiol 33:317.
8. Fincham JRS, Sastry GRK (1974). Controlling elements in maize. Ann Rev Genet 8:15.
9. Finlay KW, Wilkinson GN (1963). The analysis of adaptation in a plant breeding programme. Aust J Agric Res 14:742.
10. Fischer RA (1975). Yield potential in a dwarf spring wheat and the effect of shading. Crop Sci 15:607.
11. Fischer RA, Turner ND (1978). Plant productivity in the arid and semi-arid zones. Ann Rev Plant Physiol 29:277.
12. Fischer RA (1979). Growth and water limitation to dryland wheat yield in Australia. A physiological framework. J Aust Inst Agr Sci, p 83.
13. Hanks RJ, (1983). Yield and water-use relationships: An overview. In Taylor HH, Jordan WR, Sinclair TR (eds): "Limitations to Efficient Water Use in Crop Production", Madison: Am Soc Agro, Crop Sci Soc Am, Soil Sci Soc Am, p 393.
14. Hanson AD, Nelsen CE (1980). Water: Adaptation of crops to drought prone environments. In Carlson PS (ed): "The Biology of Crop Productivity," New York: Academic Press, p 77.
15. James CW (1981). Estimated losses of crops from plant pathogens. In Pimentel (ed): "CRC Handbook of Pest Management in Agriculture", Boca Raton, Florida: CRC Press Inc, p 79.
16. Johnson CB (1981). "Physiological Processes Limiting Plant Productivity", London: Butterworths.
17. Kadman A, Ben Yaakov A (1970). Avocado: Selection of root stocks and other work related to salinity and lime. In Oppenheimer H (ed) "The Division of Subtropical Horticulture 1960-1969", Bet-Dagan , Israel: Volcani Inst Ag Res, p 23.
18. Lange OL, Nobel PS, Osmond CB, Ziegler H (eds) (1981). Encyclopedia of Plant Physiology, New Series, Vol 12A, Berlin: Springer.

19. Lange OL, Nobel PS, Osmond CB, Ziegler H (1982). Encyclopedia of Plant Physiology, New Series Vol 12B, Berlin: Springer.
20. Loomis RS (1983). Crop manipulation for efficient use of water. In Taylor HH, Jordan WR, Sinclair TR (eds): "Limitations to Efficient Water Use in Crop Production", Madison: Am Soc Agro, Crop Sci Soc Am, Soil Sci Soc Am, p 345.
21. Loomis RS, Adams SS (1983). Integrative analysis of host-pathogen relations. Ann Rev Phytopathol 21:341.
22. Loomis RS, Rabbinge R, and Ng E (1979). Explanatory models in crop physiology. Ann Rev Plant Physiol 30:339.
23. Mizrahi, Y (1982). Effect of salinity on tomato fruit ripening. Plant Physiol 69:966.
24. Mussell H, Staples RC (eds) (1979). "Stress Physiology in Crop Plants", New York: John Wiley & Sons.
25. Parra MA, Romero GC (1980). On the dependence of salt tolerance of beans (*Phaseolus vulgaris* L.) on soil water matric potentials. Plant and Soil 56:3.
26. Passioura JB (1977). Grain yield, harvest index and water use of wheat. Aust Inst Agric Sci, p 117.
27. Penning de Vries FWT, van Laar HH (eds) (1982): "Simulation of Plant Growth and Crop Production", Wageningen: Pudoc.
28. Roberts E (1847). On the management of wheat. J R Soc Agric Soc, England 8:60.
29. Sachs MM, Freeling M, Okimoto R (1980). The anaerobic proteins in maize. Cell 20:761.
30. Shannon MC (1984). Principles and strategies in breeding more salt tolerant crops. In San Pietro A, Pastemak D (eds): "Saline Water for Agriculture and Aquaculture" Proc Intl Symp NCRD Israel (in press).
31. Tanner CB, Sinclair TR (1983). Efficient water use in crop production: Research or Re-Search? In Taylor HH, Jordan WR, Sinclair TR (eds): "Limitations to Efficient Water Use in Crop Production", Madison: Am Soc Agro, Crop Sci Soc Am, Soil Sci Soc Am, p 1.
32. Teng PS, Krupa SV (eds) (1980). "Crop Loss Assessment", Univ Minn Agric Exp Stn Misc Publ 78, 327 pp.
33. van Keulen H (1982). Crop production under semi-arid conditions as determined by moisture availability. In Penning de Vries FT, van Lar HH (eds): "Simulation of Plant Growth and Crop Production", Wageningen: Pudoc, p. 159.

34. Waggoner PE (1984). Agriculture and carbon dioxide. Am Scientist 72:179.
35. Wadleigh CH, Gauch HG, Magistad OC (1946). Growth and rubber accumulation in guayule. USDA Technical Bulletin 925.

Cellular and Molecular Biology of Plant Stress, pages 41–49
© 1985 Alan R. Liss, Inc.

LOSS IN CHLOROPLAST ACTIVITY AT LOW LEAF WATER POTENTIALS
IN SUNFLOWER: THE SIGNIFICANCE OF PHOTOINHIBITION[1]

Robert E. Sharp and John S. Boyer

Departments of Plant Biology (R.E.S., J.S.B.) and
Agronomy (J.S.B.), University of Illinois, and
USDA/ARS (J.S.B.), Urbana, Illinois 61801

ABSTRACT It has been proposed that loss in chloroplast capacity to fix CO_2 at low leaf water potentials may result from photoinhibition caused at
least in part by low intercellular CO_2 concentrations accompanying stomatal closure in the light.
We investigated these possibilities in sunflower
using gas exchange techniques. Measurements of
quantum yield and light- and CO_2-saturated photosynthesis showed that both parameters in situ were
similarly inhibited in leaves exposed to high or
very low photon flux densities during dehydration.
This occurred regardless of decreases in CO_2 availability resulting from stomatal closure. The results
show that photoinhibition is not involved in the
loss in chloroplast capacity to fix CO_2 at low leaf
water potentials in sunflower.

INTRODUCTION

At low leaf water potentials (Ψ_w), photosynthesis
is inhibited by stomatal closure and by losses in chloroplast activity (1). Stomatal closure increases the resistance to the diffusion of CO_2 into the leaf, thereby
restricting the availability of CO_2 for fixation, while
losses in chloroplast activity decrease the capacity to
fix available CO_2. In some cases, the inhibition of

---

[1]This work was supported by National Science Foundation Grant PCM 79-09790 (to J.S.B.).

chloroplast activity may be the more limiting factor (2,3).

Recently, there has been much speculation that the loss in chloroplast capacity to fix CO_2 at low Ψ_w may result from, or be enhanced by, "photoinhibition" – light dependent damage to the photosynthetic apparatus. Photoinhibition is generally believed to result from the accumulation of excessive excitation energy at the photosystem II reaction center complex (4,5). It has been proposed that at low Ψ_w the decrease in intercellular CO_2 concentration (internal CO_2) accompanying stomatal closure in the light, coupled perhaps with direct impairment of photosynthetic reactions, may predispose the chloroplasts to photoinhibition as a result of the restricted utilization of photochemical energy in photosynthesis (e.g., 6, 7). Evidence for photoinhibition at low Ψ_w was reported by Björkman et al. (8), who showed that in Nerium oleander, a plant native to arid regions, inhibition of chloroplast reactions (electron transport and variable fluorescence from photosystem II) were more severe when low Ψ_w developed in full sunlight than in the shade. However, experimental data are as yet too few to evaluate either the importance of stomatal closure in inducing photoinhibition at low Ψ_w, or the extent to which the overall loss in chloroplast capacity to fix CO_2 at low Ψ_w may be attributed to photoinhibitory effects.

In this report, we have addressed these questions in sunflower by using gas exchange techniques to investigate the influence of high light at atmospheric or reduced CO_2 concentrations on the in situ chloroplast capacity to fix CO_2 at various Ψ_w. When inhibition of photosynthesis can be attributed to damage by excess light, for example if leaves are illuminated at photon flux densities greatly in excess of light saturation (9,10), reductions in the capacity for both light-limited and light-saturated photosynthesis are observed. We have studied both of these parameters and show that, in sunflower, photoinhibition is of little consequence to the inhibition of photosynthesis at low Ψ_w regardless of decreases in CO_2 availability resulting from stomatal closure.

METHODS

Three to four week old sunflower (Helianthus annuus L. cv. IS894) plants were grown in soil as previously described (11). After withholding water, leaves received daily light pretreatments at either a high, potentially photoinhibitory photon flux density (6 h at 2000 μmol photons$\cdot$m$^{-2}\cdot$s^{-1} photosynthetically active radiation (PAR), approximating full sunlight) or a very low flux density (40 μmol photons$\cdot$m$^{-2}\cdot$s^{-1} PAR). Decreases in Ψ_w occurred predominantly during these pretreatments. The CO_2 and O_2 concentrations during pretreatment were varied for different experiments, and the internal CO_2 concentration was calculated as previously described (11). The quantum yield for CO_2 fixation (mol $CO_2 \cdot$mol photons absorbed PAR^{-1}) (11) and, in separate experiments, the light- and CO_2-saturated rate of photosynthesis (3) were measured immediately following pretreatment to determine, respectively, the light-limited and maximum capacity of the chloroplasts for CO_2 fixation in situ. Leaf water potentials (12) were determined in the experimental leaves immediately upon completion of the photosynthesis measurements.

RESULTS

Figure 1 shows that as Ψ_w decreased after withholding water, the quantum yield for CO_2 fixation was markedly inhibited when leaves received daily pretreatments at a high photon flux density at atmospheric levels of CO_2, decreasing from 0.100 at high Ψ_w to 0.035 at a Ψ_w of -21.4 bars. However, the quantum yield was similarly inhibited when leaves received only very low light pretreatments during dehydration, decreasing to 0.038 at a Ψ_w of -20.4 bars. The rate of development of low Ψ_w was similar during the high and low light experiments. When leaves received the same high or low light pretreatments but were maintained at high Ψ_w for the 5 day experimental period, no inhibition of quantum yield was observed (data not shown).

Table 1 shows that the inhibition of light- and CO_2-saturated photosynthesis at low Ψ_w was also similar regardless of whether leaves were exposed to the high or low light pretreatments during dehydration at atmospheric levels of CO_2. In both high and low light

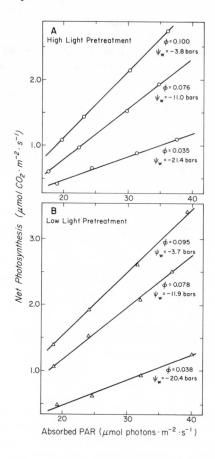

FIGURE 1. Net photosynthesis at limiting flux densities of absorbed PAR measured immediately following daily
6h light pretreatments at (A) 2000 μmol photons·m^{-2}·s^{-1}
and (B) 40 μmol photons·m^{-2}·s^{-1} as leaf water potentials
(Ψ_w) decreased after withholding water. Quantum yields
(ϕ) for CO_2 fixation are represented by the slope of the
photosynthesis-flux density relationship. Water potentials were measured in the experimental leaves immediately upon completion of the photosynthesis measurements.
Pretreatment conditions: 330 μl·l^{-1} CO_2, 21% O_2, 25°C air
temperature. Measurement conditions: 1500 μl·l^{-1} CO_2, 1%
O_2, 25°C air temperature.

TABLE 1

EFFECT OF LIGHT PRETREATMENT ON LIGHT- AND CO_2-SATURATED
PHOTOSYNTHESIS AT VARIOUS LEAF WATER POTENTIALS[a]

Water potential (bar)	Light- and CO_2-saturated photosynthesis (% of control[b])
High light pretreatment	
-8.7	95.7
-14.0	69.2
-17.5	32.0
-27.1	14.5
Low light pretreatment	
-8.5	89.6
-13.9	54.1
-16.3	25.8
-28.5	10.3

[a]After withholding water, leaves received daily 6h pretreatments at either 2000 µmol photons·m^{-2}·s^{-1} (high light) or 40 µmol photons·m^{-2}·s^{-1} (low light). Light- and CO_2-saturated photosynthesis and leaf water potential were measured in the same leaf immediately following pretreatment. Pretreatment conditions: 330 µl·l^{-1} CO_2, 21% O_2, 25°C air temperature. Measurement conditions: saturating CO_2, 21% O_2, 25°C leaf temperature.
[b]Control rates of light- and CO_2-saturated photosynthesis were measured before withholding water and prior to light pretreatment, and ranged from 61.5 to 74.8 µmol CO_2·m^{-2}·s^{-1}.

experiments, at Ψ_w between -25 and -30 bars leaves had lost more than 85% of their capacity for CO_2 fixation at high Ψ_w. Again, the rate of development of low Ψ_w was similar in high and low light experiments, and no inhibition was observed when leaves received the same light pretreatments but were maintained at high Ψ_w for the duration of the experiment (data not shown).

TABLE 2

EFFECT OF CO_2 CONCENTRATION DURING HIGH LIGHT
PRETREATMENT ON THE QUANTUM YIELD FOR CO_2 FIXATION AT
VARIOUS LEAF WATER POTENTIALS[a]

Water potential (bar)	Minimum internal CO_2 during pretreatment ($\mu l \cdot l^{-1}$)	Quantum yield
High light pretreatment at atmospheric CO_2		
−4.6	199	0.092
−14.5	100	0.055
−22.5	100	0.033
High light pretreatment at CO_2 compensation concentration		
−4.5	41.7	0.089
−12.1	49.0	0.060
−21.9	57.5	0.044

[a]After withholding water, leaves received daily
6 h pretreatments at 2000 μmol photons$\cdot m^{-2} \cdot s^{-1}$
at either atmospheric CO_2 (330 $\mu l \cdot l^{-1}$) or the
CO_2 compensation concentration. The O_2 concentration was 21% and the air temperature 25 °C.
The quantum yield for CO_2 fixation and leaf
water potential were measured in the same leaf
immediately following pretreatment. Measurement
conditions: 1500 $\mu l \cdot l^{-1}$ CO_2, 1% O_2, 25 °C air
temperature.

The internal CO_2 during the high light pretreatments at atmospheric levels of CO_2 was approximately
200 $\mu l \cdot l^{-1}$ at high Ψ_w, and, because of stomatal closure,
fell to a minimum of 100 $\mu l \cdot l^{-1}$ at lower Ψ_w. This was
approximately double the CO_2 compensation concentration
measured at low Ψ_w under the same conditions (Table 2).
The CO_2 compensation concentration presumably represents
the lowest internal CO_2 that could result from stomatal
closure in the light. Therefore, to test the possibility
that more rapid stomatal closure and hence lower internal
CO_2 availability might increase the inhibition of chloro-

plast activity in leaves exposed to high light during dehydration, we imposed the high light pretreatments while
maintaining the leaf at the CO_2 compensation concentration. Table 2 shows that the inhibition of quantum yield
at low Ψ_w was similar regardless of whether atmospheric
CO_2 or the CO_2 compensation concentration was maintained
during the high light pretreatments.

TABLE 3

EFFECT OF HIGH LIGHT PRETREATMENT AT LOW CO_2 AND O_2
ON THE QUANTUM YIELD FOR CO_2 FIXATION AND LIGHT- AND
CO_2-SATURATED PHOTOSYNTHESIS[a]

	Quantum yield	Light- and CO_2-saturated photosynthesis (μmol $CO_2 \cdot m^{-2} \cdot s^{-1}$)
Before pretreatment	0.097	59.7
After pretreatment	0.052	35.8
	(46.4% inhibition)	(40.0% inhibition)

[a]Leaves of well watered plants received a 6h pretreatment at 2000 μmol photons$\cdot m^{-2} \cdot s^{-1}$ in an atmosphere
containing 1% O_2 and less than 1 $\mu l \cdot l^{-1}$ CO_2 at an
air temperature of 25°C. The quantum yield for CO_2
fixation and light- and CO_2-saturated rate of photosynthesis were measured immediately following pretreatment. Measurement conditions were as given for
Tables 1 and 2.

To check our methods of evaluating the inhibition of
photosynthesis attributable to prior high light exposure,
leaves of well watered plants were exposed to high light
for 6 h in an atmosphere containing 1% O_2 and less than 1
$\mu l \cdot l^{-1}$ CO_2. In numerous reports (e.g., 13), comparable
pretreatments have been shown to cause photoinhibition,
manifest as reductions in subsequent measurements of both
quantum yield and light-saturated photosynthesis. Table 3
shows that the pretreatment resulted in a 40-50% inhibition of the quantum yield and the light- and CO_2-satu-
rated rate of photosynthesis. When the same low O_2 and
CO_2 conditions were imposed in the dark, no inhibition of
subsequent photosynthesis was observed (data not shown).

Therefore, our measurements at low Ψ_w would have detected any inhibition of photosynthesis resulting from photo-inhibition.

DISCUSSION

The results show that, in sunflower, photoinhibition is not involved in the loss in chloroplast capacity to fix CO_2 at low Ψ_w, regardless of decreases in internal CO_2 availability resulting from stomatal closure. Both the quantum yield and light- and CO_2-saturated rate of photosynthesis were strongly inhibited at low Ψ_w when leaves had been exposed only to very low photon flux densities during dehydration. In neither case was the extent of inhibition at any Ψ_w increased by exposure to a high photon flux density approximating that of full sunlight. Further, the inhibition of quantum yield at low Ψ_w was not increased even though the CO_2 compensation concentration, which represents the maximum decrease in CO_2 availability that could result from stomatal closure, was maintained while leaves were exposed to high light during dehydration.

Our data do not exclude the possibility that photo-inhibitory damage to primary photochemistry occurred at low Ψ_w, but suggest that any such effects were not at the rate-limiting step(s) that define either the light-limited or the maximum capacity of the chloroplasts for CO_2 fixation at low Ψ_w. The data indicate instead that, in sunflower, loss in chloroplast capacity to fix CO_2 at low Ψ_w in situ can be entirely attributed to direct effects of dehydration on chloroplast function. Further work is required to evaluate the significance of photo-inhibition at low Ψ_w in other species.

REFERENCES

1.   Boyer JS (1976). Water deficits and photosynthesis. In Kozlowski TT (ed): "Water Deficits and Plant Growth, Vol. 4," New York San Francisco London: Academic Press, p 153.
2.   Boyer JS (1971). Nonstomal inhibition of photosynthesis in sunflower at low leaf water potentials and high light intensities. Plant Physiol 48:532.

3. Matthews MA, Boyer JS (1984). Acclimation of photo-
    synthesis to low leaf water potentials. Plant
    Physiol 74:161.
4. Osmond CB (1981). Photorespiration and photoinhibi-
    tion; some implications for the energetics of photo-
    synthesis. Biochim Biophys Acta 639:77.
5. Kyle DJ, Ohad I, Arntzen CJ (1984). The molecular
    basis of light damage to chloroplast membranes. J
    Cell Biochem Suppl 8B:232.
6. Björkman O (1981). Responses to different quantum
    flux densities. In Lange OL, Nobel PS, Osmond CB,
    Ziegler H (eds): "Physiological Plant Ecology, Vol
    12A: Interactions with the Physical Environment,"
    Heidelberg Berlin New York: Springer, p 57.
7. Osmond CB, Winter K, Powles SB (1980). Adaptive sig-
    nificance of carbon dioxide cycling during photosyn-
    thesis in water-stressed plants. In Turner NC,
    Kramer PJ (eds): "Adaptation of Plants to Water and
    High Temperature Stress," New York: John Wiley and
    Sons, p 139.
8. Björkman O, Powles SB, Fork DC, Öquist G (1981).
    Interaction between high irradiance and water stress
    on photosynthetic reactions in Nerium oleander. 1981
    Carnegie Institution of Washington Yearbook: 57.
9. Powles SB, Critchley C (1980). Effect of light
    intensity during growth on photoinhibition of intact
    attached bean leaflets. Plant Physiol 65:1181.
10. Powles SB, Thorne SW (1981). Effect of high-light
    treatments in inducing photoinhibition of photosyn-
    thesis in intact leaves of low-light grown Phaseolus
    vulgaris and Lastreopis microsora. Planta 152:471.
11. Sharp RE, Matthews MA, Boyer JS (1984). Kok effect
    and the quantum yield of photosynthesis: light par-
    tially inhibits dark respiration. Plant Physiol (in
    press).
12. Boyer JS (1966). Isopiestic technique: measurement
    of accurate leaf water potentials. Science 154:1459.
13. Powles SB, Osmond CB (1978). Inhibition of the capa-
    city and efficiency of photosynthesis in bean leaf-
    lets illuminated in a CO_2-free atmosphere at low
    oxygen: a possible role for photorespiration. Aust J
    Plant Physiol 5:619.

Cellular and Molecular Biology of Plant Stress, pages 51–69
© 1985 Alan R. Liss, Inc.

MOLECULAR MECHANISMS OF COMPENSATION TO LIGHT STRESS IN CHLOROPLAST MEMBRANES[1]

David J. Kyle, Itzhak Ohad and Charles J. Arntzen

MSU-DOE PLANT RESEARCH LABORATORY, MICHIGAN STATE UNIVERSITY, EAST LANSING, MI 48824

ABSTRACT Photoinhibition of photosynthetic capacity occurs when an organism is unable to fully compensate to an imposed light stress. This stress can occur at either high photon flux densities, or at low photon flux densities if the exit of electron equivalents from the plastoquinone pool is prevented. The latter condition occurs, for example, when CO_2 fixation ceases and there is no regeneration of a terminal acceptor of electron transport. Thus, drought or temperature-induced stomatal closure will predispose leaf tissue to light sensitivity and photoinhibition. Primary photoinhibition damage is focussed on the secondary acceptor of photosystem II (Q_B) and two biochemical strategies are identified which compensate for this light-induced damage. A strategy of avoidance involves phosphorylation of certain thylakoid membrane proteins in response to a reduced plastoquinone pool, and results in a decrease in the absorptive cross-section of photosystem II and an increase in that of photosystem I. A strategy of tolerance is characterized by a highly efficient and rapid mechanism of repair in an attempt to match the accelerated rate of damage. Both mechanisms operate to minimize the impact of environmentally-imposed light stress.

[1]This work was supported by DOE contract #DE-AC02-76ERO-1338 to Charles J. Arntzen.
[2]Department of Biological Chemistry, Hebrew University, Jerusalem, Israel.

INTRODUCTION

Illumination of higher plants and algae with photon flux densities in excess of that required to saturate photosynthetic reactions will result in a loss of photosynthetic capacity due to specific damage of the thylakoid membrane at or near the position of photosystem II (PS II). This has been termed photoinhibition (1-3). The photon flux density required for photoinhibition is generally very high (3000 μE/m^2/sec) for sun-adapted, non-stressed plants (4). However, photoinhibition can also occur at very low light intensities if photosynthetic processes are limited by other environmental stresses, or if the organism has undergone an adaptive change for existence in a low light environment (i.e., shade species). The synergistic effect of light with other environmental stresses has been reported as an increased sensitivity to relatively low light intensities (photoinhibition) under conditions of low water potential (5) or low temperature stress (6). Thus, many environmental stresses may not be directly responsible for damage of the photosynthetic apparatus, but rather, may predispose the tissue to a condition of light sensitivity or vulnerability to photoinhibition.

Algal mutants deficient in the ability to fix carbon (for example, those lacking ribulose biophosphate carboxylase) exhibit severe light sensitivity (7). Furthermore, studies with algal mutants defective in photosystem I (PS I) activity, also exhibit light sensitivity (7). These results suggest that photoinhibitory damage is accelerated by illumination under conditions in which electron flow out of the plastoquinone pool is restricted.

We have characterized two molecular biochemical strategies which allow photosynthetic species to compensate for conditions of an excess light load. The first strategy is one of avoidance. Avoidance strategies such as alteration in leaf angle (8), chloroplast orientation (9) or negative phototaxis in motile algae have previously been reported. In this manuscript, however, we will discuss a biochemical strategy; reducing the absorptive cross section, or target size of PS II and thereby reducing the number of photons interacting with PS II. The second biochemical strategy is one of tolerance. Given that the avoidance response is not sufficient to prevent damage, a mechanism exists to rapidly repair the light damaged function. If the repair rate can match the rate of damage no net loss in the overall photosynthetic function should occur. The more rapid and efficient the repair mechanism, the greater the light damage that can be tolerated.

THE AVOIDANCE STRATEGY - Alterations in the Absorptive Cross-sections of PS II and PS I

The fact that higher plants and algae can control the distribution of excitation energy between PS II and PS I was recognized simultaniously by Murata (10) working with Porphyridium and Bonnaventura and Myers (11) working with Chlorella. Only recently, however, was the biochemical basis of this control mechanism identified. Bennett (12) was the first to observe the principal light-harvesting pigment protein complex in higher plants (LHC) becomes covalently phosphorylated when isolated thylakoids are incubated in the light with ATP. Furthermore, the protein kinase appears to be regulated by the redox state of the plastoquinone pool; that when the pool becomes reduced, the kinase becomes activated (13). Concomitant with the LHC phosphorylation is a decrease in the absorptive cross-section of PS II and an increase in that of PS I (14). These changes are shown in Fig. 1 as a change in the ratio of the low temperature (77K) flourescence emmisions arising from PS II and PS I [amplitude of the flourescence signal is directly pro- portional to the amount of excitation energy arriving at the reaction center], as well as changes in the relative quantum yields of PS II and PS I. How does the phosphorylation of LHC alter the absorptive cross-sections of PS II and PS I? Small particles observed on the periplasmic face (PF) following freeze-fracture analysis of chloroplast membranes have been shown to migrate from the PS II-enriched appressed grana membranes to the PS I-enriched stroma lamellae membranes in response to protein phosphorylation (15). The particles are the same size as those in an LHC-containing, PS II-deficient mutant of maize (16), and those observed in freeze-fracture replicas prepared from purified LHC incorporated into phospholipid vesicals (17). We, therefore, interpret these particles as LHC which is detached physically, although not energetically, from PS II reaction centers so that they will fracture on the PF rather than on the EF with PS II (15). As this LHC appears to be laterally mobile in the plane of the membrane in response to phosphorylation, we refer to

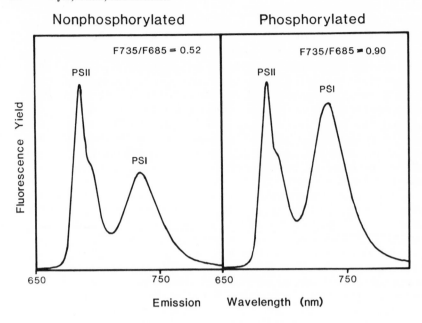

Figure 1. The effect of thylakoid protein phosphorylation on the distribution of excitation energy between PS II and PS I. Chloroplast thylakoids were incubated for 20 min in light in the presence (phosphorylated) or absence (non-phosphorylated) of 200 μ M ATP (as in ref. 5), before freezing to 77k. Flourescence spectra were scanned at 77k using an excitation wavelength of 440 nm. Values for relative quantum yield changes were obtained from the literature: a, from Steinback KE, Bose S, Kyle DJ (1982) Arch Biochem Biophys 216, 356; b, from Farchaus WJ, Widger WR, Cramer WA, Dilley RA (1982) Arch Biochem Biophys 217, 362;   c, from Horton P, Black MT (1982) Biochim Biophys Acta 680, 22.

this as "mobile LHC". The supposition that the movement of this mobile antennae from the grana to the stroma lamellae occurs in response to phosphorylation (the "mobile antennae" hypothesis) is further supported by observations of radioactive pulse/chase experiments (18), changes in the Chl a/b ratio of grana and stroma lamellae. [an increase in the concentration of LHC in the stroma lamellae is accompanied by a decrease in the Chl a/b ratio since most of the Chl b in higher plants is associated with LHC] (18,19), and by an increase in the proportion of commassie-stainable LHC polypeptide in the stroma lamellae following protein phosphorylation (18).

The above observations have led us to propose the model shown in Fig. 2. An excess light load on the photosynthetic membrane is "sensed" by an overreduction of the plastoquinone pool ($[PQH_2]>>[PQ]$). In response, a protein kinase is stimulated which phosphorylates a surface exposed threonine residue (20) of the LHC. The balance between hydrophobic attractive forces and electrostatic repulsive forces of the "mobile LHC" is altered by the additional negative charge (21) and the preferred membrane domain of the phospho-LHC is now the stroma lamellae. [The LHC "attached" to PS II is presumably also phosphorylated but, as it does not move out to the stroma lamellae, we feel that the strong hydrophobic attractive forces between the LHC and PS II core cannot be overcome by the additional negative surface charge on the "attached LHC"]. Following phosphorylation, the mobile-LHC migrates out of the PS II-enriched appressed grana membranes into the PS I-enriched stroma lamellae. The net result is a loss of antennae sensitizing PS II (decrease in absorptive cross section) and an increase in the antennae sensitizing PS I (increase in absorptive cross-section). This would effectively decrease the input of reducing equivalents into the plastoquinone pool, as well as increase their exit out of the plastoquinone pool, thereby avoiding the damage which is potentiated by the overreduction of PS II.

In addition to altering absorptive cross-sections, there is evidence to suggest that protein phosphorylation may lead to an increase in the rate of an alternative mechanism of PS II deexcitation; namely back reaction or PS II cycling (22). It should be noted that although the observed changes in absorptive cross sections of PS II and PS I correlate with the kinetics of LHC phosphorylation/ dephosphorylation, other PS II polypeptides are also

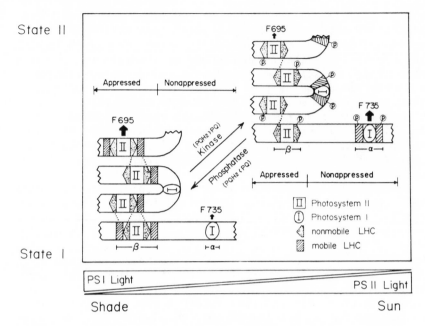

Figure 2.   The mobile antenna hypothesis for the control of
excitation energy distribution in higher plants.   Leaves
illuminated with light preferentially exciting PS I (shade)
have thylakoid membranes in the "State I" condition and are
characterized by nonphosphorylated LHC and a lesser
intensity of flourescence from PS I (F735) relative to that
of PS II (F695).   Illumination with wavelength-balanced
light (sun) which preferentially excites PS II, results in a
"State II" condition of the thylakoids.   This is
characterized by phosphorylated LHC and a higher
flourescence yield of PS I relative to that from PS II.
Sudden transfers from a State I to State II condition (as
during a sunfleck in a plant canopy), or stomatal closure
results in an increased reduction of the plastoquinone pool
($PQH_2 > PQ$) and a protein kinase is activated.   In the
nonphosphorylated condition, most of the LHC acts as an
antenna for PS II resulting in a large absorptive
cross-section (β) and high quantum yield of PS II.   Once
phosphorylated, the mobile population of LHC migrates
laterally along the thylakoid membrane from the PS II
enriched appressed membranes (grana) to the stroma lamella
where it acts to increase the absorptive cross-section (α)
and relative quantum yield of PS I.

phosphorylated by a similar light-regulated mechanism. The kinetics of dephosphorylation of the other PS II polypeptides, however are much slower. Based on delayed light measurements, we have proposed that phosphorylation of these other PS II polypeptides may be involved in stimulating a back reaction or cycling of PS II in order to enhance deexcitation. Horton (23) has recently supported this view by demonstrating a hydroxylamine enhancement of the room temperature variable flourescence yield of phosphorylated thylakoids. Both the increase in PS II deexcitation mechanisms and the decrease in the PS II absorptive cross-section reflect a function of protein phosphorylation as a biochemical mechanism for avoiding light-induced damage to PS II.

THE TOLERANCE STRATEGY - Effecient Repair of the Damage

Until recently, the primary site of photoinhibition damage to the photosynthetic membrane was only vaguely recognized as occurring at, or near, PS II (3,24). Photoinhibition studies using intact leaves of higher plants, however, are fraught with the unavoidable problem of light attenuation across the leaf. Not only is the light intensity attenuated by absorption of the cell layers on the light exposed surface, but the quality of the light transmitted into the deeper cell layers is of far-red enriched, PS I-exciting wavelengths which should be less damaging. This difficulty was recongnized by Powles and Bjorkman (5) who measured light-induced damage to cells on the upper surface of a leaf of Nerium oleander with little or no damage to the chloroplasts in cells on the shaded undersurface of the leaf. Clearly, a preparation of chloroplasts from such a light-stressed leaf will be heterogeneous with respect to primary and secondary sites of damage.
In order to obtain a homogenous sample of damaged membranes with which we could pinpoint the primary site of photodamage, we chose to study photoinhibition using the single-chloroplast-containing unicellular alga Chlamydomonas reinhardii. Using this system, we were able to identify the loss of activity of the secondary quinone acceptor [Q_B] of PS II as the first photosynthetic membrane lesion in light-stressed cells (25,26). Prolonged light stress leads ultimately to loss of PS II reaction center activity (Fig 3) and chlorophyll bleaching.

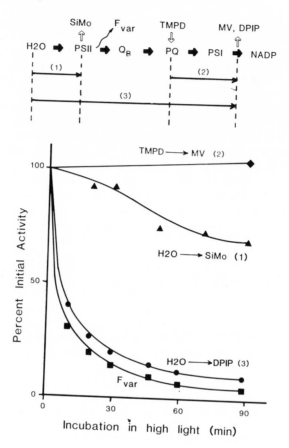

Figure 3.   The primary site of photoinhibition damage is the Q_B complex of PS II. Chlamydomonas cells were incubated at a concentration of 30 mg Chl/ml in high light (2,000 $\mu E/m^2/sec$) for various periods of time, before removal for sonication and measurement of thylakoid photochemical activities as described in ref. 16. PS I assay, O_2 uptake from the Mehler reaction of methyl viologen (mv) using N,N,N'N'-tetra methyl phenylenediamine (TMPD) as electron donor; PS II assay, spectrophotometrically monitored photoreduction of silicomolybdate (SiMo); whole chain assay (through Q_B) was 2,6-dichloroindophenol (DPIP) photoreduction using water as endogenous electron donor.

The binding of the plastoquinone moiety Q_B to the apoprotein can be affected by many photosynthetic herbicides through competition of the herbicide with the plastoquinone at the binding site (27). Using a radiolabelled azido derivative of such an herbicide (^{14}C-azido atrazine) the quinone binding protein has been identified as a polypeptide migrating with an approximate molecular weight of 32,000 daltons on SDS-polyacylamide gels (28). A protein of similar molecular weight has been identified as the product of a gene which is light-induced (a "photogene") during greening of maize (29). In addition, a PS II polypeptide of approximately 32,000 daltons has been shown to be rapidly synthesized and degraded (i.e. rapid turnover) in the light, but not in darkness, in the alga Spirodella (30) and other species (31). Steinback et al (32) have shown that proteolytic digestion patterns of the rapid-turnover protein, the light-induced photogene product and the herbicide binding protein are identical. We, therefore, conclude that the apoprotein of Q_B is rapidly turned over in the light but not in darkness.

Is the rapid turnover of the Q_B protein related to photoinhibition? Chlamydomonas cells, pulse labelled with [35]SO_4 for a short time (15 min) under light conditions for optimal growth, incorporate a large proportion of the added label into the Q_B protein (33,34). Following a certain interval of time [1-2 hr; to allow the newly synthesized Q_B protein to integrate into a functional PS II center], we have monitored the rate of loss of this label from the photosynthetic membrane at various light intensities. At nominal light intensities for growth the half-time for turnover is about 6-8 hr. Under photoinhibitory light intensities (10x growth intensities) this half time is shortened to about 90 min. Similar results have been reported for Spirodella (35). Thus, the more severe the light stress (photoinhibition), the more rapid is the turnover of the Q_B protein. When the rate of damage of this protein exceeds the rate of repair, or replacement of the damaged protein, loss of PS II function is observed.

Although high photon flux densities can increase the rate of damage, loss of PS II function can also occur by decreasing the rate of repair. We have identified three different methods by which the rate of repair can be limited (Table 1): (1) The Q_B protein is encoded by chloroplast DNA and translated on 70 S ribosomes. Thus, de novo synthesis can be inhibited by chloramphericol. (2) A reduction in the available ATP [required for protein synthesis and, perhaps, removal of the damaged protein] through the use of inhibitors such as potassium cyanide, carbonyl cyanide m-chlorophenyl hydrazone (CCCP), dark incubation [no photophosphorylation], or severe secondary damage to the photosynthetic membrane by high light and low temperatures all restrict the rate of repair of damaged protein. (3) Addition of the protease inhibitor phenylmethylsulfonylflouride (PMSF) prevents the loss of damaged Q_B protein suggesting the involvement of a serine protease in the removal of damaged Q_B protein. Recovery may be limited under these latter conditions by the inability of the newly synthesized replacement protein to integrate with PS II until the damaged protein is removed.

TABLE 1

addition	% of recovery in the light with no additions
light	100
light + CHI	80
light + CAP	10
light + KCN	3
light + CCCP	5
dark	24
light + PMSF	14

Requirement for chloroplast-directed protein synthesis, ATP and protease activity for recovery of photoinhibited Chlamydomonas cells. Cells were incubated under high light (2,000 μ E/m^2/sec) for 90 min and allowed to recover in optimal growth light conditions (200 μ E/m^2/sec). Recovery was monitored as room temperature variable flourescence (as in ref. 16) following 7 hr of recovery. Values are tabulated as a percentage of the light-recovered variable flourescence. Additions included cycloheximide (CHI; 2μg/ml); chloramphenicol (CAP; 200μg/ml); KCN (10mM); carbonyl cyanide m-chlorophenylhydrazone (CCCP; 10 μM) phenylmethylsulfonyl fluoride (PMSF; 2mM).

Pulse labelling with $^{35}SO_4$ under the various
conditions of recovery listed above, allowed us to clearly
correlate recovery of PS II activity following
photoinhibition damage with the de novo synthesis and
insertion of the Q_B protein (34). Since the mRNA from the
chloroplast gene coding for the Q_B protein is the most
abundant mRNA species in higher plant chloroplasts (29,32),
the   chloroplast is "primed" for conditions in which rapid
synthesis of this particular protein is required. Such
conditions occur following illumination.

How is this protein damaged in light? Any inhibitor or
condition which leaves the plastoquinone pool in the reduced
state accelerates the damage. Two such experiments have
been done using mutants of Chlamydomonas defective in CO_2
metabolism (kindly donated by Dr. M. Spalding, Michigan
State Univ.). The mutant pmp-1 is unable to transport bicarbonate across the plasmalemma and at atmospheric CO_2 concentrations, difusion of CO_2 cannot maintain high levels
of CO_2 fixation (36). Thus, regeneration of the terminal
acceptor of electron transport ($NADP^+$) is   limited, and
under light intensities for optimal growth of wild type
cells, the plastoquinone pool becomes reduced and photoinhibition occurs. Under CO_2-enriched conditions, however,
(5% CO_2 in air) the high diffusion of CO_2 into the cell
allows CO_2 fixation to occur normally, and the pmp-mutant
is no longer light sensitive. Similar results are observed
for a mutant deficient in carbonic anhydrase; an enzyme required for conversion of the transported bicarbonate into
CO_2 for use in carbon fixation (Spalding and Kyle, unpublished observations). The use of a specific inhibitor of
carbonic anhydrase, ethoxyzolamide (EZA), with wild type
cells further mimics the results obtained with the mutants.

Addition of inhibitors which bind to the Q_B protein
(ie. atrazine or diuron) and displace the plastoquinone
moiety, result in an oxidized PQ pool in the light and have
been shown to offer protection against photoinhibition
damage (25). DBMIB, which also blocks electron flow but
results in a reduced PQ pool, accelerates photodamage. We
therefore propose that it is the residence of a quinone
anion radical (either QB^- or $QB^=$) on the quinone binding
protein which generates the damaging condition (Fig 4). If
the damaging interaction of the radical with the protein is
directly proportional to the total residency time of the
radical on the protein, then it is clear that any increase
in the residency time should increase the probability of a

damaging interaction, resulting in turnover. This residency
time can be increased by either reduction of the plastoquinone pool resulting in an increase in the equilibrium
concentration of $Q_B^=$ (Fig 4), or by increasing the rate
of electron flow through Q_B (high light intensities). In
this latter case, the residency time for the quinone radical
of the protein will double in a given unit of time if the
electron flow rate doubles (i.e. two times as many
turnovers). We, therefore, conclude that the turnover of

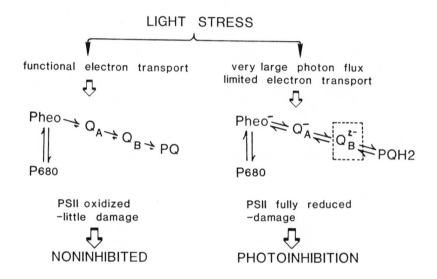

Figure 4. Summary of light stress events leading to
photoinhibition. Optimal light intensities and functional
electron transport allow the deexcitation of the energy
arriving at PS II via electron transport. Under these
conditions, avoidance and tolerance strategies operate to
prevent any loss of photosynthetic capacity. During a very
large photon flux, or if electron transport out of the
plastoquinone pool is limited in any way (ie, by cessation
of CO_2 fixation), the residency time for the reduced
quinone anion radicals on the apoprotein increases. The
damaging species (outlined in the broken line) increases the
rate of turnover of the apoprotein, blocking photosynthetic
electron transport at the position of Q_B.

the Q_B protein is a natural consequence of its enzymatic function and, as it is an integral component of electron transport, photoinhibition or light-induced loss of activity occurs when the repair mechanism cannot keep pace with the damage . Chloroplasts have the capacity to tolerate a high rate of damage to this protein due to the abundance of endogenous mRNA levels coding for replacement protein.

The biochemical mechanism of damage to the Q_B protein is presently unknown. The evidence above suggests the involvement of semiquinone anion radicals, but we are unable to exclude the possibility of an interaction of the semiquinone radical with molecular oxygen and the subsequent generation of damaging oxygen or hydroxyl radicals at the quinone binding site. Are there readily oxidizable components in the quinone biding site? The quinone binding site has not been categorically identified but comparisons of the amino acid sequences of the quinone binding proteins from organisms as phylogically diverse as Amaranthus hybridus (37), Spinacia (38), Anabaena (39), Euglena (40) and Rhodopseudomonas capsulata (41) suggest the involvement of two methionine-histidine containing regions which are totally conserved in all the above species (Fig 5). For these reasons Hirschberg et al (42) and Youvan et al (41) have suggested that these conserved regions may represent the quinone binding site(s). Although it would be highly speculative to suggest that the photodamage to this protein is due to radical attack on the methionine (forming methionine sulfoxide) or histidine (introducing a carbonyl group) in the putative quinone binding site, it is our working hypothesis. Levine (43) has recognized that the oxidative modification of the enzyme glutamine synthetase renders that protein not only nonfunctional, but susceptable to protease attack. Upon analysis of the damaged glutamine synthetase, only a single amino acid (a histidine residue) was found to be oxidatively modified. We are presently looking for a similar oxidative modification in the Q_B protein.

CONCLUSION

In this manuscript we have attempted to emphasize the importance of light stress in photosynthetic productivity. Many environmental stresses such as drought or chilling may result in a cessation of photosynthetic productivity by

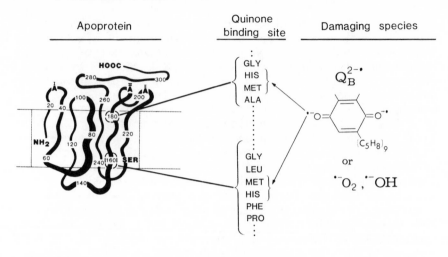

Figure 5. Hypothetical mechanism of damage of the Q_B
protein. The schematic model for the three dimensional
arrangement of the Q_B protein is an extension of that
proposed in ref. 42. This model was based on the amino acid
sequence deduced from the DNA sequence of <u>Amaranthus</u>
<u>hybridus</u> (37). We have indicated ARG-28, <u>ARG-189 and</u>
<u>ARG-202</u> as possible tripsin digestion sites. Also indicated
is the highly conserved, putative quinone binding sites
(PRO-160 to GLY-165 and ARG-177 to GLY-180) in this sequence
[note, these regions correspond to residues 172 - 177 and
189 - 192 in ref. 41]. In this model, we suggest a radical
attack on the HIS-MET pair of one of these regions either
directly by the semiquinone anion or through the involvement
of oxygen radicals. We recognize, however, that other
residues, such as SER-228 (see ref. 37) may also be involved
in the definition of this binding site.

preventing CO_2 assimilation either through a decrease in
the CO_2 substrate concentrations (following stomatal
closure) or by biochemical alterations in chloroplast
membranes (low water potential effects on chloroplast
coupling factor; J. Boyer, this volume). Although these
stresses, in moderation themselves do not damage the plant,
a condition of light sensitivity may ensue as a consequence
of the reduced ability for PS II to undergo an organized

deexcitation via electron transport when the terminal acceptor (NADP+) is no longer being regenerated (4). Under these conditions even moderate photon flux densities could result in physical damage to the photosynthetic membrane. The repair of this damage requires energy and de novo protein synthesis and will limit photosynthetic capacity when CO_2 fixation resumes (stomatal opening or alleviation of the environmental stress responsible for loss of CO_2 fixation). Photoinhibition, therefore, may represent a significant limitation to overall productivity when considering daily fluctuations in environmental parameters (especially photon flux density). By understanding the mechanisms utilized by the plant to compensate for conditions of light stress in biochemical terms, we will have greater insight into approaches toward maximizing these compensating strategies. An increase in ability to compensate for light stress should be reflected in an increased survivability in marginal agriculture land use areas.

REFERENCES

1. Myers J, Burr G (1940). Studies on photosynthesis: some effects of light at high intensity on Chlorella. J Gen Physiol 24:45.
2. Kok B (1956). On the inhibition of photosynthesis by intense light. Biochim Biophys Acta 21:234.
3. Powles S (1984). Photoinhibition of photosynthesis induced by visible light. Ann Rev Plant Physiol (in press).
4. Osmond B (1981). Photorespiration and photoinhibition: some implications for the energetics of photosynthesis. Biochim Biophys Acta 639:77.
5. Powles S, Bjorkman O (1982). High light and water stress effects on photosynthesis in Nerium oleander. Carnegie Institution of Washington Yearbook 81:75.
6. Powles S, Berry J, Bjorkman O (1983). Interaction between light and chilling temperature on the inhibition of photosynthesis in chilling-sensitive plants. Plant Cell & Environ 6:117.
7. Spreitzer RJ, Mets L (1981). Photosynthesis deficient mutants of Chlamydomonas reinhardii with associated light sensitive phenotypes. Plant Physiol 67:565.

8. Powles S, Bjorkman O (1981). Leaf movement in the shade species Oxalis oregana. II role in protection against injury by intense light. Carnegie Institution of Washington Yearbook 80:63.
9. Haupt W (1982). Light-mediated movement of chloroplasts. Ann Rev Plant Physiol 33:205.
10. Murata N (1969). Control of excitation energy transfer in photosynthesis. Biochim Biophys Acta 172:242.
11. Bonaventura C, Myers J (1969). Flourescence and oxygen evolution from Chlorella pyrenoidosa. Biochim Biophys Acta 189:366.
12. Bennett J (1977). Phosphorylation of chloroplast membrane polypeptides. Nature 269:344.
13. Allan JF, Bennett J, Steinback KE, Arntzen CJ (1981). Chloroplast protein phosphorylation couples plastoquinone redox state to distribution of excitation energy distribution between photosystems. Nature 291:21.
14. Haworth P, Kyle DJ, Horton P, Arntzen CJ (1982). Chloroplast membrane protein phosphorylation. Photochem Photobiol 36:743.
15. Kyle DJ, Staehelin LA, Arntzen CJ (1983). Lateral mobility of the light harvesting complex in chloroplast membranes controls excitation energy distribution in higher plants. Arch Biochem Biophys 222:527.
16. Wollman F-A, Olive J, Bennoun P, Recouvreur M (1980). Organization of photosystem II centers and their associated antennae in the thylakoid membranes. J Cell Biol 87:728.
17. Mullet JE, Arntzen CJ (1980). Simulation of grana stacking in a model membrane system. Biochim Biophys Acta 589:100.
18. Kyle DJ, Kuang T-Y, Watson JT, Arntzen CJ (1984). Movement of a sub-popluation of the light harvesting complex (LHCD) from grana to stroma lamellae as a consequence of its phosphorylation. Biochim Biophys Acta (in press).
19. Chow WS, Telfer A, Chapman DJ, Barber J (1981). State I - State II transition in leaves and its association with ATP-induced chlorophyll flourescence quenching. Biochim Biophys Acta 638:60.

20. Mullet JE, Baldwin TO, Arntzen CJ (1982). A mechanism for chloroplast thylakoid adhesion mediated by the Chl a/b light harvesting complex. In Akoyunoglou G (ed): "Photosynthesis III. Structure and molecular organization of the photosynthetic apparatus", Philadelphia: Baldban International Science Services, p 577.
21. Kyle DJ, Arntzen CJ (1982). Thylakoid membrane protein phosphorylation selectively alters the local membrane surface charge near the primary acceptor of photosystem II. Photochem Photobiol 5:11.
22. Jursinic P, Kyle DJ (1983). Changes in the redox state of the secondary acceptor of photosystem II associated with light-induced thylakoid protein phosphorylation. Biochim Biophys Acta 723:37.
23. Horton P, Lee P (1984). Stimulation of a cyclic electron transfer pathway around photosystem II by phosphorylation of chloroplast thylakoid proteins. FEBS 162:81.
24. Critchley C (1981). Studies on the mechanism of photoinhibition in higher plants. Plant Physiol 67:1161.
25. Arntzen CJ, Kyle DJ, Wettern M, Ohad I (1983). Photoinhibition: A consequence of the accelerated breakdown of the apoprotein of the secondary acceptor of photosystem II. In Hallick R, Staehelin LA, Thornber JP (eds): "Biosynthesis of the photosynthetic apparatus: molecular biology, development and regulation". UCLA Symposium Series (in press).
26. Kyle DJ, Ohad I, Arntzen CJ (1984). Membrane protein damage and repair. I. Selective loss of a quinone protein function in chloroplast membranes. Proc Natl Acad Sci USA (in press).
27. Vermaas WFJ, Arntzen CJ, Gu L-Q, Yu C-A (1983). Interactions of herbicides and azidoquinones at a photosystem II binding site in the thylakoid membrane. Biochem Biophys Acta 723:266.
28. Pfister K, Steinback KE, Gardner G, Arntzen CJ (1981). Photoaffinity labelling of an herbicide receptor protein in chloroplast membranes. Proc Natl Acad Sci USA 78:981.
29. Bogorad L (1981). Chloroplasts. J Cell Biol 91:256s.
30. Reisfeld A, Matoo AK, Edelman M (1982). Processing of a chloroplast translated membrane protein in vivo. Eur J Biochem 124:125.

31. Hoffman-Falk H. Matoo AK, Marcer JB, Edelman M, Ellis RJ (1982). General occurance and structural similarity of the rapidly synthesized 32,000 dalton protein of the chloroplast membrane. J Biol Chem 257:4583.
32. Steinback KE, McIntosh L, Bogorad L, Arntzen CJ (1981). Identification of the triazine receptor protein as a chloroplast gene product. Proc Natl Acad Sci USA 78:7463.
33. Wettern M, Owen GC, Ohad I (1983). Role of thylakoid polypeptide phosphorylation and turnover in the assembly and function of photosystem II. Methods Enzymol 97:554.
34. Ohad I, Kyle DJ, Arntzen CJ (1984). Membrane protein damage and repair II. Removal and replacement of inactivated 32 kilodalton polypeptides in chloroplast membranes. J Cell Biol (in press).
35. Edelman M, Matoo AK, Marder JB (1983). Three hats of the rapidly metabolized 32 kilodalton protein of the thylakoids. In Ellis RJ (ed): "Chloroplast biogenesis" Cambridge University Press (in press).
36. Spalding MH, Spreitzer RJ, Ogren WL (1983). Genetic and physiological analysis of the CO_2 concentrating system of Chlamydomonas reinhardii. Planta 159:261.
37. Hirschberg J, McIntosh L (1983). Molecular basis of herbicide resistance in Amaranthus hybridus. Science 222:1346.
38. Zurawski G, Bohnert HJ, Whitfield PR, Bottonly W (1982). Nucleotide sequence of the gene for the Mr 32,000 thylakoid membrane protein from Spinacia oleracea and Nicotiana debneyi predicts a totally conserved primary translation product of Mr 38,950. Proc Natl Acad Sci USA 79-7699.
39. Curtis SE, Haselkorn R, personal communication.
40. Hollingsworth MJ, Johanningmeier U, Karabin GD, Stiegler GL, Hallick RB (1984). Detection of multiple, unspliced presursor mRNA transcripts for the Mr 32,000 thylakoid membrane protein from Euglena gracilis chloroplasts. Nucleic Acids Res 12:2001.
41. Youvan DC, Bylina EJ, Alberti M, Begusch H, Hearst JE (1984). Nucleotide and deduced polypeptide sequences of the photosynthetic reaction center, B870 antennae and flanking polypeptides from Rps capsulata. J Cell Biol: (in press).

42. Hirschberg Y, Bleecker A, Kyle DJ, McIntosh L, Arntzen CJ (1984). The molecular basis of triazine resistance in higher plant chloroplasts. Z Naturforsch: (in press).
43. Levine RL (1983). Oxidative modification of glutamine synthase. I. inactivation is due to loss of one histidine residue. J Biol Chem 258:11823.

Cellular and Molecular Biology of Plant Stress, pages 71–92
© 1985 Alan R. Liss, Inc.

BETAINE ACCUMULATION: METABOLIC PATHWAYS AND GENETICS[1]

Andrew D. Hanson and Rebecca Grumet

MSU-DOE Plant Research Laboratory, Michigan State University
East Lansing, Michigan 48824

ABSTRACT   Advances in molecular genetics of higher
plants make it timely to identify and dissect simple
metabolic   traits   that confer stress   tolerance.   In
certain groups of plants, including the Chenopodiaceae
and the grass tribe Hordeae, water and salt stress
provoke   accumulation of betaine (glycinebetaine).
Circumstantial evidence indicates that the stress-
induced betaine accumulation is adaptive; betaine
accumulated by stressed cells may function either as a
non-toxic cytoplasmic osmoticum, as a stabilizing
compound, or as both. Betaine accumulation is thus a
potential metabolic component of stress tolerance that
may prove simple enough to manipulate by genetic
engineering.   Questions that must be addressed to
assess   the   potential worth and feasibility   of
manipulating betaine content include the following:
        1.   Is betaine accumulation really an adaptive and
heritable   trait? No direct evidence yet shows that
within a species, naturally occuring levels of betaine
are beneficial during stress,   or that raising these
levels   will   improve   stress   tolerance.   We   are
therefore using natural variation within the barley
gene   pool to develop isopopulations differing in
betaine   level in order to   test performance during
stress.   We also   investigated the inheritance of
betaine level in barley and found betaine content to
be a highly heritable trait controlled by nuclear
genes.

[1]This work was supported by the U.S. Dept. of Energy
under Contract No. DE-AC02-76-ER01338, and by the Rackham
Foundation.

2.   What are the enzymes and genes unique to the
betaine synthesis pathway?   Radiotracer experiments
demonstrate that betaine is synthesized in chenopods
and barley via a two-step oxidation of choline. There
is some evidence that the first step is regulated by
stress.   We have obtained in vitro oxidation of
choline   by   lysates of spinach and   beet   leaf
protoplasts;   choline oxidizing activity is associated
with chloroplasts and is promoted by light.

INTRODUCTION

In principle, traits conferring adaptation to water or
salt stress can be assigned to one of four levels of
organization: developmental or phenological;   structural;
physiological;   and biochemical or metabolic.   These are
shown in the inverted triangle of Figure 1. At the highest
level of organization are developmental or phenological
traits, such as the timing of flowering in relation to
seasonal rainfall patterns. Note that to date, most of the
improvement in adaptation of major crops such as wheat and
barley to dry regions has come from manipulations at the
developmental level, e.g. by breeding for earlier
flowering. (1).   Very little recourse has been had to
traits at any lower level.
At a second level of organization, and at a second
level of use in breeding for stress resistance, are
structural   traits.   These   range from gross   plant
architecture (e.g. awns on wheat ears, ref. 2) to fine
anatomy (e.g. metaxylem vessel diameter in wheat seminal
roots,   ref. 3).   Traits at these two upper levels,
developmental and structural, can be expressed only in
whole plants and are often multigenic. These traits have
been, and are being, successfully exploited by conventional
plant   breeding techniques using natural   genetic
variability.   Consequently genetic engineering has as yet
little special value for manipulating developmental and
structural stress-resistance characters.
Like the upper two trait levels, physiological traits
are capable of expression only in whole plants or whole
organs.   Such traits hold much interest for genetic
improvement of stress resistance;   promising examples
include osmotic adjustment (4) and water use efficiency
(5). However, physiological adaptations to stress occupy a
no-man's-land between traits that can be handled with

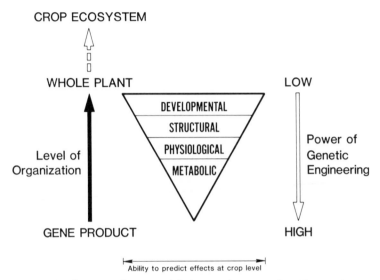

FIGURE 1. Levels at which traits conferring adaptation to stress-prone environments can be expressed.

regular screening and breeding methods, and traits simple enough to be construed in molecular terms. This unfortunately means that they are not readily suited either to conventional or to molecular genetic approaches at present.

It is at the lowest organizational level, metabolic or biochemical, that there is the best chance of defining stress-resistance traits that are governed by very small numbers of genes, and for which selection might be practiced at the cell culture level (6). So, it is with metabolic traits that molecular-genetic tools could be most effective, were it not for the following inevitable, and ironic, problem. Regardless of the level at which an adaptive trait is expressed, to be agronomically useful the trait must have a favorable effect that will be translated upwards to reach not only the level of the whole plant, but also the still higher level of the crop community. For genetic manipulations at the phenological or developmental level, our ability to predict the effect on crop performance is fairly good; at the metabolic or biochemical level--to which molecular genetics is most applicable--it is almost zero. Figure 1 restates this problem using an inverted triangle. The width of the triangle is proportional to our ability to predict the effect of manipulating traits at the various levels on crop performance in the field. Since the characters most apt for genetic engineering are the least predictable in terms of impact on crop performance, genetic engineering of stress-resistance will necessarily be a series of shots in the dark.

Despite this limitation, it seems worthwhile (a) to seek simple metabolic pathways that are associated with stress-tolerance; (b) to attempt to use conventional genetic tools to confirm their adaptive worth at the crop community level, and (c) to begin characterizing the enzyme proteins, and hence the genes involved. We can then hope to have candidate "stress-resistance" genes in hand for gene transfer experiments. Betaine accumulation in certain plant taxa may be a stress-resistance trait suitable to genetic engineering efforts, since it meets the criteria of relative metabolic simplicity and of strong general association with stress-tolerance. Our laboratory is therefore attempting to confirm the adaptive worth of betaine accumulation and to characterize the betaine pathway enzymes. Below, we review the indirect evidence for the adaptive value of betaine accumulation, and outline our physiological-genetic approach to testing directly for adaptive value. We then summarize the steps in betaine biosynthesis and describe recent progress towards isolating the biosynthetic enzymes. These enzymes appear to be chloroplastic. We end by speculating on the potential advantages of genetic engineering of chloroplast or mitochondrial metabolism via the organelle genome.

INDIRECT EVIDENCE FOR ADAPTIVE VALUE OF BETAINE ACCUMULATION

Comparative Physiology and Physiological Ecology

Among crop families, significant levels (≥ 5 umol.g^{-1} dry wt) of betaine are accumulated by chenopods, amaranths and grasses, but by few, if any, other economic taxa (7).

$$CH_3-\overset{\overset{\displaystyle CH_3}{\displaystyle \overset{\oplus}{|}}}{\underset{\underset{\displaystyle CH_3}{\displaystyle |}}{N}}-CH_2-COOH$$

Betaine
(Glycinebetaine)

Within the grasses, betaine accumulation is marked in
certain tribes (e.g. Hordeae), but far less so in others
(e.g. Maydeae, Oryzae, ref. 8). In vegetative tissues of
both chenopods and grasses there is a basal level of
betaine present in unstressed plants, and an additional
component inducible by salt- or water-stress. In salt-
stress, the stress-induced betaine accumulation is closely
and linearly correlated with external salinity and the
solute potential (ψ_S) of the tissue (e.g. 9,10), This
connotes a precisely regulated response to stress in the
environment. Further, there is a general ecological
association between betaine accumulation and adaptation to
saline environments, and possibly to dry environments (7).

Biochemistry and Bioenergetics

Betaine is accumulated by grasses during long term
field water deficits (11,12) and by grasses and chenopods
during progressive salinization (10,13); during such
gradually developing stresses, betaine is synthesized de
novo continuously over many days (12). Because betaine is
an energy rich compound, and the carbon and energy income
of the stressed plant becomes progressively limited, a
significant fraction of the restricted energy budget can be
accounted for by betaine synthesis. For example, in water
stressed barley in the field, the energy demand for betaine
synthesis approaches the energy demand for protein turnover
(12). Some value to the plant either of betaine itself, or
of the process of synthesizing betaine, is implied.

Cell Physiology

The cytoplasmic osmoticum hypothesis. In general,
quaternary ammonium compounds such as betaine are found
throughout the living world as osmolytes in organisms
subject to salinity or desiccation (14). Wyn Jones has
proposed that betaine acts in plants as a benign or
compatible osmoticum preferentially accumulated by the
cytoplasm of stressed cells to maintain cytoplasmic ψ_S in
equilibrium with vacuolar ψ_S , which in turn is lowered by
salt accumulation (7). This hypothesis requires that
betaine be primarily localized in the cytoplasm, and that
plant metabolism be far less sensitive to betaine than to
salts. Evidence supports both propositions (15,16,17).

Specific protective effects of betaine in high salt
environments.   There are indications from work with higher
plant tissues (18,19), bacteria (20) and N_2-fixing
rhizobium/legume associations (21) that exogenous betaine
may have a positive protective role during salt stress,
i.e. that it is not merely benign with respect to metabolic
function.

ISOPOPULATION DEVELOPMENT: TESTING ADAPTIVE VALUE OF BETAINE

The indirect evidence cited above constitutes a prima
facie case for some adaptive value of betaine accumulated
in response to salt-stress or water-stress.   However, it
proves none of the following points:
#1.   That the levels of betaine now present in crops like
barley are beneficial during stress;
#2.   That increasing the level in barley would enhance
overall crop stress-resistance;
#3.   That genetically engineering a high betaine-
accumulating capacity into a non-accumulating crop, e.g.
corn, would augment the relatively low salt tolerance of
this crop.
One way to test possibility #1, that the present
levels of betaine are beneficial during stress, is to seek
mutants blocked early in the betaine synthesis pathway.
This is of interest also for biochemical pathway studies
and we are currently screening an azide-mutagenized barley
population for betaine nulls.   However, even were such null
mutants recovered and found to be abnormally stress-
sensitive, this would give no information about the effects
of increasing the level, or of introducing betaine to a
species that does not normally accumulate it.   For this
reason we have undertaken a breeding program, based on
natural variation for betaine level in barley, with the aim
of developing isopopulations differing only in betaine
level.
By constructing populations genetically equivalent for
all traits other than betaine and testing them in
laboratory and field stress environments, we establish an
experimental system for examining the effects of betaine
independently of other stress-related traits.   Isolating
the effects of betaine from other stress-related traits is
crucial because of the difficulty in extrapolating
metabolic changes to a whole plant or crop community level.
Care was therefore taken when creating the isopopulations

PROCEDURE FOR
MODIFIED ISOPOPULATION DEVELOPMENT

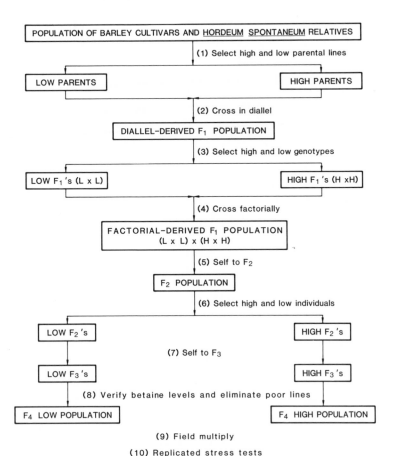

FIGURE 2. Summary of procedure used to develop high and low betaine isopopulations for stress testing.

to prevent bias due to non-betaine traits. The usual isopopulation method was modified in two ways. Many parents (13 instead of 2) were chosen to found the populations and two sets of hybridizations (a 13 x 13 diallel and a 7 x 7 factorial) were used instead of a

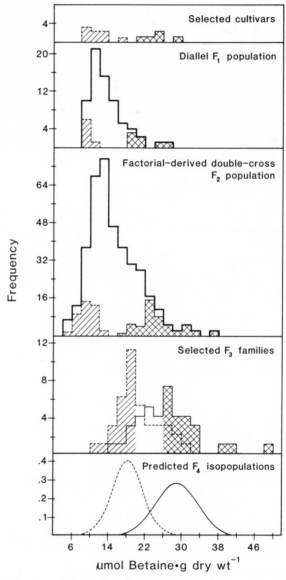

FIGURE 3. Distribution of betaine contents during isopopulation development. Distributions in the bottom frame are predicted using $R=ih^2\sigma$. Hatched areas = selected low and high genotypes. Thick lines = progeny from high x low crosses. Thin broken and solid lines = progeny from selected low and high genotypes.

single high x low cross. These changes increase the range of non-betaine traits within the populations and provide two opportunities to break linkages and recombine genes. The steps in preparation for isopopulation development are listed below; the crossing scheme is shown in Figures 2 and 3.

- Identification of genotypic variability for betaine level. Seventy-one cultivars of <u>Hordeum</u> <u>vulgare</u> and 268 lines of the wild progenitor <u>H.</u> <u>spontaneum</u> were screened in controlled environment tests using young (4-leaf) plants (22). About a three-fold range of betaine level was found in both wild and cultivated types.

- Confirmation that the genotypic differences in betaine level measured in controlled environment tests were retained in field water stress. In 13 diverse cultivars, betaine contents in the field and controlled environments were highly correlated (r=0.71**) (22).

- Choice of genotypes to found the isopopulations; five high and eight low betaine cultivars were selected.

We are now at the stage of advancing selected high and low betaine F_3 families to the F_4 and F_5 generations for stress testing. As can be seen from Figure 3, the roughly two-fold difference in betaine level between the high and low parents can be expected to persist between the final high betaine and low betaine isopopulations. In principle these populations will have a broad genetic base and should differ little in characters other than betaine accumulation. Although not large, the difference in average betaine level between the populations will come close to the extreme natural range for the species as it exists today and as it existed before domestication. This is acceptable for tests of adaptive value.

The above outline of isopopulation development has been included not only to show how adaptive value can be tested, but also to underscore three points. First, it is a long process. Second, the final populations cannot be as different in betaine content as we would like them to be. Third, both these drawbacks could be overcome were it possible to specifically manipulate the betaine trait by genetic engineering. Conceptually, genetic engineering of a normal cultivar to give low and high betaine types is jumping straight from the top frame of Figure 3 to the bottom frame, thereby saving perhaps three years.

In conjunction with the development of isopopulations, we have made a formal inheritance study of the betaine trait in barley (23). The results indicate that betaine content is a highly heritable (h ≈ 0.53), predominantly additive trait governed by nuclear genes. It is worth noting that these nuclear genes are identifiable solely via an effect on the level of a pathway end-product (betaine); they are presumably regulatory rather than structural. Although such genes are very useful in plant breeding, they are unlikely to help either with biochemical-genetic dissection of the steps in the betaine synthesis pathway, or with isolation of the genes for the betaine pathway enzymes.

BETAINE BIOSYNTHESIS: RADIOTRACER EVIDENCE FOR PATHWAY STEPS

The steps in the betaine synthesis pathway are shown in Figure 4. This figure is based on radiotracer results with leaf tissue of barley, sugarbeet and spinach. As in mammals, and also microorganisms, betaine in grasses and chenopods is derived from the oxidation of choline rather than from methylation of glycine (24). There are differences, however, among the pathways of plant taxa and between plants and other organisms; such differences suggest multiple evolutionary origins.

In mammals the stepwise methylations leading from ethanolamine to choline occur at the level of phospholipid-bound bases (24), but in plants these methylations involve water-soluble intermediates, probably phosphate esters of the bases (25,26,27). In chenopods the product of the methylation sequence, phosphorylcholine, ((P) -choline) is processed differently from the route in grasses. In sugar beet, (P) -choline is hydrolyzed directly to free choline, but in barley (P) -choline is first incorporated into the phospholipid, phosphatidylcholine (PC) before free choline is released by a phospholipase D-type reaction (26,27). Figure 5 shows the type of radiotracer data from which this distinction between chenopods and grasses is drawn. When fed [^{14}C] (P) -choline, beet leaf tissue rapidly accumulates [^{14}C]betaine; PC labels only as a minor end-product. In barley leaf tissues a large intermediate pool of [^{14}C]PC builds up before label passes on to betaine via choline.

Although they differ in how free choline is derived from (P) -choline, grasses and chenopods share the feature that the free choline so derived is tightly compartmented away from the bulk pool of free choline in the leaf, and is destined for rapid oxidation to betaine (26,27). It is possible that the PC step in grasses and (P) -choline hydrolysis in chenopods correspond to processes for transporting choline into a subcellular compartment containing the choline oxidizing system (26,27). In both barley and spinach, choline is oxidized via a small pool of betaine aldehyde to betaine (25,28); the oxidation steps are physiologically irreversible (28).

Betaine is quite metabolically inert; on a time-scale of days it is not catabolized. It is, however, readily translocated from its site of synthesis in the leaves to roots and to shoot growing points (10,29).

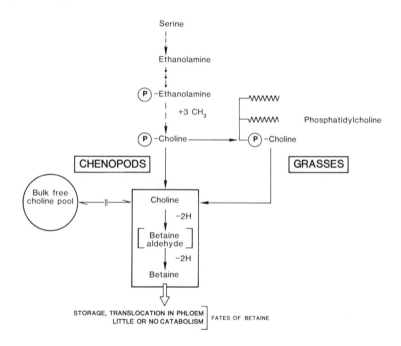

FIGURE 4. Summary of radiotracer evidence on betaine metabolism in chenopods (sugarbeet, spinach) and grasses (barley). The box surrounding the last two steps (choline oxidation to betaine via a very small pool of betaine aldehyde) denotes tight compartmentation of the choline destined for oxidation to betaine.

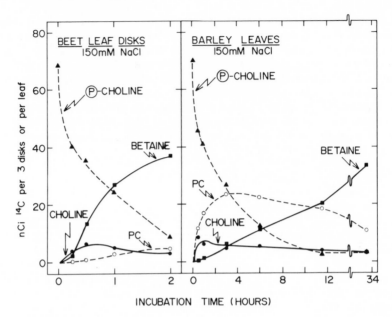

FIGURE 5. Metabolism of a 70 nCi (7.4 nmol) dose of [^{14}C] Ⓟ-choline by 11-mm disks cut from leaves of sugarbeet or 10-cm sections of barley leaves. Plants were salinized gradually to 150 mM NaCl before the experiment. Note the different time axes for the two species.

The two choline oxidation steps appear to be the only steps unique to the betaine synthesis path, inasmuch as Ⓟ-choline is a primary metabolite, and free choline is a very common, if not ubiquitous, plant constituent. Consistent with this, roots and hypocotyls of sugarbeet actively incorporate label into all the compounds in Figure 4, except betaine aldehyde and betaine, when supplied with [^{14}C]ethanolamine or [^{14}C]formate (a methyl precursor) (10). Since the choline→ betaine aldehyde reaction is the formal committed step in betaine synthesis, pathway regulation at this point is expected. There is in vivo radiotracer evidence for stress enhancement of this step in water-stressed barley leaves (28). Radiotracer experiments implicate additional regulation upstream from choline oxidation in barley and also suggest major regulatory steps upstream from choline in chenopods (25,26,28).

To summarize, the betaine synthesis path probably comprises two irreversible steps, of which the first may be modulated by stress. Since the end-product betaine is degraded slowly or not at all, control over its level within the plant must be exerted mainly via the rate of its synthesis. If betaine level is indeed positively correlated with crop stress tolerance, then the controlled expression of just two genes might affect crop performance during stress.

ENZYMES OF BETAINE SYNTHESIS

In mammalian liver, choline is oxidized to betaine in mitochondria. The choline → betaine aldehyde step is catalyzed by choline dehydrogenase, a flavoprotein of the inner membrane (30). The betaine aldehyde is then oxidized to betaine by a mitochondrial NAD-linked betaine aldehyde dehydrogenase (31). An analogous dehydrogenase system operates in Pseudomonas aeruginosa, where the choline dehydrogenase is associated with the respiratory electron transport chain (32). Other microorganisms such as Alcaligenes sp. and the fungus Cylindrocarpon have flavoprotein "choline oxidases" which can oxidize both choline and betaine aldehyde; these oxidations involve molecular oxygen and generate H_2O_2 (33,34). Comparative biochemistry would therefore indicate a mitochondrial or perhaps peroxisomal location for choline oxidation in leaves.

Until recently, in vitro choline oxidation has not been observed in plants, although low activities of an NAD-linked cytosolic betaine aldehyde dehydrogenase have been reported in spinach (35). To avoid problems of choline oxidation by leaf microflora (36) and to secure gentle cell breakage, we have used protoplasts from young spinach or sugarbeet leaves isolated under sterile conditions by standard methods (37,38). In darkness, intact protoplasts readily oxidize tracer [^{14}C]choline to betaine aldehyde and betaine (Fig. 6). Although the [^{14}C]choline oxidation rates were lower than the in vivo rate of betaine synthesis in unstressed spinach leaves (20 nmol.mgchl^{-1}.h^{-1}, ref. 25), substantial dilution of the tracer labeled substrate by endogenous choline is likely (25). Protoplast lysates obtained by passage through 15um nylon mesh were also able to oxidize [^{14}C]choline in darkness, but the proportion of [^{14}C]betaine aldehyde formed was higher than

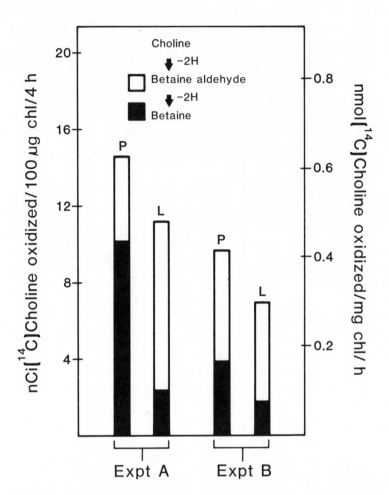

FIGURE 6. Conversion of [^{14}C]choline to betaine aldehyde and betaine by two preparations of spinach leaf protoplasts (P) and the lysates (L) derived therefrom. Reaction mixtures (50 ul) contained 0.3 mM [^{14}C]choline (58 nCi.nmol^{-1}) and approximately 100 ug chlorophyll, and were incubated for 4h at 25°C in darkness. Choline oxidation to betaine aldehyde and betaine was measured essentially according to ref. 39.

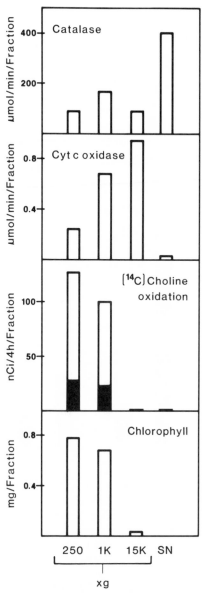

FIGURE 7. Distribution of choline-oxidizing activity and markers after differential centrifugation of the spinach protoplast lysate of Expt A, Figure 6. [^{14}C]choline oxidation was assayed in darkness. Open and solid [^{14}C]choline oxidation bars as in Figure 6.

TABLE 1
RECOVERY OF CHOLINE-OXIDIZING ACTIVITY AND MARKERS
IN PURIFIED SPINACH CHLOROPLASTS[a]

Sample characteristic	Protoplast lysate value	Purified chloroplasts	
		Value	% of lysate
Chloroplast intactness (%)	~80	>90	–
Chlorophyll (mg/sample)	0.92	0.29	31
[14C]Choline oxidation (nCi/4h/sample)			
To betaine aldehyde	48	60	125
To betaine	17	18	106
Catalase (μmol/min/sample)	445	7	1.6
Cytochrome c oxidase (nmol/min/sample)	1370	14	1.0

[a]Chloroplasts were isolated from a spinach protoplast lysate by sedimentation (1 min, 2500g) through a 40% Percoll medium. The lysate was from Expt B, Figure 6. [14C]Choline oxidation was assayed in darkness. Chloroplast intactness was estimated by light-dependent ferricyanide reduction. Note that >100% recoveries of [14C]choline oxidizing activities in purified chloroplasts can be accounted for by depletion of the endogenous choline present in the lysate.

TABLE 2

EFFECT OF LIGHT ON [^{14}C]CHOLINE OXIDATION BY PROTOPLASTS AND CHLOROPLASTS[a]

Conditions		Pea		Spinach		Sugarbeet	
		Betaine aldehyde	Betaine	Betaine aldehyde	Betaine	Betaine aldehyde	Betaine
				--nCi/100µgchl--			
Protoplasts	Dark	0	0.02	4.8	1.9	13.7	3.3
	Light	0	0.03	6.9	46.0	56.7	30.5
Chloroplasts	Dark	0	0	12.2	0.4	14.4	5.6
	Light	0	0.08	32.3	5.1	9.3	19.2

[a]Illumination was with red light ($2-3 \times 10^{-3}$E.m^{-2}.s^{-1}). Chloroplast fractions were washed 500g pellets. For protoplasts, incubations were for 1h in light or darkness except for pea in darkness, which was for 4h. For chloroplast fractions, incubations were for 30 min in light or darkness except for pea in darkness, which was for 4h.

in intact protoplasts (Figure 6).   The choline-oxidizing
activity of the lysates sedimented with intact chloroplasts
in both differential centrifugation (Figure 7) and   through
a 40% Percoll layer (Table 1).   No choline-oxidizing
activity   could   be   detected   in   intact protoplasts   of
chloroplasts prepared under similar conditions from pea,   a
species that does not synthesize betaine in vivo.

These results indicate that the chloroplast,   not   the
mitochondrion   or   peroxisome,   is   the   site   of   choline
oxidation in chenopods.   Note that this is consistent with
the   finding   that in sugarbeet only the leaves   synthesize
betaine,   and with radiotracer   evidence   favoring   tight
compartmentation   of   the   choline-oxidizing   steps.   The
electron   acceptors   are not yet known for   either   of   the
oxidation   steps,   but   it   is   noteworthy   that   choline
oxidation   was   enhanced   by high-intensity light   in   both
intact   protoplasts   and   chloroplasts   obtained   from
protoplasts (Table 2).

Now that in vitro choline oxidation has been obtained,
it   is   possible   to   search   for   the   enzyme   proteins
responsible, and for the genes specifying them.

CONCLUDING SPECULATION:
ENGINEERING OF CYTOPLASMICALLY INHERITED METABOLIC TRAITS.

Clearly,   the fact that betaine synthesis in chenopods
is   a   chloroplast   activity does not imply that   the   genes
involved are in the plastid genome.   However, were that the
case,   or were it possible for genetic engineering to   make
it   so,   some   interesting practical advantages   for   plant
breeding   could accrue.   Taking betaine biosynthesis as   a
specific instance,   suppose that it is desired to introduce
a   metabolic   pathway into a crop species from which it   is
lacking, or to accentuate the pathway in a species in which
it is weakly expressed.   Suppose also that the crop species
is   self-pollinated   and   shows   maternal   transmission   of
organelle   genomes,   that   keeping track of   the   metabolic
trait requires chemical analyses, that there is a technical
choice   of   incorporating   the relevant   gene(s)   into   the
nuclear   or   the cytoplasmic genome of a   current   cultivar,
and that in the nuclear case,   multiple insertion sites are
possible   (40).   If we wish only to exploit the   newly
engineered   cultivar directly (i.e. to actually grow it as
the   crop) neither cytoplasmic nor nuclear inheritance   has
any great advantage.   This predicates that in each instance

where an engineered trait is to be incorporated, it must be introduced de novo.

However, suppose we wish to use the transformed cultivar as a source of the engineered metabolic trait in a regular breeding program--for example, in a different climatic region--much as non-adapted sources of disease resistance are commonly used. Here, a typical procedure would be to use the transformed line as the exotic parent in a backcrossing program with a desirable (locally adapted) type as the recurrent parent. In this case a trait carried in the chloroplast or mitochondrial genome has three advantages over a nuclear-encoded trait.

#1.   The trait can be tracked in the breeding program by pedigree alone; chemical analyses as selection criteria are eliminated.

#2.   Since cytoplasmic and nuclear genes cannot be linked, heavy selection can be applied in the backcross generations for the recurrent parent phenotype, enabling rapid recovery of desirable agronomic traits with no risk of losing the engineered metabolic trait.

#3.   With engineered nuclear genes, the problem of breaking unfavorable linkages between their sites of insertion and adjacent chromosomal genes becomes increasingly severe as the number of insertion sites increases. If there were, for example, five such sites in linkage groups that required breaking to obtain an acceptable backcross progeny, then the enterprise would be effectively impossible.

Perhaps the potential benefits for disseminating genetically-engineered cytoplasmic traits via normal plant breeding make the technical obstacles worth attacking.

ACKNOWLEDGMENTS

Michigan Agricultural Experiment Station Journal Article No. 11233.

REFERENCES

1.   Fischer RA, Turner NC (1978). Plant productivity in the arid and semiarid zones. Annu Rev Plant Physiol 29:277.

2.   Evans LT, Bingham J, Jackson P, Sutherland J (1972). Effect of awns and drought on the supply of photosynthate and its distribution within wheat ears. Ann Appl Biol 70:67.

3. Richards RA, Passioura JB (1981). Seminal root morphology and water use of wheat II. Genetic variation. Crop Sci 21:253.
4. Morgan JM (1977). Differences in osmoregulation between wheat genotypes. Nature 270:234.
5. Farquhar GD, O°Leary OH, Berry JA (1982). On the relationship between carbon isotope discrimination and the intercellular carbon dioxide concentration in leaves. Aust J Plant Physiol 9:121.
6. Chaleff RS (1983). Isolation of agronomically useful mutants from plant cell cultures. Science 219:676.
7. Wyn Jones RG, Storey R (1981). Betaines. In Paleg LG, Aspinall D (eds): "Physiology and Biochemistry of Drought Resistance in Plants," Sydney: Academic Press, p 171.
8. Hitz WD, Hanson AD (1980). Determination of glycine betaine by pyrolysis-gas chromatography in cereals and grasses. Phytochemistry 19:2371.
9. Coughlan SJ, Wyn Jones RG (1980). Some responses of Spinacea oleracea to salt stress. J of Exp Bot 31:883.
10. Hanson AD, Wyse R (1982). Biosynthesis, translocation, and accumulation of betaine in sugar beet and its progenitors in relation to salinity. Plant Physiol 70:1191.
11. Ford CW, Wilson JR (1981). Changes in levels of solutes during osmotic adjustment to water stress in leaves of four tropical pasture species. Aust J Plant Physiol 8:77.
12. Hitz WD, Ladyman JAR, Hanson AD (1982). Betaine synthesis and accumulation in barley during field water-stress. Crop Sci 22:47.
13. Storey R, Wyn Jones RG (1978). Salt stress and comparative physiology in the Gramineae. III. Effect of salinity upon ion relations and glycinebetaine and proline levels in Spartina x townsendii. Aust J Plant Physiol 5:817.
14. Yancey PH, Clark ME, Hand SC, Bowlus RD, Somero GN (1982). Living with water stress: evolution of osmolyte systems. Science 217:1214.
15. Hall JL, Harvey DMR, Flowers TJ (1978). Evidence for the cytoplasmic localization of betaine in leaf cells of Suaeda maritima. Planta 140:59.
16. Leigh RA, Ahmad N, Wyn Jones RG (1981). Assessment of glycinebetaine and proline compartmentation by analysis of isolated beet vacuoles. Planta 153:34.

17. Pollard A, Wyn Jones RG (1979). Enzyme activities in concentrated solutions of glycinebetaine and other solutes. Planta 144:291.
18. Jolivet Y, Hamelin J, Larher F (1983). Osmoregulation in halophytic higher plants: The protective effects of glycine betaine and other related solutes against the oxalate destabilization of membranes in beet root cells. Z Pflanzenphysiol 109:171.
19. Wyn Jones RG, Gorham J (1983). Aspects of salt and drought tolerance in higher plants. In Kosuge T, Meredith CP, Hollaender A (eds): "Genetic Engineering of Plants - An Agricultural Perspective," New York & London: Plenum Press, p 355.
20. Strom AR, LeRudulier D, Jakowec MW, Bunnell RC, Valentine RC (1983). Osmoregulatory (Osm) genes and osmoprotective compounds. In Kosuge T, Meredith CP, Hollaender A (eds): "Genetic Engineering of Plants - An Agricultural Perspective," New York & London: Plenum Press, p 39.
21. LeRudulier D, Bernard T, Pocard J-A, Goas G (1983). Accroissement de l'osmotolerance chez Rhizobium meliloti par la glycine betaine et la proline betaine. CR Acad Sci 297:155.
22. Ladyman JAR, Ditz KM, Grumet R, Hanson AD (1983). Genotypic variation for glycinebetaine accumulation by cultivated and wild barley in relation to water stress. Crop Sci 23:465.
23. Grumet R, Isleib TG, Hanson AD (1984). Genetic control of glycinebetaine level in barley. Crop Sci (in press).
24. Hanson AD, Hitz WD (1982). Metabolic responses of mesophytes to plant water deficits. Annu Rev Plant Physiol 33:163.
25. Coughlan SJ, Wyn Jones RG (1982). Glycinebetaine biosynthesis and its control in detached secondary leaves of spinach. Planta 154:6.
26. Hanson AD, Rhodes D (1983). ^{14}C Tracer evidence for synthesis of choline and betaine via phosphoryl base intermediates in salinized sugarbeet leaves. Plant Physiol 71:692.
27. Hitz WD, Rhodes D, Hanson AD (1981). Radiotracer evidence implicating phosphoryl and phosphatidyl bases as intermediates in betaine synthesis by water-stressed barley leaves. Plant Physiol 68:814.
28. Hanson AD, Scott NA (1980). Betaine synthesis from radioactive precursors in attached water-stressed barley leaves. Plant Physiol 66:342.

29. Ladyman JAR, Hitz WD, Hanson AD (1980). Translocation and metabolism of glycine betaine by barley plants in relation to water stress. Planta 150:191.

30. Tsuge H, Nakano Y, Onishi H, Futamura Y, Ohashi K (1980). A novel purification and some properties of rat liver mitochondrial choline dehydrogenase. Biochim Biophys Acta 614:274.

31. Wilken DR, McMacken ML, Rodriquez A (1970). Choline and betaine aldehyde oxidation by rat liver mitochondria. Biochim Biophys Acta 216:305.

32. Nagasawa T, Kawabata Y, Tani Y, Ogata K (1975). Choline dehydrogenase of Pseudomonas aeruginosa A-16. Agr Biol Chem 39:1513.

33. Ohta-Fukuyama M, Miyake Y, Emi S, Yamano T (1980). Identification and properties of the prosthetic group of choline oxidase from Alcaligenes sp. J Biochem 88:197.

34. Yamada H, Mori N, Tani Y, Properties of choline oxidase of Cylindrocarpon didymum M-1. Agric Biol Chem 43:2173.

35. Pan S, Moreau RA, Yu C, Huang AHC (1981). Betaine accumulation and betaine-aldehyde dehydrogenase in spinach leaves. Plant Physiol 67:1105.

36. Kortstee GJJ, (1970) The aerobic decomposition of choline by microorganisms I. The ability of aerobic organisms, particularly coryneform bacteria, to utilize choline as the sole carbon and nitrogen source, Arch Mikrobiol 71:235.

37. Nishumura M, Douce R, Akazawa T (1982). Isolation and characterization of metabolically competent mitochondria from spinach leaf protoplasts. Plant Physiol 69:916.

38. Edwards GE, Robinson SP, Tyler NJC, Walker DA (1978). Photosynthesis by isolated protoplasts, protoplast extracts, and chloroplasts of wheat. Influence of orthophosphate, pyrophosphate and adenylates. Plant Physiol 62:313.

39. Haubrich DR, Gerber NH (1981). Choline dehydrogenase - Assay, properties and inhibitors. Biochem Pharmacol 30:2993.

40. Barton KA, Binns AN, Matzke AJM, Chilton M-D (1983). Regeneration of intact tobacco plants containing full length copies of genetically engineered T-DNA, and transmission of T-DNA to R1 progeny. Cell 32:1033.

Cellular and Molecular Biology of Plant Stress, pages 93–114
© 1985 Alan R. Liss, Inc.

POLYAMINE METABOLISM AND PLANT STRESS[1]

Hector E. Flores[2], Nevin D. Young,[3] and Arthur W. Galston[4]

[2]ARCO Plant Cell Research Institute, Dublin, CA. 94568,
[3]Department of Plant Pathology,Cornell University
Ithaca, N. Y. 14853, and
[4]Biology Department, Yale University
New Haven, CT.06511

ABSTRACT   The diamine putrescine (1,4-diamino-butane), and to a lesser extent the polyamine spermidine, accumulate in higher plants in response to various stresses. Cereal leaf segments exposed to osmotic stress (0.4 - 0.6 M sorbitol) or low pH ($<$5.0) show up to 60-fold increases in putrescine titer within 6 hr of exposure. The rise in putrescine is paralleled by an increase in the activity of arginine decarboxylase, while the alternative putrescine biosynthetic enzyme, ornithine decarboxylase, is unaffected. DL-α-difluoromethylarginine, the specific enzyme-activated irreversible inhibitor of arginine decarboxylase, completely prevents the stress response, whereas α-difluoromethylornithine, the corresponding inhibitor of ornithine decarboxylase, is ineffective. Putrescine increase is prevented by cycloheximide, suggesting that protein synthesis is involved in the response. Other stress situations, involving K^+-deficiency, ammonium nutrition, water and salt stress, also result in putrescine accumulation and arginine decarboxylase increase.The first leaf of 18 day old K^+-deficient oat seedlings shows 30-fold higher ADC specific activity than control leaves. The enzyme has been purified by ammonium sulfate and acetone fractionation, gel filtration and ion exchange chromatography, resulting in a 650-fold enrichment. The purified ADC migrates as a single band with M_r 39,000 upon SDS-polyacrylamide gel electrophoresis. In addition to free polyamines, bound forms conjugated to hydroxycinnamic acids are known to accumulate in response to viral and fungal infection.

[1] Supported by NIH grant 1RO1AG-02742 to A.W.G.

INTRODUCTION

The diamine putrescine (1,4-diaminobutane) (Put) and
the polyamines spermidine (Spd) and spermine (Spm), are
ubiquitous in bacteria, animals and plants (1). Spd may also
be found as an intrinsic component of animal and plant
viruses (2). Spd and Spm are related biosynthetically to Put,
through the transfer of one and two aminopropyl groups
(derived from decarboxylated S-adenosylmethionine), respectively, to the 1,4-diaminobutyl moiety (Figure 1).
Although polyamines have been known since the classic
studies of Leewenhoek on human semen, their chemical
structures were not determined until the mid 1920's, and it
was only 3 decades ago that interest started in their
physiology and biochemistry.

In animal and bacterial cells (1), a close correlation was
found between high polyamine titers (Spd, Spm) and rapidly
dividing cells. With the availability of synchronous animal
cells cultures and polyamine biosynthesis inhibitors, it was
shown that polyamine-depleted cells were arrested in the G_1-
S phase of the cell cycle (3). Upon addition of polyamines (Put,
Spd, Spm), cycling was resumed. The recent development of
suicide inhibitors of polyamine biosynthesis (4) provided clear
evidence for a causal link between polyamine function and
cell division and differentiation (5). These results have been
recently confirmed in plant systems. In tomato ovaries,
polyamines and their biosynthetic enzyme ornithine
decarboxylase, have been shown to be essential for the stage
of rapid cell proliferation following fertilization (6). In carrot
cell suspensions, embryogenesis is blocked by an irreversible
inhibitor of arginine decarboxylase, and the effect is reversed
by addition of Put or Spd to the culture medium (7).

Although the basic findings relating polyamines to cell
division have been confirmed in higher plants, recent
observations suggest that unique functions, not directly
related to the above, may have evolved for polyamines in
plant cells (8). For example, polyamines are effective
retardants of senescence in mono- and dicotyledonous plants
(9). In etiolated pea seedlings, ADC activity is regulated by
phytochrome (10). The induction of Crassulacean acid
metabolism in *Bryophyllum*, results in a rhythmic oscillation
in polyamine content in the leaves which parallels that of
organic acids (11).

A more neglected aspect of polyamine physiology in
plants concerns the response to various stress conditions. In
1952, Richards and Coleman (12) found that putrescine is the

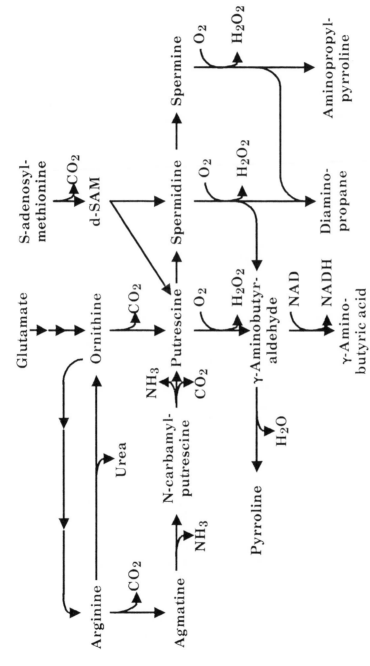

Figure 1. Pathways of polyamine biosynthesis and degradation in higher plants.

main nitrogenous compound accumulating in potassium-deficient barley seedlings, and this was subsequently shown to be a universal response (13). In this paper, we describe our recent findings concerning changes in polyamine metabolism in response to osmotic and water stress, low pH stress, and K+-deficiency. We discuss approaches to understanding the physiological significance of such responses, and speculate on the.underlying mechanism(s).

RESULTS

Cereal Leaves and Osmotic Stress

While leaf mesophyll protoplasts of numerous dicotyledonous species can be succesfully cultured and regenerated *in vitro*, cereal protoplasts have remained refractory in spite of attempts at culture by a number of groups (14). The reason for this lack of success is not clear, but recent work suggests it may be at least in part physiological. Freshly isolated cereal protoplasts are very labile, showing a decreased incorporation of precursors into macromolecules and an increase in ribonuclease and protease activity, both part of the complex of senescence related changes in cereal leaves (9). In the course of these studies we found that Put, Spd and Spm added to the protoplast culture medium were effective against protoplast lysis, and promoted incorporation of ^{3}H-uridine and ^{3}H-leucine into RNA and protein, respectively. Polyamine addition also increased thymidine incorporation into DNA and inception of mitotic activity (15). While continued cell division did not result from such treatments, these observations prompted a study of polyamine metabolism in cereal protoplasts.
      The first surprising result was that the polyamine pattern of freshly isolated oat mesophyll protoplasts, analyzed by TLC and HPLC (16), is strikingly different from that of the oat mesophyll cells from which they are derived. While spermidine and spermidine remain unchanged on a per cell basis, the putrescine content is 5 to 10-fold higher in protoplasts, and this change occurs in the 2-3 hr period of digestion of peeled oat leaves with cellulolytic enzymes (0.5% Cellulysin, Calbiochem, in 0.4 M sorbitol at 30 C in darkness) (17, 18). The increase in putrescine is not a direct result of the toxic, low molecular weight components present in crude Cellulysin, since the same result is obtained with enzyme that has been desalted through a Sephadex G-50 column. It is

also not dependent on the source of cellulase, since Put titer is high in protoplasts isolated with Onozuka Cellulase, which has widely different ratios of proteolytic:cellulolytic:pectinolytic activity (18).On the other hand, it is well known that the osmoticum present in the protoplast isolation medium triggers increases in RNAase and protease activity, causes depolarization of cell membranes, and inhibits photosynthesis (8, 14). Therefore, we tested the possibility that the increased putrescine content of protoplasts is a direct result of osmotic stress.

The lower epidermis of primary leaves of 7 to 9 day old cv. Victory oat seedlings was removed, and the leaves floated under fluorescent light (ca. 12,000 lux) in the presence or absence of 0.6 M sorbitol in 1 mM phosphate at pH 5.8. The experimental setup was identical to that used for protoplast isolation, except for the absence of cellulase. After 2 hr of such incubation in the presence of the osmoticum, a significant increase in Put was evident, and a titer 60-fold higher than the initial value was reached by 6 hours (Figure 2 A). The rise in Put reached a plateau after 8-10 hr. In contrast, Put titer remained constant during incubation in control phosphate buffer. The polyamines Spd and Spm showed a gradual decline during the initial 4 hr of osmotic treatment, but an eventual increase of 30-50% in Spd titer became apparent by 6-8 hr of stress (data not shown). No change was observed in Spm titer.

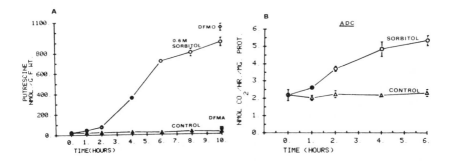

Figure 2. Time course of putrescine (A) and ADC (B) increase in oat leaf segments floated on sorbitol or buffer control (1 mM phosphate, pH 5.8) under light at 25⁰ C.

That the increase in Put is in fact a result of osmotic stress is supported by our observation (18) that both relatively non-permeating (sorbitol, mannitol, proline, betaine) as well as permeating osmotica (sucrose) can induce similar increases in Put. The response is also not restricted to oat leaves. Leaf segments of wild oat (*Avena fatua*), barley, corn and wheat floated over 0.4 M sorbitol respond with significant increases in Put after 4 hr (17) The Put concentration at the start of the experiment was not signifcantly different from that of control leaves incubated in buffer for 4 hr (19) However, there is a 2-fold increase in Put in corn leaf segments floated in buffer, possibly because of the wounding induced by peeling. Wheat leaf segments, whch cannot be peeled and do not show obvious signs of wilting in the presence of osmoticum, show a 4 to 6-fold increase in Put titer. Thus, injury caused by peeling is not essential for the increase in Put. The threshold levels of osmoticum necessary for the response are probably much lower than the concentrations used in these experiments, since 0.2 M sorbitol can induce a 2.5-fold rise in Put after 4 hr (17, 19).

It is unlikely that the osmoticum-induced increase in Put and Spd is due to release from a bound form. Acid hydrolysis of the perchloric acid-soluble and insoluble fractions of leaf extracts causes little or no Put or Spd release (19). We thus conclude that Put biosynthesis is in fact induced by osmotic treatment.

In animal cells, Put is synthesized solely through ornithine decarboxylase (ODC) (Figure 1). In contrast, it is now accepted that most, if not all, higher plants, share with bacterial cells the existence of an alternate Put biosynthetic pathway, involving arginine decarboxylase (ADC), via agmatine and N-carbamylputrescine (20). As shown in Figure 2B, ADC activity declines slightly in oat leaf segments incubated in control buffer, but increases 2 to 3-fold in the presence of osmoticum. The ADC increase closely parallels the rise in Put.

The correlation between ADC increase and Put accumulation is of similar magnitude to that observed under K^+-deficiency (21, 22), but is not by itself proof that only ADC mediates the stress-induced rise in polyamines. The availability of enzyme activated specific inhibitors of polyamine biosynthesis allowed us to established a causal link (17). Difluoromethylarginine (DFMA) and α-difluoro-methylornithine (DFMO) are the suicide inhibitors of ADC and ODC, respectively. In contrast to the classic irreversible inhibitors, compounds such as DFMO and DFMA are chemi-

TABLE 1
EFFECT OF IRREVERSIBLE POLYAMINE BIOSYNTHETIC INHIBITORS ON THE RESPONSE OF OAT LEAF SEGMENTS TO OSMOTIC STRESS

Treatment	Polyamine Titer nmol/g. fresh wt.		Enzyme Activity (nmol CO_2/hr. mg protein)	
	Put	Spd	Arginine decarboxylase	Ornithine decarboxylase
Oh Control	16 ± 1	161 ± 18	2.27 (115)	4.24 (93)
4h Control	13 ± 2	152 ± 10	1.98 (100)	4.55 (100)
0.6 M Sorbitol	102 ± 5**	120 ± 13	3.81 (192)**	4.65 (102)
0.6 M Sorbitol + 10^{-3} M DFMO	$129\pm$***	186 ± 10*	5.31 (268)**	4.68 (103)
0.6 M Sorbitol + 10^{-4} M DFMA	20 ± 2	123 ± 7	1.24 (63)	4.64 (102)
0.6 M Sorbitol + 10^{-3} M DFMA	10 ± 2	128 ± 8	0.41 (21)	4.55 (100)

Numbers represent the mean ± S.E.M.

*, ** Significantly different from the 4 hr control at $P < 0.05$ and $P < 0.01$, respectively. Adapted from Ref. 17.

TABLE 2

PREVENTION BY CYCLOHEXIMIDE OF THE RESPONSE TO OSMOTIC STRESS

Treatment	Time of CH addition	Putrescine	Spermidine nmol/gfresh wt	Spermine	ADC CO$_2$/h.mg protein	ODC
0 h Control	---------	19	158	28	2.09	8.14
4 h Control	---------	19	149	25	2.27	8.14
0.4 M Sorbitol	---------	466**	210*	35	4.20**	7.60
Sorbitol + CH 10 µg/ml	0 h	85*	97*	30	2.54	7.8g
Sorbitol + CH 25 µg/ml	'''	55*	125	28	1.98	7.91
Sorbitol + CH 50 µg/ml	'''	63*	91*	37	2.05	7.75
Sorbitol + CH 50 µg/ml + 1 mM L-Arginine	'''	103**	100*	41	2.72	7.97
Sorbitol + CH 25 µg/ml	1 h	341**	210*	32	3.38**	6.35*
Sorbitol + CH 25 µg/ml	2 h	502**	257**	29	3.53**	6.00*
Sorbitol + CH 25 µg/ml	3 h	561**	311**	30	4.11**	4.79**

*, ** Significantly different from the 4 hr control at $P < 0.05$ and $P < 0.01$, respectively. Adapted from Ref. 18

cally inert substrates, that when acted upon by the enzyme generate electrophilic forms inside the active site. The covalent link formed between the reactive intermediate and a nucleophilic residue in the active site leads to the irreversible inactivation of the enzyme (23).

As shown in Table 1, DFMA (0.1 to 1.0 mM) completely prevents the osmotic stress-induced rise in Put, in parallel with the inhibition of ADC activity. In contrast, DFMO potentiates the increase in Put and Spd in response to sorbitol treatment, while having no effect on ODC activity. No cross inhibitions were observed in *in vitro* assays, supporting the specificity of DFMO and DFMA (18). The preferential activation of ADC by osmotic stress is supported by the following additional observations (18): a) α-methylornithine, a competitive inhibitor of ODC which *is* effective both *in vivo* and *in vitro*, has similar effects to DFMO; it increases the Put and Spd titer above the sorbitol treatment; b) D-arginine and L-canavanine, which at 1 mM inhibit oat ADC activity by 60 and 50% , respectively (24), also reverse the increase in Put when added at the same concentration to osmotically stressed leaves.

The increase in Put and ADC in cells exposed to osmotic shock (17, 18) or high external acidity (25) is seen after a lag time of 1 to 2 hr, which is consistent with a requirement for protein synthesis. Table 2 shows the effect of the translation inhiibitor cycloheximide on the response to osmotic stress. CH (10-50 µg/ml) completely prevents the rise in Put and Spd. The effect of transcription inhibitors is not as clear. Cordycepin (5-10 µg/ml) and Actinomycin D (10-20 µg/ml) inhibit the rise in Put by 11% and 28%, respectively (not shown). If transcription of the mRNA for ADC mediates the stress response, these results may reflect incomplete penetration of the inhibitors. If added at the start of the osmotic treatment, CH effectively prevents the ADC and Put rise, but at progressively later times there is an escape from prevention by CH which is already completed by 1-2 hr (Table 1) This supports the view that a protein synthesis-dependent event occurring during the first hr of exposure to osmotic stress is required for the rise in ADC and Put accumulation.

Are the short term changes oberved in osmotically stressed leaves comparable to a water stress condition in the whole plant? If 11-day old oat seedlings are allowed to wilt by witholding the water supply, polyamine titer remains unchanged for 2 days before rising significantly at the time when water stress symptoms become obvious. In parallel, ADC specific activity goes up almost 4-fold, while ODC is not affected (18). These changes have been followed in the first leaf, which at this stage is more susceptible to wilting than the emergent second leaf. Therefore, the results obtained with the osmotically treated leaf segments are of direct relevance to the changes that occur at the whole plant level.

Polyamines and Low pH Stress

Since most of the stress situations which induce Put biosynthesis may have a direct effect on the ionic balance of the cell (12, 17, 25-27), the response to changes in the external pH was studied. The close monitoring of environmental conditions that is possible with the *in vitro* peeled leaf system (9) make it appropriate for approaching this problem. Oat and pea leaf segments were incubated over buffers covering a pH range of 3.0 to 8.0 (Figure 3).

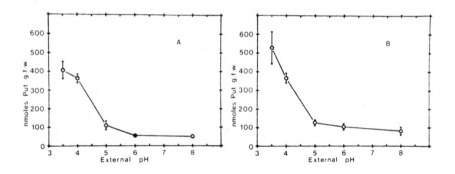

Figure 3.   Response of putrescine titer to a range of external pH in oat leaf segments (A) and pea leaf discs (B) incubated for 8 hr under light at 25°C. Adapted from Ref. 25.

Both species showed remarkably similar responses. Put titer did not change during 8 hr incubation at pH > 5.5., while incubation on media of pH 5.0 or below led to a rapid and massive increase in Put (25). Spd and Spm did not show any response to changes in external pH. The observed results were not related to the buffer system used. Several buffer combinations (Succinate/Tris, Succinate/NaOH, Mes/Tris, Kphthalate/Tris) used at low (4.0-4.8) and high (6.0) pH values, invariably caused Put buildup under the more acid conditions. The time course of increase in Put titer in response to low pH was similar to that observed in osmotically shocked tissue. While remaining low and constant at pH 6.0, Put titer was significantly higher after 3 hr incubation at pH 4.0, reaching a peak after 9 hr, after which it declined. As in the osmotically stressed leaves, protein and chlorophyll levels remained equivalent in the control and stressed tissues, and no loss in cell viability was observed within the time frame of the experiments (18, 25).

The response to low pH shows other striking similarities to the response to osmotic stress: 1) ADC activity increases about 50% at pH 4.0, while ODC activity declines at both pH 4.0 and 6.0; 2) DFMA completely inhibits the response to low pH, while DFMO is ineffective; 3) Cycloheximide (5 µg/ml) inhibits the rise in Put. Upon withdrawal of CH, Put accumulation is preceded by a lag of at least 3 hr, consistent with the requirement for resumption of protein synthesis. There are, however, some differences that must be pointed out. In the acid-treated tissues, the rise in Put seems to be the most significant change in organic nitrogen pools. Under osmotic stress, however, notable changes occur in the amino acid content, in addition to the diamine accumulation (28). We observed a 3 to 5-fold increase in ammonia under osmotic stress, most likely reflecting an increase in titer of acid-labile amides. In fact, glutamic acid declines sharply, paralleled by an increase in glutamine.There is an overall increase in the level of the other amino acids, probably as a result of the proteolysis known to be caused by osmotic stress (29). Osmotic stress seems to induce a more dramatic change in polyamine metabolism, at least in part due to increased availability of precursors (Arg, Orn) derived from protein breakdown.

Arginine Decarboxylase and Potassium Deficiency

The most generalized stress- induced changes in poly-amine metabolism have been observed in K-deficient plants (12,13,21,24). However, little information has been available until recently concerning the development of Put buildup and ADC increase in the whole plant. The effect of K-deficiency was investigated by growing oat seedlings in a low-K (6µM KCl) but otherwise complete nutrient medium (29). No difference in growth rate became apparent in the K-deficient plants as compared to the controls (6mM KCl)until 9-12 days.

By this time, white necrotic spots and chlorotic regions were visible in the first leaf of the -K seedlings.

Potassium starvation had a dramatic effect on Put long before visible symptoms became apparent. Put titer in -K seedlings was 15-fold higher than in the controls only 6 days after germination (Figure 4A). Diamine accumulation leveled off after 12 days, at which time the Put titer had reached 6300 nmoles/g fresh wt. As in the case of osmotic and low pH stress, Put rise was paralleled by increased ADC activity (Figure 4B). Both Put and ADC rise were at least partially reversed by addition of 6 mM KCl nutrient solution from days 12 to 18. Appearance of chlorosis and necrotic lesions on first and second leaves was also inhibited. The increase in ADC activity in potassium starved plants was even more dramatic in the first leaf, in which the earliest symptoms of deficiency are observed. From 6 to 18 days, ADC specific activity in the first leaf increased 30-fold.

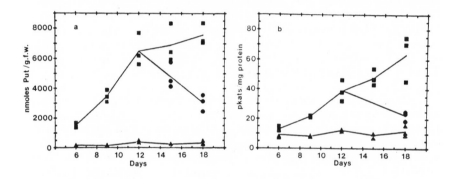

Figure 4. Effect of K-deficiency on the putrescine content (A) and ADC activity (B) of oat seedlings. Δ, grown in + K medium; □, grown in -K medium; ●, grown in -K medium until day 12, and in + K medium thereafter. Solid lines connect the means of the triplicate data points shown.

Several monovalent cations can reverse the effect of K-deficiency on ADC activity (29). Addition of RbCl or NaCl (6.0 mM) to the K-deficient plants improved growth partially. On the other hand, plants fed LiCl grew less than the -K plants. However, both Li and Na could supress the increase in ADC normally observed in -K oats, while Rb caused ADC to drop to 1/3 of the activity found in the +K controls.

Potassium-deficient plants provided an aprropriate starting material for purification of ADC (Young and Galston, in preparation). About 50 g of first leaves of 18 day old -K cv. Victory oats were extracted on ice in 0.05 M Na_2HPO_4. The crude supernatant fraction obtained after centrifugation at 20,000g for 10 min was purified through ammonium sulfate (70%) fractionation, precipitation with acetone, gel filtration through Bio-gel A (100-200 mesh, Bio-Rad), and DEAE-cellulose (DE 32, Whatman) ion exchange chromatography. While the final recovery was only 5%, the purified oat ADC had a 650-fold higher specific activity than the crude extract, and upon SDS-gel electrophoresis migrated as a single, major band of M_r 39,000. This is similar to the value of M_r 36,000 obtained for the subunit of ADC purified from Lathyrus (30). Purified oat ADC also migrated as a major band in non-denaturing gels, showed a K_m of 30 μM for L-arginine and was effectively inhibited in vitro by D-arginine (96%, 5 mM) and DFMA (88%, 1 mM). Activity was significantly enhanced by pyridoxal phosphate and dithiothreitol (24, 30).

The Di- and Polyamine Conjugates

The responses discussed in the previous sections concern stress-induced changes in the levels of free polyamines. However, bound forms of these compounds are of widespread occurrence in higher plants, and are mostly found as the basic and neutral amides of caffeic, coumaric and ferulic acid. The potential significance of these compounds has been recognized only recently (see Ref. 31 for review) Hydroxycinnamic acid amides (HCAs) of di- and polyamines may represent in some cases over 80% of the total amine pool, and are found at levels of 1 μmol/g (or higher) in the inflorescences of all mono- and dicotyledonous species studied to date (32, 33).

Conditions known to cause increase in free polyamines, such as deficiencies of K, Mg and Ca (34) also result in the accumulation of mono- and dicaffeoylspermidine in tobacco leaves (35). Tobacco cell suspensions grown on urea, γ-aminobutyric acid and ammonium as sole nitrogen sources, show an elevated titer of free Put, but also accumulate cinnamoyl-Put to levels of about 5% dry wt (H. Flores, unpublished). NH_4^+ nutrition is known to result in Put accumulation in mono- and dicot species.(36). In addition, there are several situations in which HCAs appear to be specifically involved, namely, response of plant cells to pathogen infection.

In a study of the phenolic compounds formed in tobacco *cv. Xanthi* leaves upon infection with tobacco mosaic virus (TMV) (37), mono- and diferuloylputrescine were found in large amounts. Similarly, *Nicotiana sylvestris* formed diferuloylputrescine when infected with TMV strain Aucuba. The highest concentration of HCAs was reached when the synthesis of new viral particles had been completed, and they accumulated at particularly high levels in the living cells surrounding the hypersensitive zone. Coumaroyl-, dicoumaroyl- and caffeoylPut, when applied to tobacco leaf discs at 0.25-0.75 mg/ml, cause a reduction of up to 90% in the number of local lesions caused by TMV infection (38).Very suggestively, a mutant of *N. tabacum cv. Xanthi* named RMB7, which cannot flower and does not synthesize HCAs, does not show a hypersensitive reaction to TMV. Infection of this mutant with TMV is lethal (39).

HCAs also appear to be involved in the response to fungal infection. A water-soluble elicitor from *Phytophthora parasitica* can induce either resistance of the basal ends of carnation (*Dianthus caryophyllus*) cuttings, or biosynthesis of fungitoxic phytoalexins (40). When cuttings become resistant to the fungus, a set of highly fluorescent compounds appears, some of which have been identified as HCAs. The resistance of young barley seedlings to infection by *Helminthosporium sativum* is attributed to the occurrence in the shoots of dimers of 4-coumaroylagmatine. These compounds, known colectively as hordatines, inhibit the germination of a wide variety of fungal spores at low concentrations (1-10μM) (see Ref. 31).

DISCUSSION

We have described several situations which lead to an activation of the ADC-mediated pathway of Put biosynthesis, and to a fast, massive and reversible increase in diamine titer (17, 25, 29). This occurs whether the response takes place within days (K deficiency) or hours (osmotic shock, low pH stress).Such striking similarities suggest the existence of a common mechanism.

Since the early studies on K+ deficiency (41), it was suggested that alkali metal deficiency shifts the internal balance between the organic anions and cations in the direction of increased acidity. Amine accumulation would function as an homeostatic mechanism to keep intracellular pH at a constant value. This phenomenon may be the counterpart of the well known balancing of excess inorganic cations by the production of organic acids. Consistent with this hypothesis, the conditions of ion imbalance upon K deficiency have been mimicked by feeding barley seedlings with 0.025 N HCl or H_2SO_4 (42), and *Cucumis sativus* cotyledons with 5-10 mM HCl (43). Similarly, NH_4^+ nutrition, the uptake and assimilation of which is expected to lead to the production of H^+ ions in the cytoplasm, causes Put and ADC increase in pea, corn and wheat seedlings (36). SO_2 fumigation of pea seedlings induces accumulation of free and bound Put and Spd (44). Since SO_2 absorption by cells may also result in higher acidity as a result of the formation of HSO_3^-, $SO_3^=$ and $SO_4^=$ ions, amine accumulation could also in this case compensate for the relative cation deficit. More suggestively, acid pH induces amino acid decarboxylation in bacteria (45). In *E. , coli*, 2 forms of ADC exist, showing high (7.8) and low (4.5) pH optima, respectively. The latter is induced by growth at low pH (46).

The obvious possibilities of there being changes in internal pH and/or alterations in K^+ fluxes can be readily tested in the *in vitro* detached leaf system. NMR techniques have been developed for the close monitoring of cytoplasmic pH changes in living plant tissues, relying on the strong pH dependence of chemical shifts of ^{31}P-labeled phosphates (47). Also, specific ionophores which disrupt charge and/or proton gradients across cell membranes (48) should prove useful as probes, as well as the specific inhibitors of Put synthesis mentioned above (17, 18, 25, 29).

Is *de novo* ADC synthesis involved in the stress-induced Put increase? The prevention of the response by cycloheximide suggests that this is in fact the case (18, 25). Furthermore, the labeling with ^{35}S-methionine of a band comigrating with purified ADC is enhanced 28% in osmotic stressed leaf segments (Young and Galston, in preparation). While the enhancement in labeling is relatively small, it is consistent with the 2-fold rise in ADC activity within 4 hr of stress (18). Alternatively, the imposition of stress may result in the conversion of ADC to an active form by post-translational modification. ODC activity is known to be regulated post-translationally, either through the action of an antizyme (49) or via phosphorylation (50). While the question is still open in the case of ADC, a straightforward approach to distinguish between these alternatives involves the use of DFMA. Since the inhibitor is actually bound specifically and covalently to the active site, the availability of labeled DFMA would make it possible to follow directly the changes in the actual amount of the enzyme, upon labeling of cells and electrophoresis of the labeled extract. This approach has been used successfully to localize DFMO in animal cells (51), and to monitor ODC purification (52).

Enzyme-activated irreversible inhibitors may be particularly useful as probes in the case of salt and osmotic stress, which induce dramatic metabolic changes. For example, complex changes in amino acid pools occur in osmotically stressed oat leaves. DFMA treatment specifically causes arginine to accumulate 3 to 10-fold, at the same time preventing Put buildup, while having no effect on the other amino acids (28). Thus, specific inhibitors may allow us to dissect specific components of stress responses. We would expect this approach to be useful in other instances of stress-induced accumulation of metabolites, as in the case of proline and betaine, provided information becomes available on the enzymes involved and on their mechanism of action. We cannot underestimate the impact that specific inhibitors made by design can have on our understanding of the regulation of plant metabolism. The widespread use of aminooxy-phenylpropionic acid (53), an inhibitor of phenylalanine ammonia-lyase, is a case in point.

Although the physiology of stress-induced changes in polyamine metabolism is now well documented at the whole plant level and in *in vitro* systems, the question of the significance of such changes remains unanswered. The hypothesis (22, 41) that synthesis of Put is a response to acidification of the cytoplasm resulting from cation/anion imbalance, and serves to regulate cellular pH, is certainly compelling, and the observations described above are consistent with it. However, if in fact Put is acting as an organic cation, one sees no apparent reason why Spd and Spm should not be effective in the same way. The fact is, in every case where stress-induced Put accumulation occurs, its increase is disproportionately larger than the rise in Spd (17, 18), and in some cases there is no change in Spd or Spm (13, 36). Thus, in addition to the induction of Put synthesis, stress results either in the compartmentation of Put such that it is not made available for polyamine synthesis, or in the direct inhibition of Spd/Spm synthase.

The above differences may be significant in light of what is known about di- and polyamine function. It is widely accepted that Spd and Spm are essential for continued growth and division of cells (1, 8). Put is required inasmuch as it is the obligate precursor for Spd and Spm. Differences in the distribution of polyamines are suggestive of different functions for Put and the polyamines. In shoots and roots of pea and corn seedlings (19, 54), Spd and Spm are always higher in the meristem and decline sharply towards the region of cell elongation. Put shows exactly the opposite gradient. The stimulation of growth of pea internodes by gibberellic acid (55), of tomato ovary development after fertilization (6), and the induction of embryogenesis in carrot cell suspensions (56), always result in rapid synthesis of Spd and Spm following a transient rise in Put, regardless of whether Put synthesis is mediated by ADC (7, 55, 57) or ODC (6). If the apparent block in polyamine synthesis induced by stress is long lived, it may eventually lead to Spd and Spm depletion, with the expected adverse effects on cell growth and division. Thus, if the pH hypothesis is correct, although the accumulation of Put may have a role in the maintenance of cellular homeostasis, the response as a whole could be a metabolic *cul de sac.*

Some final considerations are relevant to the previous discussion. Even if Put by itself is not directly toxic to the cell, as has been suggested for salt stressed and K-deficient plants (27, 41), or if it is not involved in the regulation of cellular pH, its synthesis under stress conditions may impose a significant drain on the pools of amino acid precursors. We have shown (28) that arginine availability limits the rise of Put under osmotic stress. It is obvious that, when considering stress responses that involve dramatic changes in metabolism, no alteration in any single compoment is without effect on the rest. This is particularly true for polyamines, considering their metabolic pathways (Figure 1), and because Put may have a key position in nitrogen metabolism which has not been fully realized yet. Besides being a precursor for Spd, Put gives rise to hydroxycinnamic acids (31), which may be involved in plant-pathogen interactions, to γ-aminobutyric acid (58), which accumulates under hypoxia (59), and to alkaloids such as nicotine and atropine. While much remains to be known about the regulation of polyamine pools and turnover in higher plants, the systems and approaches we have outlined in this paper may give in the near future clear answers concerning the function of polyamines in stress situations, and their significance for plant growth and development.

ACKNOWLEDGEMENTS

We are grateful to Kent F. McCue for his advice on word processing and graphics, and to Mark Staebell and Jaen Andrews for critical comments on the manuscript.

REFERENCES

1. Bachrach U (1973). "Function of Naturally Occurring Polyamines." New York: Academic Press, p 46.
2. Cohen SS, Greenberg ML (1981). Spermidine, an intrinsic component of turnip yellow mosaic virus. Proc Natl Acad USA 78:5470.
3. Rupniak HT, Paul D (1978)Inhibition of spermidine and spermine synthesis leads to growth arrest of rat embryo fibroblasts in G_1. J Gen Physiol 94:161.
4. Metcalf BW, Bey P, Danzin C, Jung MJ, Vevert JP (1978). Catalytic irreversible inhibition of mammalian ornithine decarboxylase (E.C.4.1.1.17) by substrate and product analogs. J Amer Chem Soc 100:2551.
5. Fozard JR, Part ML, Prakash NJ, Grove J, Schechter PJ, Sjoerdsma A, Koch-Weser J (1980) L-ornithine decarboxylase: an essential role in early mammalian embryogenesis. Science 208:505.
6. Cohen E, (Malis)Arad S, Heimer YM, Mizrahi Y (1982). Participation of ornithine decarboxylase in early stages of tomato fruit development. Plant Physiol 70:540.
7. Feirer RP, Mignon G, Litvay JD (1984). Arginine decarboxylase and polyamines required for embryogenesis in the wild carrot. Science 223:1433.
8. Galston AW (1983). Polyamines as modulators of plant development. BioScience 33:382.
9. Kaur-Sawhney R, Galston AW (1979). Interaction of polyamines and light on biochemical processes involved in leaf senescence. Plant, Cell & Env 2:189.
10. Dai YR, Galston AW (1981). Simultaneous phytochrome-controlled promotion and inhibition of arginine decarboxylase activity in buds and epicotyls of etiolated peas. Plant Physiol 67:266.
11. Morel C, Villanueva VR, Queiroz O (1980) Are polyamines involved in the induction and regulation of the Crassulacean acid metabolism? Planta 149:440.
12. Richards FJ, Coleman RG (1952). Occurrence of putrescine in potassium deficient barley. Nature 170:460.
13. Smith TA (1970). Putrescine, spermidine and spermine in higher plants. Phytochem 9:1479.
14. Flores HE, Kaur-Sawhney R, Galston AW (1981). Protoplasts as vehicles for plant propagation and improvement. In Maramorosch K (ed): "Advances in Cell Culture Vol 1," New York, Academic Press, p 241.

15. Kaur-Sawhney R, Flores HE, Galston AW (1980). Polyamine-induced DNA synthesis and mitosis in oat leaf protoplasts. Plant Physiol 65:368.
16. Flores HE, Galston AW (1982) Analysis of polyamines in higher plants by high performance liquid chromatography. Plant Physiol 69:701.
17. Flores HE, Galston AW (1982). Polyamines and plant stress: activation of putrescine biosynthesis by osmotic shock. Science 217:1259.
18. Flores HE, Galston AW (1984). Osmotic stress-induced polyamine accumulation in cereal leaves. I. Physiological parameters of the response. Plant Physiol 75(in press).
19. Flores HE (1983). "Studies on the Physiology and Biochemistry of Polyamines in Higher Plants." Ph. D. Dissertation, Yale University.
20. Smith TA (1984) Putrescine and inorganic ions. In "Recent Advances in Phytochemistry Vol. 18." New York: Plenum Press (in press).
21. Smith TA (1963). L-arginine carboxy-lyase of higher plants and its relation to potassium nutrition. Phytochem 2:241.
22. Smith TA (1970). The biosynthesis and metabolism of putrescine in higher plants. Ann N Y Acad Sci 171:988.
23. Mamont PS, Bey P, Koch-Weser J (1981). Biochemical consequences of drug-induced polyamine deficiency in mammalian cells. In Gaugas JM (ed) "Polyamines in Biomedical Research." New York: Wiley, p 147.
24. Smith TA (1979). Arginine decarboxylase of oat seedlings. Phtyochem 18:1447.
25. Young ND, Galston AW (1983). Putrescine and acid stress. Induction of arginine decarboxylase activity and putrescine accumulation by low pH. Plant Physiol 71:767.
26. Le Rudulier D, Goas G (1971). Mise en evidence et dosage de quelques amines dans les plantules de *Soja hispida* Moench. privées de leurs cotyledons et cultivées en présence de nitrates, d'uree et de chlorure d'ammonium. C R Acad Sci (Paris) Sér D 273:1108.
27. Strogonov BP (1973) "Structure and Function of Plant Cells in Saline Habitats." Jerusalem: Israel Program for Scientific Translations, p 284.
28. Flores HE, Galston AW (1984). Osmotic stress-induced polyamine accumulation in cereal leaves. II. Relation to amino acid pools. Plant Physiol. 75 (in press)

29. Young ND, Galston AW (1984). Physiological control of arginine decarboxylase activity in K-deficient oat shoots. Plant Physiol (in press).

30. Ramakrishna S, Adiga PR (1975). Arginine decarboxylase from *Lathyrus sativus* seedlings. Eur J Biochem 59:377.

31. Smith TA, Negrel J, Bird CR (1983) The cinnamic acid amides of the di- and polyamines. In Bachrach U, Kaye A, Chayen R (eds) "Advances in Polyamine Research Vol. 4." New York: Raven Press, p 347.

32. Martin-Tanguy J, Cabanne F, Perdrizet E, Martin C (1978) The distribution of hydroxycinnamic amides in flowering plants. Phytochem 17:1927.

33. Ponchet M, Martin-Tanguy J, Marais A, Martin C (1982).Hydroxycinnamic acid amides and aromatic amines in the inflorescences of some *Araceae* species. Phytochem 21:2865.

34. Basso LC, TA Smith (1974). Effect of mineral deficiency on amine formation in higher plants. Phytochem 13:875.

35. Delétang J (1974). Présence de caffeoyl putrescine, de caffeoyl spermidine et de dicaffeoyl spermidine chez *Nicotiana tabacum*. Ann Tabac (Sect 2) 11:123.

36. Le Rudulier D, Goas G (1979). Contribution a l'étude de l'accumulation de putrescine chez des plantes cultivées en nutrition strictment ammoniacale. C R Acad Sci (Paris) Sér D 288:1387.

37. Martin-Tanguy J, Martin C, Gallet M (1973). Présence de composés aromatiques liés à la putrescine dans divers *Nicotiana* virosés. C R Acad Sci (Paris) Sér D 276:1433.

38. Martin-Tanguy J, Martin C, Gallet M, Vernoy R (1976). Sur de puissants inhibiteurs naturels de multiplication du virus de la mosaique du tabac. C R Acad Sci (Pars) Sér D 282:2231.

39. Martin C, Martin-Tanguy J (1981). Polyamines conjuguées et limitation de l'expansion virale chez les végétaux. C R Acad Sci (Paris) Sér D 293:249.

40. Ponchet M, Martin-Tanguy J, Andréoli C, Martin C (1982). Apparition de substances de type phénolamide lors de l'interaction *Dianthus caryophyllus* L. var. «*Scania*» - *Phytophthora parasitica* Dastur. Agronomie 2:37.

41. Coleman RG, Richards FL (1956). Physiological studies in plant nutrition. XVIII. Some aspects of nitrogen metabolism in barley and other plants in relation to potassium deficiency. Ann Bot NS 20:393.

42. Smith TA, Sinclair C (1967). The effect of acid feeding on amine formation in barley. Ann Bot NS 31:103.
43. Suresh MR, Ramakrishna S, Adiga PR (1978). Regulation of arginine decarboxylase and putrescine levels in *Cucumis* cotyledons. Phytochem 17:57.
44. Priebe A, Klein H, Jager HJ (1978). Role of polyamines in SO_2-polluted pea plants. J Exp Bot 29:1045.
45. Gale EF (1946) The bacterial amino acid decarboxylases. Adv Enzymol 6:1.
46. Wu WH, Morris DR (1973). Biosyntetic arginine decarboxylase from *Escherichia coli*. J Biol Chem 248:1687.
47. Roberts JKM, Calllis J, Wemmer D, Walbot V, Jardetzky O (1984). The mechanism of cytoplasmic pH regulation in hypoxic maize root tips and its role in survival under hypoxia. Proc Natl Acad Sci USA (in press).
48. Nicholls DG (1982). "Bioenergetics. An Introduction to the Chemiosmotic Theory." London: Academic Press, p 190.
49. Kyriakidis DA (1983). Macromolecular effectors of ornithine decarboxylase activity in germinating barley seeds. In Bachrach U, Kaye A, Chayen R (eds) "Advances in Polyamine Research Vol. 4." New York, Raven Press, p 427.
50. Kuehn GD, Atmar VJ (1983). New perspectives on polyamine-dependent protein kinase and the regulation of ormithine decarboxylase by reversible phosphorylation. In Bachrach U, Kaye A, Chayen R(eds) "Advances in Polyamine Research Vol. 4." New York: Raven Press, p 615.
51. Pegg AE, Seeley JE, Zagon IS, (1982). Autoradiographic identification of ornithine decarboxylase in mouse kidney by means of α-[5-14C]difluoromethylornithine. Science 217:68.
52. Pösö H, Pegg AE (1983). Measurement of ornithine decarboxylase in *Saccharomyces cerevisiae* and *Saccharomyces uvarum* by using α-[5-14C]difluoromethylornithine. Biochim Biophys Acta 747:209.
53. Amrhein N, Gödeke KH (1977). α-aminooxy-ß-phenylpropionic acid- a potent inhibitor of L-phenylalanine ammonia-lyase in vitro and in vivo. Plant Sci Lett 8:313.
54. Dumortier FM, Flores HE, Shekhawat NS, Galston AW (1983). Gradients of polyamines and their biosynthetic enzymes in coleoptiles and roots of corn. Plant Physiol 72:915.

55. Dai YR, Kaur-Sawhney R, Galston AW (1982). Promotion by gibberellic acid of polyamine biosynthesis in internodes of light-grown dwarf peas. Plant Physiol 69:103.
56. Montague M. Armstrong T, Jaworski E (1978). Polyamine metabolism in embryogenic cells of *Daucus carota*. I. Changes in intracellular contents and rates of synthesis. Plant Physiol 62:430.
57. Montague M, Armstrong T, Jaworski E (1979). Polyamine metabolism in embryogenic cells of *Daucus carota*. II. Changes in arginine decarboxylase activity. Plant Physiol 63:341.
58. Wielgat B, Kleckowski K (1971). Putrescine metabolism in pea seedlings. Acta Soc Bot Bot Pol 40:197.
59. Streeter JG, Thompson JF (1972). Anaerobic accumulation of γ-aminobutyric acid and alanine in radish leaves (*Raphanus sativus* L.). Plant Physiol 49:572.

Cellular and Molecular Biology of Plant Stress, pages 115–128
© 1985 Alan R. Liss, Inc.

REGULATION OF THE OSMOTICALLY STIMULATED
TRANSPORT OF PROLINE AND GLYCINEBETAINE
IN *SALMONELLA TYPHIMURIUM*[1]

Virginia Joan Dunlap and Laszlo N. Csonka

Department of Biological Sciences, Purdue University
West Lafayette, Indiana 47907

ABSTRACT   We are studying the osmoregulatory mechanisms
of the bacterium *S. typhimurium* because the results
with this simple organism might help to elucidate the
osmoregulatory mechanisms in higher plants. Exogenous
proline or glycinebetaine stimulates the growth rate of
S. typhimurium under conditions of osmotic stress.
Assays of the rate of transport of ^{3}H-proline showed
that there are two distinct proline transport systems
in *S. typhimurium* whose activity is stimulated by osmotic stress. One of the systems is stimulated because
osmotic stress causes at least a 10-fold induction of
the transcription of the gene (proU) that encodes one
of its constituent proteins. The other system, encoded
by the proP gene, is not regulated at the level of
transcription. Its stimulation by osmotic stress is
probably the result of a conformational change or covalent modification of a transport protein. We isolated
regulatory mutants that express the former (proU) system at a high level even in the absence of osmotic
stress. We also constructed mutants that do not respond
to stimulation by glycinebetaine in media of inhibitory
osmotic strength. These mutants probably lack a glycinebetaine transport system, that might be analogous to one
of the osmotically stimulated proline transport systems.

[1]This work was supported by the Public Health Service
Grant GM 3194401 from the National Institutes of
Health.

INTRODUCTION

Organisms generally respond to increases in the osmotic strength of their environment by elevating the internal concentrations of a number of inorganic ions and low molecular weight organic molecules (1-3). Prominent among the organic metabolites that are accumulated by a wide variety of organisms, ranging from bacteria to higher plants and even some animals, is the imino acid, L-proline (4-6), or the quaternary amine, glycinebetaine (7,8). These two compounds, at high intracellular concentrations, can somehow alleviate osmotic inhibition in bacteria, because exogenous proline or glycinebetaine stimulate the growth rate of some species under conditions of osmotic stress (9-13). This observation has been the rationale for the construction of proline overproducing mutants of *Salmonella typhimurium*, some of which as a consequence acquired increased tolerance of osmotic stress (11).

The enhanced accumulation of proline or glycinebetaine can be caused by two mechanisms: an increase in the net rate of synthesis, or an increase of uptake from the environment. In higher plants, this increased accumulation is accomplished by the former mechanism. Although increased proline accumulation in bacteria under osmotic stress can be brought about by increased synthesis, as evidenced by the proline overproducing mutants, normally the synthesis of proline or glycinebetaine is not enhanced during osmotic stress; rather the transport from the medium is stimulated (11-14). Britten and McClure noted that when *E. coli* was supplied with exogenous proline, the intracellular concentration of this metabolite was directly proportional to the osmotic strength of the medium (15). Kaback and Deuel reported that the rate of proline transport into cell-free vesicles of *E. coli* membranes was stimulated upon increasing the osmolarity of the assay buffer (16). Similar results were obtained by Weaver with whole *E. coli* cells (17).

The enhancement of proline transport in response to osmotic stress could be the consequence either of a direct modification of the transport protein(s) endowing them with elevated activity, or the consequence of induction of the gene(s) of the transport proteins, resulting in elevated synthesis of these proteins. The analysis of the stimulation of proline transport by osmotic stress in *S. typhimurium* is complicated by the presence of three distinct transport systems for proline. The first proline transport system, PPI, which is encoded by the putP⁺ gene, is not affected by osmotic stress (14), but rather it is induced by growth of the organism on proline as a nitrogen or carbon source (18). The other two proline transport systems, PPII and PPIII, might be subject to osmotic regulation. Mutants that lack either PPII or PPIII display a partial reduction in the ability of proline to stimulate growth under conditions of osmotic stress, and double mutants which lack both PII and PPIII are not stimulated by proline at all (14). PPII seems to be active in cells grown in media of normal and elevated osmotic strength, whereas PPIII functions only in cells grown in media of elevated osmotic strength (14).

In order to distinguish between the two possible modes of stimulation of proline transport, we constructed mutant strains in which the gene for PPII (the proP⁺ gene), or the gene for PPIII (proU⁺) was fused to the E. coli lacZ⁺ gene, encoding the enzyme β-galactosidase. In strains carrying such gene fusions, the synthesis of β-galactosidase is dependent on transcription initiated at the promoters of the respective genes to which lacZ has been fused (Cf. 19). Thus, by measuring the β-galactosidase activity in these strains we can easily monitor the regulation of the expression of the proline transport genes, as the osmotic strength of the medium is varied. We also determined the transport activities of PPII and PPIII by assaying the rates of uptake of ³H-proline in appropriate mutants under various conditions of osmotic stress. The results were that the transport of ³H-proline via either PPII or PPIII, as determined by the direct measurements, was enhanced by osmotic stress. However, the β-galactosidase activity in strains carrying lacZ fusions to proP were invariant, whereas in strains carrying lacZ fusions to proU, the β-galactosidase levels were regulated by the osmolarity of the medium. These results indicate that both of the two possible mechanisms of stimulation of proline transport occur in S. typhimurium: PPII is activated by increased "catalytic" function, and PPIII is activated as a consequence of increased transcription.

At present, we do not have comparable results with the glycinebetaine transport system. We have isolated mutants which no longer respond to stimulation by glycinebetaine under conditions of osmotic stress. In this communication, we also present a preliminary description of a number of these mutants.

METHODS

Bacterial culture conditions. The composition of the
minimal medium 63 (M63) and the culture conditions were described previously (11).
Bacterial strains. The derivation of *S. typhimurium*
LT2 strains TL106, TL179, TL185, TL192, TL195, and TL199 has
been published (14). Strains carrying genetic fusions of
lacZ to proP or proU have been constructed using phage Mud-1
(Ap^R lac) developed by Casadaban and Cohen (20). Strains
lacking PPI (putP$^-$) but possessing PPII (proP$^+$) are sensitive to the proline analogue 3,4-dehydro-D,L-proline (Dhp),
but derivatives lacking PPII (putP$^-$ proP$^-$) are resistant
(14). The proP-lacZ fusions were generated by infecting (21)
strain TL212 (ΔputPA proP$^+$ galE$^-$) with Mud-1 and selecting
derivatives resistant to ampicillin and Dhp in medium 63.
Strains lacking PPI and PPII but possessing PPIII (putP$^-$
proP$^-$ proU$^+$) are resistant to Dhp in medium 63, but they are
sensitive to it in medium 63 containing 0.3 M NaCl (14);
strains lacking all three transport systems (putP$^-$ proP$^-$
proU$^-$) are resistant to Dhp even in the presence of 0.3 M
NaCl. To select proU-lacZ fusions, strain TL200 (ΔputPA proP$^-$
proU$^+$ galE$^-$) was infected with Mud-1, and derivatives that
were resistant to ampicillin and Dhp in the presence of 0.3
M NaCl were selected. The putative proP- or proU-lacZ fusions
were confirmed by genetic mapping and proline transport assays (V. J. Dunlap, A. Korty, and L. N. Csonka, unpublished
data). Strains that were not stimulated by glycinebetaine in
media of elevated osmolarity were constructed by subjecting
mutagenized cells of strain TL179 (ΔputPA proP$^-$) to ampicillin selection for mutants (22) that were unable to grow
in medium 63 containing 1.0 M NaCl and 1 mM glycinebetaine.
Proline transport assays. The measurement of uptake of
^{3}H-proline was by the method of equilibrium dialysis (23),
according to the procedures of Dankert et al. (24). The assays were done at 24°, with 20 µM proline as the substrate,
and in the presence of 100 µg/ml chloramphenicol.
β-galactosidase assays were performed according to the
procedure of Miller (22).

RESULTS

The basis of all the experiments we present is the fact
that proline or glycinebetaine are stimulatory to some bacteria under conditions of osmotic stress. This phenomenon is
illustrated with *Salmonella typhimurium* LT2 in Figure 1.

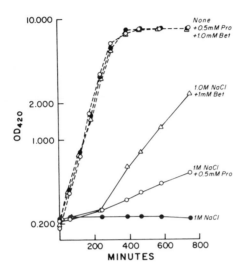

FIGURE 1. The stimulation of growth rate of *Salmonella typhimurium* by proline (Pro) or glycinebetaine (Bet) under conditions of osmotic stress. The optical density of the culture is plotted v.s. time of incubation.

As can be seen, the growth of *S. typhimurium* is completely inhibited by 1 M NaCl. However, proline and glycinebetaine can overcome this inhibition, so that the doubling time with 1 M NaCl in the presence of these two compounds is 7 h and 3 h respectively. The figure shows the results obtained with NaCl, but similar results were obtained with KH_2PO_4, $(NH_4)_2SO_4$, or sucrose as the inhibitory substance (data not shown).

Stimulation of Proline Transport by Osmotic Stress.

As we stated in the INTRODUCTION, the transport of proline into the cells is enhanced under conditions of osmotic stress, and it is probably the accumulation of proline to high levels that results in the stimulation of the growth rate. In order to determine the effect of osmotic stress on proline transport, we measured the rate of uptake of ^{3}H-proline in strains carrying mutations that inactivated one or more of the proline transport systems.

TABLE 1

THE STIMULATION OF PROLINE TRANSPORT ACTIVITY
BY OSMOTIC STRESS

Strain	Proline Transport System[a]			Transport Activity in M63 plus[b]	
	PPI	PPII	PPIII	Nothing	0.3 M NaCl
TL106	+	+	+	0.78	4.78
TL192	–	+	–	0.66	3.46
TL179	–	–	+	0.50	1.93
TL199	+	–	–	0.61	0.63
TL195	–	–	–	0.42	0.29

[a]+ or - indicates a wild type or mutant transport system, respectively.

[b]The transport activity, expressed as nmoles/min mg protein, was determined as described in METHODS. The cells were grown and assayed in the indicated media.

In a strain which is wild type for all the three transport systems (TL106), there is a 6-fold stimulation of proline uptake by growth in medium whose osmolarity has been increased by NaCl, even at a concentration as low as 0.3 M. The rate proline transport is stimulated by osmotic stress in strains that are wild type for either PPII (TL192) or PPIII (TL179). However, the residual transport activity in the strain that has only PPI (TL199) is not affected by exposure to 0.3 M NaCl. Thus, the stimulation of proline transport in wild-type *S. typhimurium* that occurs upon osmotic stress is due to the activation of two independent transport systems, PPII and PPIII. The third transport system, PPI, is not subject to osmotic regulation.

Transcriptional Regulation of PPII and PPIII.

On the basis of the above experiments alone, we cannot distinguish between the two alternatives whether the stimulation of proline transport is due to an activation of pre-existing transport proteins, or to an increased synthesis of the transport proteins. Therefore, we constructed strains that carry fusions of the lacZ gene to the proP or proU, the genes for PPII and PPIII respectively. Since the β-galactosidase activity in these strains is a measure of the rates of transcription of the proP and proU genes, we can analyze the genetic regulation of these genes by assaying β-galactosidase. The results obtained with three independent proP-lacZ fusion strains are in Table 2, and results with three independent proU-lacZ fusion strains are in Table 3.

TABLE 2
THE EXPRESSION OF β-GALACTOSIDASE ACTIVITY
IN proP-lacZ FUSION STRAINS

Strain	β-galactosidase (nmoles/min mg) in cells grown in M63 plus	
	Nothing	0.3 M NaCl
TL423	38	50
TL424	65	129
TL425	34	60

TABLE 3
THE EXPRESSION OF β-GALACTOSIDASE ACTIVITY
IN proU-lacZ FUSION STRAINS

Strain	β-galactosidase (nmoles/min mg) in cells grown in M63 plus		
	Nothing	0.3 M NaCl	0.54 M Sucrose
TL391	21	232	Not determined
TL393	26	455	316
TL390	11	140	Not determined

These data show that in proP-lacZ fusion strains the
β-galactosidase activity, hence the expression of the proP
gene, is virtually unaffected as the osmolarity of the medium
is increased. However, in the proU-lacZ fusion strains, 0.3 M
NaCl causes an 11- to 17-fold induction in the synthesis of
β-galactosidase, hence an 11- to 17-fold induction of the
proU gene. This induction can also be elicited by 0.54 M
sucrose (Table 3), or other solutes, ionic or non-polar, provided they are not freely permeable across the membrane (V.
J. Dunlap and L. N. Csonka, manuscript in preparation). The
induction is rapid (<10 min), and the levels of expression of
the proU gene increase linearly with the osmotic strength of
the medium from 0 to 0.5 M NaCl.

Mutations Resulting in Constitutive Expression of PPIII.

One of our long-term objectives is to analyze the mechanisms by which osmotic stress is sensed by *S. typhimurium*.
In an attempt to generate strains with possible alterations
in the osmoregulatory mechanisms, we selected mutants which
express PPIII at an elevated level in the absence of osmotic
stress. The starting strain for this construction was TL185,
which carries a mutation (ΔproBA), causing a loss of the
first and second enzymes of proline biosynthesis. Because of
this lesion, the strain must be supplemented with exogenous
proline for growth. The strain has two additional mutations,
putP⁻ and proP⁻ which inactivate PPI and PPII, respectively.
Because the latter two mutations block proline transport, the
strain needs abnormally high levels of proline (>1 mM) for
growth on the normal medium. However, because PPIII can be
induced by growth on media of elevated osmolarity, the strain
can grow optimally with only 0.1 mM proline, in the presence
of an extra 0.3 M NaCl. So we speculated that derivatives of
strain TL185 which could grow with 0.1 mM proline in the normal medium might be strains that express PPIII in the absence
of osmotic stress.

We obtained approximately 20 derivatives of strain TL185
with the desired phenotype (J. Druger-Liotta and L. N. Csonka,
unpublished data). As a preliminary test of whether the mutation might elevate the synthesis of PPIII, we introduced a
proU-lacZ fusion into one of these strains. The results of
measurement of the β-galactosidase activity in the latter
strain (TL430), and in a control strain with normal regulation
of PPIII (TL428) are in Table 4.

TABLE 4

THE β-GALACTOSIDASE ACTIVITY IN A proU-lacZ
FUSION STRAIN CARRYING A REGULATORY MUTATION
FOR THE EXPRESSION OF PPIII

Strain	β-galactosidase (nmoles/min mg) in cells grown in M63 plus	
	Nothing	0.3 M NaCl
TL430	71	230
TL428	11	117

Thus, the mutation in strain TL430 elevates the expression of the proU gene 6-fold in the absence of osmotic stress,
as compared with that seen in the control strain, TL428. But,
the expression of the gene in strain TL430 still responds to
regulation by osmotic stress. Mapping experiments indicate
that the regulatory mutation is very closely linked to the
proU gene (J. Druger-Liotta, D. Overdier and L. N. Csonka,
unpublished data). We currently have no further information
about the mutation; it might be an alteration in the promoter
of proU, or in some regulatory gene.

Mutants That Are Not Stimulated by Glycinebetaine during
Osmotic Stress.

Our search for mutations affecting the osmoregulatory
mechanisms of the cell included the selection of strains
that were no longer stimulated by glycinebetaine. We have
initially obtained eight such strains (Cf. METHODS).
The growth rates of two of these strains, in media of
elevated osmotic strength containing glycinebetaine, are
listed in Table 5.

TABLE 5
THE GROWTH CHARACTERISTICS OF MUTANTS THAT DO NOT
RESPOND TO STIMULATION BY GLYCINEBETAINE
DURING OSMOTIC STRESS

Strain[a]	Growth Rate (generation/h) in M63 plus		
	Nothing	0.5 M NaCl	0.5 M NaCl + 1 mM glycinebetaine
TL241 (bet-8)	1.05	0.45	0.28
TL242 (bet-11)	1.00	0.42	0.47
TL179 (bet[+])	1.11	0.44	0.71
TL195 (bet[+])	1.11	0.47	0.80

[a]Strains TL241 and TL242 were selected from
strain TL179 (putP⁻ proP⁻) as strains that
are not stimulated by glycinebetaine under
conditions of osmotic stress, and like the
parental strain TL179, they lack PPI and
PPII. The control strain TL195 lacks PPI,
PPII, and PPIII.

Strains TL241 and TL242 which are not stimulated by
glycinebetaine during osmotic stress, probably lack a transport system for this metabolite. Strain TL195, which is devoid of all three proline transport systems, hence is not at
all stimulated by proline during osmotic inhibition (14), is
nevertheless stimulated by glycinebetaine. Conversely, strains
TL241 and TL242 still respond to stimulation by proline under
conditions of osmotic inhibition (L. N. Csonka, unpublished
data). These results indicate that the transport system which
takes up glycinebetaine is probably distinct from the osmotically activated proline transport systems. We have not yet
characterized the glycinebetaine transport system more extensively, but we anticipate that the regulation of this system might be analogous to the regulation of PPII or PPIII.

DISCUSSION

The major new finding presented in this manuscript is that two of the three proline transport systems of *S. typhimurium*, PPII and PPIII, are stimulated by osmotic stress. However, the mechanism of stimulation of the two systems is different. The rate of transcription of the gene for PPII is invariant under conditions of osmotic stress, and therefore the stimulation of proline uptake via this system is probably due to some modification of a component of the system that confers on it higher activity. This modification might be a conformational change, chemical modification, or an altered interaction with other membrane constituents. The stimulation of PPIII, on the other hand, is accomplished by a genetic induction, resulting in a higher level of transcription of it structural gene(s).

In *S. typhimurium* and other enteric bacteria a number of cellular phenomena other than proline and glycinebetaine transport are known to be subject to osmotic regulation. The intracellular levels of glutamate, glutamine, and putrescine are regulated in response to osmotic stress (11,25). Osmotic shock stimulates the uptake of potassium by a mechanism that entails the induction of the transcription of the genes for one of the potassium transport systems (26). There are two outer-membrane "porin" proteins, OmpC and OmpF, whose synthesis is regulated such that the ratio of OmpF to OmpC increases with increasing external osmotic strength (27). The region between the cytoplasmic membrane and the cell wall, known as the periplasmic space, contains a membrane-derived oligosaccharide, MDO, whose function is thought to be to balance the osmotic strength of the periplasm to match that of the cytoplasm. The synthesis of MDO is inhibited under conditions of high external osmolarity (28). It is not clear whether all these phenomena are controlled by a common regulatory element. Our results, that the two proline transport systems are activated in response to osmotic stress by different mechanisms points out the probability that several osmotic regulatory mechanisms exist in bacteria.

In light of the fact that it is possible to generate bacterial mutants with enhanced osmotic stress tolerance by selecting proline over-producing strains, it is curious that wild-type *S. typhimurium* does not elevate the synthesis of proline, or glycinebetaine in response to osmotic stress. However, it does accumulate high levels of these metabolites during osmotic stress by enhanced transport from the exterior (13,14). Our analysis of the regulation of the osmotically stimulated transport systems for proline and glycinebetaine should shed light on the central osmoregulatory mechanisms of *S. typhimurium*. Because proline and glycinebetaine are also important osmoregulatory molecules in plants, the results obtained during the analysis of the function of these metabolites under osmotic stress in bacteria might help in the elucidation of the mechanisms of osmoregulation in higher plants.

ACKNOWLEDGMENTS

We thank W. A. Cramer for teaching us the use of the equilibrium dialysis apparatus for proline transport assays. We also thank S. B. Gelvin and C. A. Lindquist for helpful critique of the manuscript.

REFERENCES

1. Yancey PH, Clark ME, Hand SC, Bowlus RD, Somero GN (1982). Living with water stress: evolution of osmolyte systems. Science 217:1214.
2. Flowers TJ, Troke PF, Yeo AR (1977). The mechanisms of salt tolerance in halophytes. Ann Rev Plant Physiol 28:89.
3. Jeffries RL (1980). The role of organic solutes in osmoregulation in halophytic higher plants. In Rains DW, Valentine RC, Hollaender, A (eds): "Genetic Engineering of Osmoregulation," New York: Plenum Press, p 135.
4. Measures JC (1975). Role of amino acids in osmoregulation in nonhalophilic organisms. Nature 257:398.
5. Stewart GR, Lee JA (1974). The role of proline accumulation in halophytes. Planta 120:279.
6. Fyhn HJ (1976). Haloeuryhalinity and its mechanisms in a cirriped crustacean, *Balanus improvisus*. Comp Bioch Physiol 53A:19.
7. Wyn Jones RG (1980). An assessment of quaternary ammonium and related compounds as osmotic effectors. In Rains DW, Valentine RC, and Hollaender A (eds): "Genetic Engineering of Osmoregulation," New York: Plenum Press, p 155.
8. Hitz WD, Landyman JAR, Hansen AD (1982). Betaine synthesis and accumulation in barley during field water stress. Crop Sci 22:47.

9.   Christian JHB (1955). The influence of nutrition on the water relations of *Salmonella orianenburg*. Aust J Biol Sci 8:75.

10. Shkedy-Vinkler C, Avi-dor Y (1975). Betaine-induced stimulation of respiration at high osmolarities in a halotolerant bacterium. Bioch J 150:219.

11. Csonka LN (1981). Proline over-production results in enhanced osmotolerance in *Salmonella typhimurium*. Molec Gen Genet 182:82.

12. LeRudulier D, Yang SS, Csonka LN (1982). Nitrogen fixation in *Klebsiella pneumoniae* during osmotic stress: effect of exogenous proline or a proline over-producing plasmid. Bioch Bioph Acta 719:723.

13. LeRudulier D, Boulliard L (1983). Glycine betaine, an osmotic effector in *Klebsiella pneumoniae* and other Enterobactericeae. Appl Env Microbiol 46:152.

14. Csonka LN (1982). A third L-proline permease in *Salmonella typhimurium* which functions in media of elevated osmotic strength. J Bacteriol 151:1433.

15. Britten RJ, McClure FT (1962). The amino acid pool in *Escherichia coli*. Bacteriol Rev 26:292.

16. Kaback HR, Deuel TF (1969). Proline uptake by disrupted membrane preparation from *Escherichia coli*. Arch Bioch Bioph 132:118.

17. Weaver CA (1981). "Interaction of Colicin Ia with *Escherichia coli*: I. Mechanisms of Colicin Sensitivity. II. Inhibition of Colicin Binding by Phenolic Un-couplers." University of Illinois at Champaign-Urbana: Ph.D. Thesis.

18. Ratzkin B, Grabnar M, Roth J (1978). Regulation of the major proline permease in *Salmonella typhimurium*. J.Bacteriol 133:737.

19. Bassford P, Beckwith J, Berman M, Brickman E, Casadaban M, Guerente L, Saint-Girons I, Sarthy A, Schwartz M, Shuman H, Silhavy T (1978). Genetic fusions of the lac operon: a new approach to the study of bio-logical processes. In "The Operon," Cold Spring Harbor: Cold Spring Harbor Laboratory, p 245.

20. Casadaban MJ, Cohen SN (1979). Lactose genes fused to exogenous promoters in one step using a Mu-lac bacteriophage: *in vivo* probe for transcriptional control sequences. Proc Natl Acad Sci USA 76:4530.

21. Csonka LN, Howe MM, Ingraham JL, Pierson III LS, Turnbough Jr CL (1981). Infection of *Salmonella typhimurium* with coliphage Mud-1 (Ap[r] lac): construction of pyr::lac gene fusions. J Bacteriol 145:299.

22. Miller JH (1972). "Experiments in Molecular Genetics."
     Cold Spring Harbor: Cold Spring Harbor Laboratory,
     p 230.
23. Kaback HR (1977). Molecular biology and energetics of
     membrane transport. In Semenza G, Carafoli E (eds):
     "Biochemistry of Membrane Transport," Berlin: Springer-
     Verlag, p 598.
24. Dankert J, Hammond SM, Cramer WA (1980). Reversal by
     trypsin of the inhibition of active transport by
     Colicin E1. J Bacteriol 143:594.
25. Munro GF, Hercules K, Morgan J, Sauerbier W (1972).
     Dependence of the putrescine content of *Escherichia
     coli* on the osmotic strength of the medium. J Biol
     Chem 247:1272.
26. Laimins LA, Rhoads DB, Epstein W (1981). Osmotic
     control of kdp operon expression in *Escherichia coli*.
     Proc Natl Acad Sci USA 78:464.
27. Ozawa Y, Mizishima S (1983). Regulation of outer
     membrane porin protein synthesis in *Escherichia coli*
     K-12: ompF regulates the expression of ompC. J
     Bacteriol 154:669.
28. Kennedy EP (1982). Osmotic regulation and biosynthesis
     of membrane-derived oligosaccharides in *Escherichia
     coli*. Proc Natl Acad Sci USA 79:1092.

Cellular and Molecular Biology of Plant Stress, pages 129–143
© 1985 Alan R. Liss, Inc.

EFFECT OF SALINITY ON GROWTH AND MAINTENANCE
COSTS OF PLANT CELLS

Suzan J. Stavarek and D. William Rains

Department of Agronomy & Range Science
University of California
Davis, California 95616

ABSTRACT   The widespread occurrence of salinity
imposes a limitation to irrigated agriculture. The
development of salt tolerant crop plants through
genetic improvement provides a means to reduce the
detrimental effects of salinity on productivity. The
knowledge of salt tolerance in biological systems
would enhance breeding efforts. Understanding the
processes related to salt tolerance could provide
potential biological markers and characteristics that
could be used to identify possible economically
useful, salt tolerant crop plants.
    Plants exposed to salinity show alterations in
physiological processes. This can involve a diversion of metabolic energy to enhance the tolerance to
the salt stress. Understanding the changes in energy
costs associated with the stress will provide insight
into the potential productivity of growing crops in
saline environments.
    Through the use of a cell culture system, the
effects of salt on the growth and maintenance costs
of plant cells was determined. The plant cells were
grown in media containing precise concentrations of
sugar and salt, thereby exposing the cells to quantified amounts of energy and stress. The proportion of
glucose used for growth and maintenance was determined and the effect of NaCl was evaluated. By
comparing NaCl selected cells to nonselected cells,
changes in the cost of metabolic processes which
allow the NaCl selected cells to be productive were
evaluated.

INTRODUCTION

Plants exposed to environmental stress show alterations in physiological processes. This involves diversion
of metabolic energy to cellular processes which provides
tolerance to the stress. An understanding of the molecular basis of tolerance mechanisms by which plants and
plant cells adjust to a saline environment is important to
all programs attempting to establish salt tolerant crop
species. The identification of specific characteristics
related to salt tolerance will provide potential biological markers useful in the identification and genetic
manipulation of salt tolerant plants and plant cells. An
understanding of the changes in energy costs associated
with salinity and the specific characteristics related to
salt tolerance can provide insight into the potential
productivity of growing crops in a saline environment.
Plants exposed to saline environments encounter three
basic problems: a) a reduction in water potential of the
surrounding environment results in water becoming less
available; b) toxic ions can interfere with the physiological and biochemical processes of the organism; and c)
required nutrient ions must be obtained despite the predominance of other ions (1).
Plants have evolved several strategies to overcome
the problems of saline environments. Plants may avoid the
stress by growing during months of more favorable conditions (i.e., high rainfall months). Exclusion is another
strategy used by plants. The exclusion may be at a whole
plant level in which salts are not allowed to enter the
shoots. This may be accomplished by active pumping of
salts back into the roots. Exclusion mechanisms also may
involve salt glands located on the surface of leaves which
extrude salts that reach the shoots. Cellular level
exclusions can be accomplished by either not permitting
ions into the cells or pumping them out once they are in.
A highly significant strategy is a physiological tolerance
which may involve compartmentation and osmotic adjustment
using inorganic and organic constituents. All these
mechanisms used by plants have been discussed in detail in
many reviews (2-10).
Whichever mechanisms account for salt tolerance,
there must be an energy cost. The energy required for
the carbon skeletons to make the organic compounds used
for osmotic regulation may limit the growth of the

organism. The energy required for regulation of inorganic ions may limit the potential for plant productivity. Respiratory energy for maintenance of processes competing with growth may increase when the organism is stressed, so that a greater portion of the energy is used for processes resulting in salt tolerance rather than for growth. Yeo (11) recently reviewed how the costs of different mechanisms used by plants under a saline stress can vary.

The synthesis of dry matter requires the input of energy. Energy is also required by the plant for processes which are considered maintenance processes. The maintenance component includes such functions as protein turnover and continuous maintenance of ionic gradients across membranes. Individuals have attempted to separate the proportion of respiration which can be attributed to maintenance processes from that which is needed for growth. The growth rate and respiration rate of a plant have been found not to be directly related. Non-growing tissue also requires a level of respiration needed for maintenance.

Several methods have been developed for evaluating the maintenance costs and efficiency of dry matter production in plants (12-17). Pirt (18) developed a model for determining the maintenance energy of bacteria growing in cultures. He determined that:

$$\frac{1}{Y} = \frac{m}{\mu} + \frac{1}{Yg} \qquad \text{where} \qquad (1)$$

Y = observed growth yield,
Yg = true growth yield,
m = maintenance coefficient, and
μ = specific growth rate.

If m and Yg are constant, then the plot of $1/Y$ against $1/\mu$ will give a straight line, the slope will be m, and the y-axis intercept will equal $1/Yg$. Using data derived from other researchers, Pirt calculated maintenance coefficients (m) and true growth yields (Yg) for a number of bacteria (Aerobacter aerogenes, A. cloacae and Lipolytic bacterium). Aerobic conditions gave maintenance coefficients in the range of 0.076-0.094 g substrate·g dry wt^{-1}· hr^{-1}, while anaerobic conditions gave much higher maintenance coefficients (0.300-0.473 g substrate·g dry wt^{-1}· hr^{-1}). The true growth yields he calculated varied from

0.083 to 0.55 g dry wt·g substrate^{-1}. Selenomonad, another type of bacteria, did not follow a linear relationship when $1/Y$ was plotted against $1/\mu$. Pirt felt that this organism had a peculiar type of energy metabolism, and that other factors such as methods of determining growth could account for the nonlinearity. Calculations for E. coli gave variable results, however, Pirt suggests that this may be due to differences in growth and temperature conditions.

Several factors must be considered in using Pirt's model for calculating maintenance coefficients and true growth yields. The methods used to determine the parameters, e.g., growth (direct dry weight vs. optical density ratio) must be evaluated as this may cause variation in the coefficients calculated. The maintenance coefficient and true growth yield is assumed to be constant during measurements of specific growth rate and actual yield. This, however may not be the case if measurements are taken over different stages of the growth cycle. The maintenance energy is associated with the growing state. Cells in a stationary phase of growth may change the maintenance requirements. A criticism with this method has been that the efficiency of the respiratory processes may change during ontogeny, thereby influencing the maintenance coefficient (19). It is therefore important to evaluate the efficiency of the respiratory process at all stages in the growth cycle.

Another criticism has been that the respiration for ion uptake is included in growth respiration and assumed to be a constant portion of that respiration. However, this may not be a true assumption, especially under ionic stress (e.g., salinity), and therefore Yg would not be constant.

Cell culture provides an excellent model for studies of growth and maintenance costs. The amount of energy and stress can be precisely controlled and easily determined. Alterations in respiration can be directly measured.

In whole plant systems, measurements of available energy and consumption rates are complex. Many structures (leaves, roots, flowers, etc.) and cell types are involved in the measurements. Determinations of m and Yg will therefore be an average of the whole plant. These values would vary dramatically with different stages of growth due to changes in the proportion of each structure and cell type.

In a cell culture, alterations do not have to be made in the normal growth conditions to obtain the needed measurements. These have typically been done in whole plant studies causing concern in the validity of the m and Yg values obtained.

Pirt's model has been used in studies with plant cell cultures. Equation (1) was rearranged to give:

$$\mu \, \frac{1}{Y} \; = \; m \; + \; \frac{1}{Yg} \, \mu \; . \qquad (2)$$

Pirt showed that $-ds/dt = -\mu x/Y$, which can be rearranged to give $-ds/dt \; x = \mu 1/Y$, where ds is the amount of substrate, dt is a small time interval and x is the amount of organism. The rate of substrate consumption, ν, is $-ds/dt$ x, so equation (2) becomes:

$$\nu \; = \; m \; + \; \frac{1}{Yg} \, \mu \qquad (3)$$

Plotting ν against μ gives a linear relationship, with the y-axis intercept equal to m and the inverse of the slope equal to Yg.

Kato and Nagai (20) using <u>Nicotiana</u> <u>tabacum</u> cells and sucrose as the carbon source calculated values for m and Yg. They determined a maintenance coefficient of 0.02 mmol glucose•gram dry wt^{-1}•hr^{-1} (or 3.6 mg glucose•g dry wt^{-1}•hr^{-1}) and Yg = 107 g dry wt•mole glucose^{-1} (or 0.59 g dry wt•g glucose^{-1}). The maintenance coefficient is much less than that determined for bacteria (18), however, the Yg is relatively large. This may reflect differences in composition and complexity of the organism as well as differences in metabolic pathways and respiratory efficiency.

Parellioux and Chaubet (21) also used Pirt's model to determine m and Yg for <u>Medicago</u> <u>sativa</u> cells growing on lactose. They calculated a maintenance coefficient of 4.7 mg lactose•g dry wt^{-1}•hr^{-1} and Yg = 0.765 g dry wt•g lactose^{-1}. These values are similar to those obtained by Kato and Nagai (20) and the differences may reflect specific species characteristics or efficiency differences due to the carbon source.

In both of the plant cell culture studies described above, the disappearance of the carbohydrate (sucrose or lactose) from the media was used to determine the specific consumption rate (ν). This method has a serious error associated with it. It is assumed that the carbohydrate removed from the media is consumed for growth and maintenance purposes. However, these cells are capable of storing the carbohydrate taken up and also utilizing previously stored carbohydrates (sugars and starch). This fact makes it important that tissue sugar and starch concentrations be determined and changes be accounted for when calculating ν.

MATERIALS AND METHODS

Alfalfa seeds were obtained from Dr. E. T. Bingham, Department of Agronomy, University of Wisconsin. The variety, W74RS, was developed for its increased ability to regenerate plants from callus in culture (22). Cell cultures were initiated and selected as described previously (23).

Callus tissue was aseptically weighed into foil-capped 125 ml Erlenmeyer flasks containing 25 ml modified Blaydes growth medium. The modified medium contained glucose instead of sucrose as the carbon source and no agar. Various levels of NaCl were also added. The flasks were maintained at 26°C in the dark on a rotary shaker at 100 rpm. Periodically suspensions were removed for destructive harvest.

To harvest, the cells were gravity filtered through Whatman #4 (11 cm) filter paper. The cells were then removed from the filter paper with forceps and weighed into a tared plastic petri dish. The fresh tissue was dried 2 days at 70°C. Dried tissue was weighed, then ground with mortar and pestle.

For tissue starch and sugar determinations, approximately 50 mg dried tissue was weighed into a test tube and 2 ml double distilled, deionized water added. This was shaken for 15 minutes then filtered through a Millipore 0.45 μm filter (HVLP 02500). A 20 μl sample of the filtrate was injected into a Waters Associate high pressure liquid chromatograph (HPLC). Sugars were separated at 85°C on a BioRad carbohydrate analysis column (Aminex HPX-87C, 300 mm x 7.8 mm), with 2 prefilters: anion/OH micro-guard (Aminex A-25, 40 mm x 4.6 mm) and ion exclusion micro-guard (Aminex HPX-85 H, 40 mm x 4.6 mm). Sugars were detected with a differential refractometer.

The filter paper and plant material were placed into 15 ml centrifuge tubes and 2 ml double distilled, deionized water added. The tubes were put into a water bath at 100°C for 10 minutes, then samples were cooled to room temperature. Two ml of 9.2 N perchloric acid was added, and the tubes were stirred occasionally for 15 minutes. The volume was brought to 10 ml with water and the tubes centrifuged 7 minutes in a desk top centrifuge. The supernatant was collected in a 50 ml digestion tube. Two ml of 4.6 N perchloric acid was added to the residue in the centrifuge tube, and stirred occasionally for 15 minutes. Tubes were then centrifuged 7 minutes, and the supernatant collected in the 50 ml digestion tube. The volume was brought to 50 ml with double distilled, deionized water. The amount of glucose in the sample was then determined colorimetrically. Samples were diluted as needed, and 5 ml placed in a test tube. Tubes with samples and standards were placed into an ice bath. To each tube, 10 ml of an anthrone reagent was added slowly. The tubes were placed in a boiling water bath for 7.5 minutes, and then cooled to room temperature. When cool, the absorbance was measured at 630 mµ.

The media from which the cells were filtered was filtered through a 0.45 µm Millipore filter (HAWP 02500). A 20 µl sample of the filtrate was injected into the Waters HPLC and the amount of sugars remaining determined as described above.

Method of Calculation

Substrate was the amount of sugar in the media plus the amount of sugars in the tissue. Total dry weight and substrate was determined for each harvest, and then entered into a PDP11 computer and a polynomial regression line was fitted with a BMDP-P5R program. Daily predicted dry weight was used to calculate daily specific growth rates (μ):

$$\mu = \frac{\ln \text{dry weight}_{t_2} - \ln \text{dry weight}_{t_1}}{t_2 - t_1}$$

The daily predicted substrate values were used to determine daily specific consumption rates (ν):

$$\nu = \frac{(\text{substrate}_{t_2} - \text{substrate}_{t_1})}{(t_2 - t_1) \cdot \left(\dfrac{\text{dry wt}_{t_2} + \text{dry wt}_{t_1}}{2}\right)}$$

The ν was plotted against μ for each cell line and salt treatment. Linear regressions were run on an HP97 and intercepts and slopes determined.

RESULTS

The dry weight of the nonselected and NaCl selected cells as a function of time is shown in Figure 1 for the three NaCl treatments. The growth of nonselected cells is inhibited when salt is present and essentially no growth occurred at 1% NaCl. There is little effect of NaCl on the growth of the NaCl selected cells, except the cells have a longer lag period when grown on 1% NaCl.

The disappearance of glucose from the media is shown in Figure 2. At 0.5% NaCl the uptake of glucose was inhibited in the nonselected cells and very little glucose was removed when 1% NaCl was present in the growth media. The NaCl selected cells show little reduction of glucose uptake when NaCl is present. At 1% NaCl, the removal of glucose is slightly slower due to the longer lag phase in the growth cycle of the cells.

The concentration of sugars in the tissue has been previously described (24). The concentration of fructose in the tissue is extremely low for both cell lines and does not change with salt treatment or time.

For the nonselected cells on 0% NaCl, and for the NaCl selected cells on all salt treatments, the concentration of glucose drops to a low level within 4 days and remains low the rest of the growth cycle. The sucrose concentration initially decreases, but then increases to a high level by the end of the growth cycle.

At 0.5% NaCl, the nonselected cells show very little change in their glucose concentration over the growth cycle. The concentration of sucrose decreases slightly at first, but then increases to the initial concentration by the end of the growth cycle. When grown on 1% NaCl, the

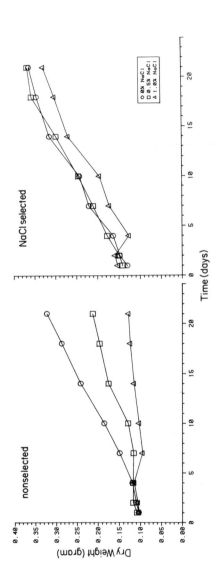

Figure 1.   The dry weight (gram) of the nonselected and NaCl selected alfalfa cells as a function of time (day), at three salt levels: 0% NaCl (O), 0.5% NaCl (□), and 1% NaCl (△).

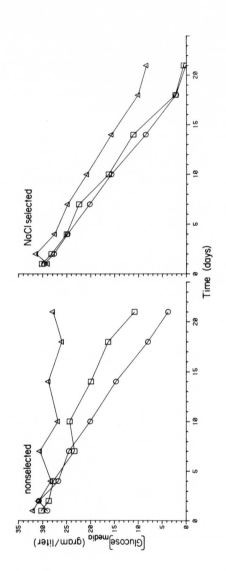

Figure 2. The glucose concentration (grams/liter) as a function of time (day) of the nonselected and NaCl selected alfalfa cells, at three salt levels: 0% NaCl (○), 0.5% NaCl (□), and 1% NaCl (△).

concentration of glucose in the nonselected cells in-creases dramatically, while the sucrose concentration drops to a low level.

The starch concentration for both the nonselected and NaCl selected cells did not change significantly during the growth cycle. The concentration of NaCl in the media did not effect the concentration of starch in the tissue.

Calculations of specific consumption rate (ν) and specific growth rate (μ) were performed using predicted daily values of dry weight and substrate obtained from either first or second degree polynomials (all curves were fitted with $r^2 > 0.82$). A linear relationship was obtained in all plots of ν as a function of μ, as predicted from Pirt's model. (All linear regression had a $r^2 \geq 0.990$.) No calculations were performed for the nonselected tissue grown on 1% NaCl because of the lack of growth.

Table 1 lists the coefficients m and Yg, obtained for both cell lines at each salt level.

TABLE 1

MAINTENANCE (m) AND GROWTH YIELD (Yg) COEFFICIENTS
FOR THE NONSELECTED AND NaCl SELECTED ALFALFA
CELL LINES AT THREE CONCENTRATIONS OF NaCl

	m mg substrate/g dry wt/hr	Yg g dry wt/g substrate
Nonselected cells		
0% NaCl	1.2667	0.4759
0.5% NaCl	5.8125	negative value
1.0% NaCl	____a	____a
NaCl selected cells		
0% NaCl	2.2375	0.5574
0.5% NaCl	1.1167	0.4262
1.0% NaCl	0.5708	0.3822

[a] No calculations performed due to lack of growth.

For the nonselected cell line the maintenance coefficient increases with increasing levels of NaCl, while the opposite is observed for the NaCl selected cell line. The nonselected cell line has a true growth yield of 0.4759 g substrate·g dry wt^{-1}. When under NaCl stress, a negative slope is obtained when ν is plotted against μ. For the NaCl selected cells the true growth yield decreases with increasing levels of NaCl.

DISCUSSION

The uptake of glucose from the media and growth of the nonselected cell line is inhibited when NaCl is present in the medium (Figs. 1 and 2). The utilization of stored sugars, sucrose and glucose, is also altered by NaCl (24). For the NaCl selected cells, however, there is no effect of NaCl on these processes at 0.5% NaCl, and only a slight reduction at 1% NaCl.

The values obtained for the maintenance coefficients (m) are in the range observed by others; 1.5 mg·g^{-1}·hr^{-1} for whole plants (13), 3.6 and 4.7 mg·g^{-1}·hr^{-1} for cell cultures (20, 21). The maintenance cost of the nonselected cells increases dramatically when under a saline stress. The increase in maintenance cost is associated with adjustments the cells must make to maintain their metabolic processes under the stress, such as increased ion regulation, compartmentation or exclusion of the toxic ions and maintenance of membrane integrity. Concomitant with the large increases in the maintenance costs to the nonselected cells, a dramatic decrease is observed in their true (potential) growth yield (Yg). The nonselected cells when grown in 0.5% NaCl actually have a negative slope when ν is plotted against μ. This suggests that the NaCl is toxic to the cells, and the cells are not productive when grown under a saline stress.

The maintenance cost of the NaCl selected cells decreases with increasing stress. These cells are selected for growth and maintained on medium with 1% NaCl. Their internal osmoticum is regulated to maintain ideal metabolic conditions efficiently at 1% NaCl. When placed in medium with a lower level of NaCl, the cells must adjust and then maintain osmotic potential under these new conditions. It appears that this increases the costs to the cells almost 2-fold at 0.5% NaCl and 4-fold at 0%

NaCl. Possible alterations in the regulation of compartmentation and types of osmoticum (ionic vs. organic) used by the cells could cause increases in the maintenance costs.

When maintenance costs are lower, more of the substrate could be available for other energy processes. A lower true growth yield (Yg), however, suggests that less growth occurs per unit of substrate consumed. The NaCl selected cells have a low maintenance cost, low true growth yield (Yg) and also a slower substrate consumption rate when grown in 1% NaCl. Rates of respiration (measured as O_2 consumption) for the NaCl selected cells are not significantly different between the three salt levels.

Lower true growth yields could result from a) possible toxic ion effects, b) reduce efficiency of substrate utilization, and c) diversion of energy substrate to osmotic regulation. Since NaCl does not effect respiration in the NaCl selected cells, this would suggest that the salt stress reduced the coupling between respiration and energy-requiring metabolic processes. Further experiments are necessary to address the effect of NaCl on the coupling between energy-requiring processes and respiration.

The use of Pirt's model (18) provided values for maintenance costs and true growth yields of alfalfa cells. The values were similar to those obtained in whole plant studies. Few studies, however, have shown how these coefficients change under a saline stress.

The ability to measure maintenance coefficients provides a method to determine the costs of different pathways and mechanisms in a cell culture system. For example, two cell lines which differ in the type of osmoticum used under saline stress for adjustments of water potential could be compared. Yeo (11) has reviewed how the cost of different mechanisms can vary. This could provide insight into the type of mechanism which should be selected for in crop improvement programs.

REFERENCES

1. Rains DW (1979) Salt tolerance of plants: strategies of biological systems. In Hollaender A (ed): "The Biosaline Concept: An Approach to the

Utilization of Saline Environments," New York: Plenum Publishing, p 47.
2. Flowers TJ, Troke PF, Yeo AR (1977) The mechanism of salt tolerance in halophytes. Ann Rev Plant Physiol 28:89.
3. Greenway H, Munns R (1980) Mechanisms of salt tolerance in non-halophytes. Ann Rev Plant Physiol 31:149.
4. Hellebust JA (1976) Osmoregulation. Ann Rev Plant Physiol 27:485.
5. Jefferies RL (1980) The role of organic solutes in osmoregulation in halophytic higher plants. In Rains DW, Valentine RC, Hollaender A (eds): "Genetic Engineering of Osmoregulation; Impact on Plant Productivity for Food, Chemicals, and Energy," New York: Plenum Publishing, p 135.
6. Levitt J (1982) "Responses of Plants to Environmental Stresses." New York: Academic Press, Vol II.
7. Rains DW (1972) Salt transport by plants in relation to salinity. Ann Rev Plant Physiol 2:367.
8. Stavarek SJ, Rains DW (1983) Mechanisms for salinity tolerance in plants. Iowa State J Res 57(4):457.
9. Stewart GR, Larher F (1980) Accumulation of amino acids and related compounds in relation to environmental stress. In Miflin BJ (ed): "The Biochemistry of Plants, Vol. 5." New York: Academic Press, Inc, p 609.
10. Wyn Jones RG (1980) An assessment of quaternary ammonium and related compounds as osmotic effectors in crop plants. In Rains DW, Valentine RC, Hollaender A, (eds): "Genetic Engineering of Osmoregulation; Impact on Plant Productivity for Food, Chemicals, and Energy." New York: Plenum Publ, p 155.
11. Yeo AR (1983) Salinity resistance: physiologies and prices. Physiol Plant 58:214.
12. McCree KJ, Silsbury JH (1978) Growth and maintenance requirements of subterranean clover. Crop Sci 18:13.
13. McCree KJ (1982) Maintenance requirements of white clover at high and low growth rates. Crop Sci 22:345.
14. McDermitt DK, Loomis RS (1981) Elemental composition of biomass and its relation to energy content, growth efficiency and growth yield. Ann Bot 48:275.

15. Penning de Vries FWT, Brunsting AHM, van Laar HH (1974) Products, requirements and efficiency of biosynthesis: A quantitative approach. J Theor Biol 45:339.
16. Penning de Vries FWT (1975) The cost of maintenance processes in plant cells. Ann Bot 39:77.
17. Thornley JHM (1976) "Mathematical Models in Plant Physiology." London: Academic Press, p 123.
18. Pirt SJ (1965) The maintenance energy of bacteria in growing cultures. Proc R Soc Lond Biol 163:224.
19. Lambers H, Szaniawski RK, de Visser R (1983) Respiration for growth, maintenance and ion uptake. An evaluation of concepts, methods, values and their significance. Physiol Plant 58:556.
20. Kato A, Nagai S (1979) Energetics of tobacco cells, Nicotiana tabacum L., growing on sucrose medium. European J Appl Microbiol Biotechnol 7:219.
21. Pareilleux A, Chaubet N (1981) Mass cultivation of Medicago sativa growing on lactose: kinetic aspects. Eur J Appl Micro Biotec 11:222.
22. Bingham ET, Hurley LV, Kaatz DM, Saunders JW (1975) Breeding alfalfa which regenerates from callus tissue in culture. Crop Sci 15:719.
23. Croughan TP, Stavarek SJ, Rains DW (1978) Selection of a NaCl tolerant line of cultured alfalfa cells. Crop Sci. 18:959.
24. Stavarek SJ, Rains DW (1983) The development of tolerance to mineral stress. HortSci (in press).

Cellular and Molecular Biology of Plant Stress, pages 145–160
© 1985 Alan R. Liss, Inc.

THE MOLECULAR RESPONSE OF CADMIUM RESISTANT DATURA INNOXIA CELLS TO HEAVY METAL STRESS[1]

Paul J. Jackson, Cleo M. Naranjo,
Peter R. McClure[2], and E. Jill Roth[3]

Genetics Group, Life Sciences Division
Los Alamos National Laboratory
Los Alamos, New Mexico 87545
and
[3]Department of Biology, The University of Utah
Salt Lake City, Utah 84112

ABSTRACT Datura innoxia suspension culture cells can be selected for the ability to grow rapidly for extended periods of time in normally toxic concentrations of cadmium ion. Resistance to this toxic heavy metal is correlated with the ability to rapidly synthesize large amounts of one or more low molecular weight, cysteine-rich, metal binding proteins. Resistance to increasing concentrations of cadmium is correlated with both the rate of de novo synthesis and the maximum accumulation of these binding proteins within cadmium resistant cells. The ability of one metal binding protein to tightly bind copper ion suggests that these proteins may normally play a role in trace metal metabolism and transport. Metal binding proteins are not easily detectable in the cadmium sensitive cells from which resistant cells were derived.

[1]This work was performed under the auspices of the U.S. Department of Energy.
[2]Present address: Allied Corporation, Syracuse Research Laboratory, P.O. Box 6, Solvay, New York 13209.

INTRODUCTION

Industrial, mining, and sewage disposal operations have resulted in the contamination of certain environments with cadmium, a group IIB heavy metal known to be harmful to human health (1). Plants grown in such environments may accumulate cadmium, subject to both the genetic characteristics of the plant (2,3) and environmental factors such as soil pH, cation exchange capacity and the concentration of cadmium versus other competing ions (4). Accumulation of cadmium in plants adversely affects different metabolic activities (5,6,7,8), yet plants can be selected for the ability to grow on normally toxic concentrations of this ion (9,10). The molecular mechanisms behind such growth are not yet understood.

Low molecular weight, cysteine rich, soluble metal binding proteins have been studied extensively in animals and are thought to function as scavengers of group IIB heavy metal ions (11,12). Similar low molecular weight, soluble metal binding proteins have also been found in several plant species (13,14,15). The role these proteins play in metal ion metabolism has not been determined. We report here the existence of low molecular weight, metal binding proteins in cadmium resistant Datura innoxia cells. De novo synthesis of these proteins is induced by cadmium, and different levels of cadmium resistance are directly correlated with both the initial rate of de novo synthesis and the maximum accumulation of the metal binding proteins within resistant cells. We also report the effects of increased copper concentrations on cadmium resistant cells and demonstrate a molecular response which explains these effects.

These results, taken together, suggest that the synthesis of cadmium binding proteins in cadmium resistant D. innoxia cells is the primary molecular mechanism of cadmium resistance. Moreover, they also suggest that cadmium binding proteins play a role in copper metabolism or transport in both resistant and sensitive cells.

MATERIALS AND METHODS

Maintenance of Plant Suspension Cultures and Selection and Cloning of Cadmium Resistant Cells.

Cadmium sensitive Datura innoxia suspension cultures

were originally obtained from Dr. O.L. Gamborg. Cultures
were maintained in darkness as 50 ml batch suspensions in
250 ml Delong flasks. Cells were grown on a gyrotory
shaker (120 rpm) at 33°C in Gamborg's 1B5 medium (16)
supplemented with 2 gm/l NZ amine A (Sheffield Products,
Memphis, TN), vitamins (17), and 1 µg/ml 2,4-dichloro-
phenoxyacetic acid. The pH of the medium was adjusted to
5.5 prior to sterilization. Cells were diluted as
necessary to maintain concentrations of 2 x 10^5 to 2 x 10^6
exponentially growing cells/ml. Cell numbers were
estimated by packed cell volume and protoplast counting
(17). Cells were grown for seven to ten generations in 1B5
medium minus NZ amine A prior to labeling with radioactive
amino acids.

The methods of initial selection of cadmium resistant
cells from cadmium sensitive cell cultures and the methods
of stepwise increase in selective pressure to isolate cells
of higher resistance from cell cultures resistant to lower
cadmium concentrations are modifications of methods
originally utilized by Schimke et al., Hildebrand et al.,
and Beach and Palmiter (18,19,20) to select drug and metal
resistant mammalian cells. These procedures have been
described previously (21). The method of single cell
cloning, utilized to derive cadmium resistant cell cultures
has also been described (21).

Isotopic Labeling of Cells.

Carrier-free [^{67}Cu] was provide by the Los Alamos
Medical Radioisotopes Research Group (INC-3), Los Alamos
National Laboratory, Los Alamos, NM. Carrier-free [^{109}Cd];
L-[^{35}S]-cystine (>300 Ci/mmol); and L-[3,4,5 ^{3}H(N)]-leucine
(>110 Ci/mmol) were purchased from New England Nuclear,
Inc., Boston, MA.

Metal binding protein induction experiments utilized
20 ml cultures containing approximately 10^6 exponentially
growing cells/ml. Cells were grown for 24 hours in medium
containing 0.15 µCi/ml carrier-free [^{109}Cd] (as CdCl$_2$)
prior to addition of unlabeled cadmium (2.5 to 250 µM
depending upon the cells utilized).

Prior to radioactive cysteine-leucine double labeling
experiments, cells were grown in 1B5 medium minus NZ amine
A.   Cells were then exposed to medium containing [^{3}H]-
leucine (10 µCi/0.1 µg/ml) plus [^{35}S]-cysteine (1 µCi/0.03
µg/ml) for different periods of time in either the presence
or absence of cadmium. Amino acid labeling was initiated

two hours prior to addition of cadmium in those cultures induced to synthesize metal binding proteins. [^{35}S]-cystine was reduced to [^{35}S]-cysteine just prior to use as previously described (21).

Copper labeling of induced metal binding proteins was accomplished by addition of 50 μM CdCl$_2$ plus 1.5 μCi/ml carrier-free [^{67}Cu], 24 hours prior to extraction and analysis of metal binding proteins. D. innoxia-Cd^r 250 cells grown in 250 μM CdCl$_2$ for 10 generations were transferred to medium containing the same concentration of CdCl$_2$ plus 0.15 μCi/ml carrier-free [^{109}Cd] for 24 hours prior to addition of 100 μM CuSO$_4$ plus 1.5 μCi/ml carrier-free [^{67}Cu]. Protein was then extracted at the times indicated.

Protein Extraction and Gel Filtration of Extracts.

The methods of protein extraction and gel filtration have been described previously (21). Under these conditions, recovery of cadmium and copper from filtration columns is 95 and 90% respectively. Absorbance of gel filtration fractions at 254 and 280 nm was measured with a Perkin-Elmer model 552 spectrophotometer. Assays of [^{3}H], [^{35}S], and [^{109}Cd] content were by scintillation spectrometry after addition of Formula 963 Aqueous Counting Cocktail (New England Nuclear, Boston, MA). Assays of [^{67}Cu] were by gamma counting. Samples containing both [^{67}Cu] and [^{109}Cd] were assayed for [^{67}Cu] content then stored for 21 days prior to assay for [^{109}Cd] content to allow for decay of [^{67}Cu] (half-life = 61 hours).

RESULTS

Growth Characteristics of Cadmium Resistant D. innoxia Suspension Cultures.

Five different D. innoxia cell lines, all derived from one cadmium resistant cell culture by stepwise increases in selective pressure followed by single cell cloning of the more resistant cells, have been isolated (21). They are: D. innoxia-Cd^r 50, D. innoxia-Cd^r 120, D. innoxia-Cd^r 160, D. innoxia-Cd^r 200, and D. innoxia-Cd^r 250, resistant to 50, 120, 160, 200, and 250 μM cadmium respectively. They will be referred to in the text as Cd^r 50, Cd^r 120, Cd^r 160, Cd^r 200, and Cd^r 250 respectively. Resistance is defined as

the ability of an exponentially growing culture to continue
rapid growth and division immediately following a change in
cadmium concentration from zero to the maximum
concentration of cadmium specified. All five cadmium
resistant cell lines grow and divided rapidly in either the
absence or presence of the toxic metal ion (data not
shown).

Accumulation of Cadmium in Cadmium Binding Complexes of
Cadmium Resistant D. innoxia Cells.

It has been demonstrated that both cadmium resistant
and cadmium sensitive cells accumulate cadmium at the same
rate, given equal concentrations of this ion (21). It has
also been demonstrated that higher concentrations of

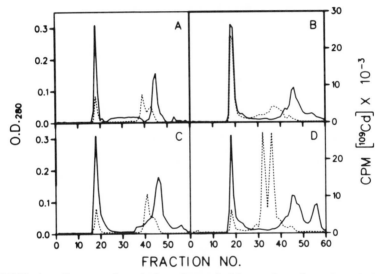

FIGURE 1. Forms of cadmium in soluble extracts of cadmium
sensitive and cadmium resistant D. innoxia cells. Panels A
and B: G-50 gel filtration profiles of the soluble portion
of extracts from cadmium sensitive cells grown in either
the absence (A) or presence (B) of 2.5 μM cadmium for 24
hours. Panels C and D: G-50 gel filtration profiles of
the soluble portion of extracts from cadmium resistant
Cd^r50 cells grown in either the absence (C) or presence (D)
of 2.5 μM cadmium for 24 hours. Fractions were assayed for
radioactive [^{109}Cd] content (----) and OD_{280} (——).
resistant cells.

cadmium result in a greater accumulation of cadmium in
resistant cells.   Therefore exclusion of the toxic ion is
not a potential mechanism of cadmium resistance in these
cells.

    After growth for 24 hours in cadmium, two apparent low
molecular weight cadmium binding complexes can be
identified in gel filtration fractions of extracts of all
five cadmium resistant cell lines (see Figure 1D for an
example).   These complexes are not found in cadmium
sensitive cells grown under the same conditions (Figure
1B).   They are also absent in extracts of both cadmium
resistant and sensitive cells following extended growth in
the absence of all but trace amounts of carrier-free
[^{109}Cd] (Figures 1A and C respectively).   Treatment of
partially purified, [^{109}Cd]-labeled cadmium binding
molecules with proteinase K followed by refiltration of the
digested material demonstrates a complete loss of metal

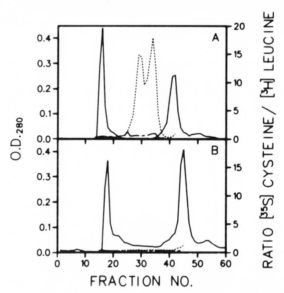

FIGURE 2.   Gel filtration profiles of soluble extracts
from cells exposed to [^{35}S]-cysteine and [^{3}H]-leucine for
six hours in the presence or absence of cadmium. <u>Panel A</u>:
Cadmium resistant Cd^{r} 50 cells.   <u>Panel B</u>:   Cadmium sensitive
cells.   OD$_{280}$, (———); [^{35}S]:[^{3}H] ratio for cells grown in
the presence of cadmium, (-----); and [^{35}S]:[^{3}H] ratio for
cells grown in the absence of cadmium, (— - —).

binding activity (21) suggesting that the metal binding
molecules are proteins.

Cysteine can account for 30 mole.percent of the amino
acid residues in metallothioneins, metal binding proteins
from other species (12). Since these residues are active
in the binding of metal ions (12), it might be expected
that Datura metal binding proteins also have a high
cysteine content. Cysteine content can be measured by
determining the OD_{254} (14), or by measuring the change in
$[^{35}S]$-cysteine:$[^3H]$-leucine ratio of gel filtration
fractions of extracts from cells grown in the presence of
these two radioactive amino acids. Figure 2 demonstrates
that, following growth of Cd^r50 cells in medium containing
50 μM cadmium plus $[^{35}S]$-cysteine and $[^3H]$-leucine, there
is a distinct increase in the $[^{35}S]$:$[^3H]$ ratio in gel
filtration fractions containing metal binding proteins.
This change is entirely the result of increased $[^{35}S]$
content. While measurement of the optical density at 254
nm compared to the optical density at 280 nm suggests no
reproducible increase in the cysteine content of metal

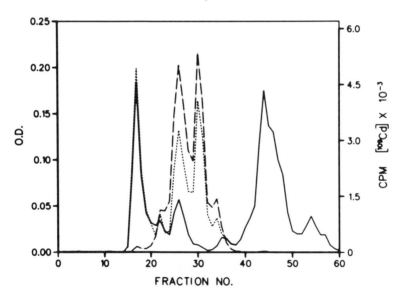

FIGURE 3. Changes in OD_{254} in gel filtration profiles
of soluble extracts from cadmium resistant Cd^r250 cells
following growth in 250 μM cadmium for 24 hours. OD_{280},
(———); OD_{254}, (— . —); and $[^{109}Cd]$ content, (----).

binding fractions of Cd[r]50 cells, measurement of this
parameter demonstrates a distinct increase in cysteine
content in metal binding gel filtration fractions of
extracts from more resistant Cd[r]250 cells following growth
for extended periods in cadmium (Figure 3), suggesting a
greater accumulation of metal binding proteins in cells
resistant to higher concentrations of cadmium.

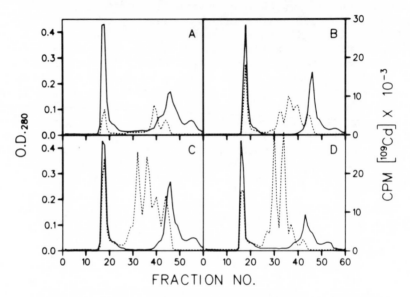

FIGURE 4. The synthesis of cadmium binding proteins in
cadmium resistant Cd[r]50 cells upon exposure to cadmium.
Resistant cells grown for at least 50 generations in the
absence of cadmium were exposed to 50 μM CdCl$_2$ labeled with
0.15 μCi/ml carrier-free [^{109}Cd] for 30, 60, 90, and 120
minutes (panels A,B,C, and D respectively). Protein
extraction and gel filtration were as previously described
(21). Addition of carrier-free [^{109}Cd] either 24 hours
prior to induction or at the time of induction gave
identical results indicating that carrier-free [^{109}Cd]
concentrations are insufficient to induce metal binding
protein synthesis and that the appearance of [^{109}Cd]
labeled metal binding proteins is not an artifact of
radioactive cadmium uptake. Symbols are (----) for [^{109}Cd]
and (———) for OD$_{280}$.

Induction of De Novo Metal Binding Protein Synthesis and
its Relationship to Different Levels of Cadmium Resistance.

Figure 1 suggests that, since metal binding proteins
are not found in cells grown in the absence of cadmium, the
synthesis of these proteins must be induced in response to
increased concentrations of cadmium in the medium. Cadmium
resistant cells were therefore grown for extended periods
of time in the absence of cadmium then were exposed to 50
μM cadmium plus 0.15 μCi/ml carrier-free [^{109}Cd]. Metal

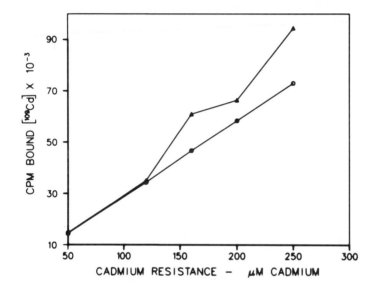

FIGURE 5. Maximum accumulation of cadmium binding
proteins in D. innoxia cells resistant to different
concentrations of cadmium. Cd^{r}50, Cd^{r}120, Cd^{r}160, Cd^{r}200,
and Cd^{r}250 cells were grown for 24 hours in the presence of
50, 120, 160, 200, and 250 μM [^{109}Cd] respectively. Metal
binding proteins extracted from these cells were then
assayed for bound [^{109}Cd]. Given the amount of cadmium
bound to metal binding proteins from Cd^{r}50, the amount to
be bound by metal binding proteins in the other more
resistant cell lines was calculated and plotted (-o-).
These values were then compared to values obtained from
each of the more resistant cell lines (-Δ-). Previous
experiments demonstrate a direct correlation between the
amount of cadmium bound and the total amount of metal
binding proteins present (21).

binding proteins with appropriate chromatographic behavior
were detectable sixty to ninety minutes following addition
of cadmium (see Figure 4 for an example).

The rate of accumulation of metal binding proteins
within the cells as measured by either [^{109}Cd] binding to
metal binding proteins or by [^{35}S]-cysteine incorporation
into newly synthesized proteins was proportional to the
maximum concentration of cadmium tolerated by the cell
line. Induced metal binding proteins were detectable
sooner and reached a maximum earlier in cells resistant to
higher concentrations of cadmium when compared to their
less resistant counterparts. Moreover, the maximum level
of metal binding proteins accumulated in resistant cells
was directly proportional to the maximum concentration of
cadmium tolerated by the resistant cell line (Figure 5).

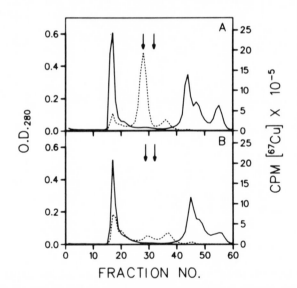

FIGURE 6. Binding of copper by cadmium induced metal
binding proteins. Cadmium resistant Cd^{r}50 cells grown in
the presence of absence of 50 µM cadmium were grown for one
generation in medium containing carrier-free [^{67}Cu]. Panel
A: Gel filtration profile of soluble extracts from
resistant cells grown in the presence of cadmium. Panel B:
Gel filtration profile of soluble extracts from resistant
cells grown in the absence of cadmium. Counts per minute
[^{67}Cu], (----) and OD$_{280}$, (——). Arrows represent the
locations of the cadmium binding proteins.

Cd^{r}250 cells accumulate more than five times as much metal binding protein as Cd^{r}50 cells.

Copper Binding by a Cadmium Induced Metal Binding Protein.

Extracts of cadmium resistant cells grown in only trace amounts of carrier-free [^{67}Cu] demonstrate that copper is not bound by any low molecular weight metal binding protein not induced by cadmium (Figure 6B). However, gel filtration profiles of extracts from the same cells grown in the presence of cadmium plus [^{67}Cu] demonstrate that one of the two metal binding proteins has a significantly higher affinity for copper than for cadmium

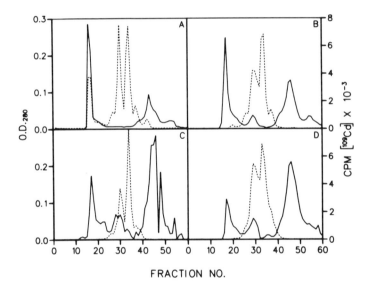

FRACTION NO.

FIGURE 7.   Removal of cadmium from metal binding proteins following growth of cadmium resistant cells in 250 µM cadmium plus 100 µM copper.   Cd^{r}250 cells grown for two generations in 250 µM cadmium were transferred to medium containing the same concentration of cadmium plus 100 µM CuSO$_4$.   Proteins were extracted 0,4,8, and 12 hours following exposure to copper (panels A,B,C, and D respectively).   Soluble portions of extracts were then passed through Sephadex G-50 gel filtration columns and fractions were assayed for [^{109}Cd] content (••••) and OD$_{280}$ (——).

(Figure 6A). The other metal binding protein either does not bind copper or has a higher affinity for cadmium.

If synthesis and accumulation of large amounts of metal binding proteins are a major factor contributing to the cadmium resistance phenotype, then growth of cadmium resistant cells in cadmium followed by addition of normally non-toxic concentrations of copper should release cadmium from one of the two metal binding proteins, killing the cells. Cd^r 250 cells grown in 250 µM cadmium are killed by addition of 100 µM copper although these cells grow rapidly in 500 µM copper in the absence of cadmium.

Addition of 100 µM CuSO$_4$ to cells growing rapidly in 250 µM [^{109}Cd] results in a marked reduction in cadmium binding to the larger of the two metal binding proteins

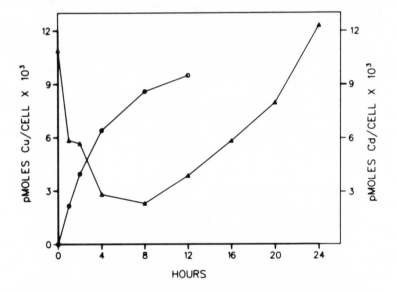

FIGURE 8. Substitution of copper for cadmium in one metal binding protein following exposure of cadmium induced Cd^r 250 cells to copper. Cd^r 250 cells grown for two generations in 250 µM [^{109}Cd] were transferred to medium containing the same concentration of cadmium plus 100 µM [^{67}Cu]. Proteins were extracted and fractions of gel filtrations of the soluble extracts were assayed for radioactive cadmium and copper content. The number of ions per cell of each metal ion was determined utilizing the number of cells extracted and the specific activity of each isotope. Symbols are cadmium, (-Δ-) and copper, (-o-).

(Figure 7). This occurs in combination with an increase in copper binding to these proteins. Measurement of the absolute number of cadmium and copper ions bound demonstrates that, up to eight hours after addition of copper, one cadmium ion is displaced by one copper ion (Figure 8). Beyond eight hours, both cadmium and copper binding increases. Since this is accompanied by an increase in OD_{254} (data not shown), this can be attibuted to continued metal binding protein synthesis.

DISCUSSION

Cadmium ions are toxic to D. innoxia cells and have been utilized to select cadmium resistant cell cultures (21). Application of stepwise increases in cadmium concentrations to cadmium resistant cells has allowed the isolation of five different cell lines, each resistant to a different maximum concentration of this heavy metal ion. Different levels of resistance suggest the presence of either a combination of mechanisms or further modification of a single mechanism which results in the more resistant phenotypes. Results presented here demonstrate that different levels of cadmium resistance are directly correlated with both the rate of accumulation of de novo synthesized metal binding proteins and the maximum accumulation of metal binding proteins within the resistant cells. That is, resistance to normally toxic concentrations of cadmium is first conferred upon Cd^r50 cells by the capacity to synthesize detectable amounts of metal binding proteins and increased resistance in the subsequently isolated resistant lines is the result of the capacity to synthesize even greater amounts of these same proteins at a faster rate.

It is probable that the ability to synthesize large amounts of cadmium binding proteins is the primary molecular mechanism responsible for cadmium resistance since blocking of this capacity by addition of normally non-toxic concentrations of copper results in cell death.

It is unlikely that the ability to synthesize cadmium binding proteins arose de novo within resistant cells. Studies of metal binding proteins in intact plants demonstrate that such proteins are expressed under certain developmental and environmental conditions (10,13,14). Resistance to cadmium is probably the result of these cells' capacity to overproduce, relative to sensitive cells, metal binding proteins.

The presence of these proteins suggests some role in normal metabolism. The capacity of one cadmium binding protein to bind copper suggests that this protein may, in the absence of cadmium, play a role in copper metabolism and transport within the cells or within intact plants. Such high affinities for copper suggest that, if this protein plays such a role, it must have some means of direct transfer of copper to copper-requiring proteins within the cells.

Changes in the genome, if any, which are responsible for the cadmium resistance phenotype have not yet been elucidated. However, overproduction of metal binding proteins is accompanied by overproduction of certain mRNA sequences (data not shown). This is also seen in cadmium resistant mammalian cells where resistance is known to sometimes, but not always, be associated with amplification of the genes responsible for metallothionein synthesis (18,23). We are currently investigating these possibilities in cadmium resistant Datura cells.

ACKNOWLEDGMENTS

The authors would like to thank Dr. K.G. Lark in whose laboratory this work was initiated. They would also like to thank Ms. Catherine Tallerico and Mr. Robert Alexander for technical assistance.

REFERENCES

1. Friberg L, Piscator M, Nordberg GF, Kjellstrom T (1974). Cadmium: occurrence, possible routes of exposure, and daily intake. In Friberg C, Piscator M, Norberg GF, Kjellstrom T (eds): "Cadmium in the Environment," Cleveland: CRC Press, Inc., p 9.
2. Foy CD, Chaney RL, White MC (1978). The physiology of metal toxicity in plants. Ann Rev Plant Physiol 29:511.
3. Turner RG (1969). Heavy metal tolerance in plants. In Rorison IH (ed): "Ecological Aspects of the Mineral Nutrition of Plants," Oxford: Blackwell, Inc., p 399.

4. Cataldo DA, Wildung RE (1978). Soil and plant factors influencing the accumulation of heavy metals by plants. Environ Health Perspect 27:149.
5. Bazzaz FA, Rolfe GL, Carlson RW (1974). Effects of cadmium on photosynthesis and transpiration of excised leaves of corn and sunflower. Physiol Plant 32:43.
6. Huang C-Y, Bazzaz FA, Vanderholf CN (1974). The inhibition of soybean metabolism by cadmium and lead. Plant Physiol 54:122.
7. Root RA, Miller RJ, Koeppe DE (1975). Uptake of cadmium - its toxicity and effect on the iron ratio in hydroponically grown corn. J Environ Qual 4:473.
8. Weigel HJ, Jager HJ (1980). Different effects of cadmium in vitro and in vivo on enzyme activities in bean plants (Phaseolum vulgaris L. cv Sankt Andreas). Z Pfanzenphysiol 97:103.
9. Page A, Bingham F, Nelson C (1972). Cadmium absorption and growth of various plant species as influenced by solution cadmium concentration. J Environ Qual 1:288.
10. Rauser WE, Curvetto NR (1980). Metallothionein occurs in roots of Agrostis tolerant to excess copper. Nature 287:563.
11. Cousins RJ (1979). Metallothionein synthesis and degradation: Relation to cadmium metabolism. Environ Health Perspect 29:131.
12. Webb M (1979). The metallothioneins. In Webb M (ed): "The Chemistry, Biochemistry, and Biology of Cadmium," Amsterdam: Elsevier/North-Holland, p 195.
13. Bartolf M, Brennan E, Price CA (1980). Partial characterization of a cadmium binding protein from roots of cadmium treated tomato. Plant Physiol 66:438.
14. Wagner GJ, Trotter MM (1982). Inducible cadmium binding complexes of cabbage and tobacco. Plant Physiol 69:804.
15. Weigel HJ, Jager HJ (1980). Subcellular distribution and chemical form of cadmium in bean plants. Plant Physiol 65:480.
16. Gamborg OL, Miller RA, Ojemos K (1968). Nutritional requirements of suspension cultures of soybean root cells. Exptl Cell Res 50:151.
17. Chu Y-E, Lark KG (1976). Cell-cycle parameters of soybean (Glycine max L.) cells growing in suspension cultures: Suitability of the system for genetic studies. Planta 132:259.

18. Schimke RI, Alt FW, Kellems RE, Kaufman R, Bertino JR
    (1978). Amplification of dihydrofolate reductase genes
    in methothrexate-resistant cultured mouse cells. Cold
    Spring Harbor Symp Quant Biol 42:649.
19. Hildebrand CE, Tobey RA, Campbell EW, Enger MD (1979).
    A cadmium-resistant variant of the chinese hamster
    (CHO) cell with increased metallothionein induction
    capacity. Exptl Cell Res 124:237.
20. Beach LR, Palmiter RD (1981). Amplification of the
    metallothionein-I gene in cadmium resistant mouse
    cells. Proc Nat Acad Sci USA 78:2110.
21. Jackson PJ, Roth EJ, McClure PR, Naranjo CM (1984).
    Selection, isolation, and characterization of cadmium
    resistant Datura innoxia suspension cultures. Plant
    Physiol, In Press.
22. Walters RA, Enger MD, Hildebrand CE, Griffith JK
    (1981). Genes coding for metal-induced synthesis of
    RNA sequences are differentially amplified and
    regulated in mammalian cells. In Brown DD (ed):
    "Developmental Biology Using Purified Genes," New York:
    Academic Press, p 229.

Cellular and Molecular Biology of Plant Stress, pages 161-179
© 1985 Alan R. Liss, Inc.

THE HEAT SHOCK RESPONSE IN SOYBEAN

Joe L. Key, Janice A. Kimpel, Chu Yung Lin,
Ronald T. Nagao, Elizabeth Vierling, Eva Czarnecka,
William B. Gurley, James K. Roberts, Michael A. Mansfield,
and Leonard Edelman

Botany Department, University of Georgia,
Athens, GA 30602

ABSTRACT When seedlings are subjected to an
immediate shift from 30° to 40° or to a gradual
(e.g. 3° per hr) increase to 40° to 45° there is
a dramatic and rapid shift in the pattern of
protein synthesis. Most normal protein synthesis
rapidly declines at a temperature of about 40°
and a new set of proteins (hsps) is rapidly
induced. The synthesis of hsps correlates with
the rapid induction, detectable within 3 to 5
minutes, of hs-specific RNAs encoding a complex
set of 30 to 40 polypeptides. These sequences
subsequently accumulate to as many as 20,000
copies per cell. The system appears to be self-
regulated since it "turns off" after 6 to 8 hr
of continuous hs, indicated by the depletion of
hs mRNAs, cessation of hsp synthesis, and the
resumption of normal protein synthesis. This
"turn off" occurs more slowly during prolonged
high temperature treatment than when the
seedlings are returned to 30° after a short, 2
to 4 hr hs at 40°.
    One putative function of hs proteins is a
role in the development of tolerance to high
temperatures through a conditioning treatment at
a lower, hsp inducing temperature. An initial
hs treatment (e.g. 2 hr at 40° or 5 to 10 min
at 45° followed by 2 hr at 30°) provides
thermoprotection to a subsequent, otherwise
lethal temperature (e.g. 2 hr at 45°). This

short-term thermal adaptation correlates with
the accumulation of hsps and their selective
localization during hs. A wide range of plant
species are known to synthesize a complex group
of hsps. There is considerable variation in the
electrophoretic pattern of the 15 to 27 kD hsps,
while the high molecular weight hsps are highly
conserved. Although many stress agents induce
low levels of hs mRNAs based on Northern
hybridization analyses using hs cDNA clones
to probe poly(A)RNAs from tissues stressed by
other agents, only arsenite treatment closely
mimics hs, as evidenced by intense radiolabeling
of hs proteins. Cadmium also induces a small
subset of the hs proteins and hs mRNAs. The
hs genes comprise several multigene families,
probably with over-lap among these families,
as analyzed by Southern hybridization, 2D gel
analyses of hybrid-selected translation products,
and DNA sequence analysis of cDNA and λ genomic
clones. Potential regulatory sequences have
been identified in the soybean genes, similar
to those identified in other organisms, particularly
Drosophila, where gene structure and expression
have been analyzed in considerable detail.

INTRODUCTION

Since 1962, it has been known that Drosophila
responds to a heat shock (hs) treatment by dramatically
altering the pattern of polytene chromosomes puffs
(reviewed in Ashburner and Bonner, 1979). During hs,
preexistent puffs, which are sites of active RNA synthesis,
regress, and there is a rapid induction of puffing at a
few new gene loci. A number of other stress agents
similarly alter the pattern of chromosome puffing. It has
been demonstrated by biochemical and molecular techniques
that most normal RNA synthesis ceases upon hs treatment,
and that a minimum of seven new mRNAs appear. These
changes correlate with the loss of normal protein
synthesis and the appearance and predominant synthesis of
seven new proteins. Most normal mRNAs persist during hs
despite the dramatic decline in synthesis of their
corresponding proteins, demonstrating that translational
in addition to transcriptional switches function during

the onset of and recovery from hs. While the regulatory
phenomena are not yet understood, there is some recent
evidence which addresses both the transcriptional (Parker
and Topol, 1984; Wu, 1984) and translational (Ballinger
and Pardue, 1983; Reiter and Penman, 1983; DiDomenico
et al., 1982B) control mechanisms operative in the hs
response of Drosophila. The hs response therefore affords
a powerful system for the study of gene regulation at many
levels, and in addition is very likely of major importance
to the organism's survival to stress.

It is clear from studies on organisms including from
bacteria (Yamamori and Yura, 1982; Neidhardt and Bogelen,
1981; Daniels et al., 1984; Tilley et al., 1983), fungi
(McAlister and Finkelstein, 1980 and 1980B), Tetrahymena
(Glover et al., 1981), Dictyostelium (Loomis and Wheeler,
1980 and 1982), insects (Ashburner and Bonner, 1979),
plants (Barnett et al., 1980; Key et al., 1981), mammals
(Li and Werb, 1982; Wang et al., 1981) and avian cells
(Schlesinger et al., 1982B) that the hs response is
ubiquitous. Although specific structural or enzymatic
functions for the heat-shock proteins (hsps) have not been
identified, synthesis of hsps has been strongly correlated
with the development of thermal tolerance. Thermal
tolerance is defined as the ability of organisms to
tolerate normally lethal temperatures after an initial
exposure to a sublethal, hsp-inducing temperature. Some
hsps are observed to associate with specific organelles
during the heat treatment and it is now widely suggested
that this localization is important to hsp function. The
question of how this set of proteins might mediate the
development of thermal tolerance in soybean is the ultimate
focus of much of the work in our laboratory.

In the subsequent discussion we assess the current
state of knowledge of the hs response in plants, including
its possible physiological significance, describe the
complexity of the multigene families involved, and present
a comparative analysis of the plant and Drosophila hs
genes. Finally, we shall address some of the major
unresolved issues relating to the hs response in plants.

Induction of Heat Shock Proteins in Plants.

A number of plant species including soybean (Barnett
et al., 1980; Key et al., 1981), corn (Baszczynski et al.,
1982; Cooper and Ho, 1983; Key et al., 1983A; Kelley and
Freeling, 1982), tomato (e.g. Scharf and Nover, 1982;

Nover and Scharf, 1984), tobacco (e.g. Barnett et al.,
1980; Meyer and Chartier, 1983), cotton, pea, millet and
sunflower (Key et al., 1983A) have been shown to undergo
the transition during a hs treatment from normal protein
synthesis to predominantly hs protein synthesis, usually
maximal at about 10° above the normal growing temperature.
The level of translational control of normal mRNAs during
hs varies considerably among plants; and even with soybean
(e.g. Key et al., 1981) the level of translational control
may not be as great as in Drosophila. While the response
has been shown to occur in this wide range of plants by a
number of laboratories, we shall focus our discussion on
the soybean system and primarily on work done in our own
laboratory.

When the temperature of incubation of 2-day-old
etiolated soybean seedlings is increased from 30° up to
about 40° (either abruptly or gradually), there is little
change in the level of externally supplied radioactive
amino acid incorporated into proteins. Above this
temperature, denoted "breakpoint" in our studies, there is
a precipitous drop in incorporation (e.g. Key et al.,
1982, 1983A). Other factors, such as amino acid uptake,
translocation and/or activation in addition to decreased
rates of protein synthesis may contribute to this abrupt
drop in incorporation; however, there is clearly a
dramatic change in the pattern of protein synthesis (Key
et al., 1981). At 35° to 37.5° hsp synthesis appears
against a background of normal proteins, and at 40° or
above, the synthesis of hsps predominates with minimal
synthesis of most normal proteins. The induction of hsp
synthesis is rapid; their synthesis becomes predominant
within the first 30 min at 40°-41°. This pattern persists
for 6 to 8 hr at which time hsp synthesis declines
dramatically, and the normal pattern of protein synthesis
returns. It is not known if this recovery of the normal
pattern of protein labeling by amino acids results from
the direct decline in hsp synthesis or an increase in the
rate of amino acid incorporation into normal proteins.
Only an analysis of the actual rates of synthesis under
these various conditions will permit a resolution of this
question. The observed "shut-down" of hsp synthesis with
increasing time at hs temperatures is consistent with the
experiments of Lindquist and coworkers (DiDomenico et al.,
1982A) demonstrating autoregulation of hs protein
synthesis in Drosophila. Their results indicate that
accumulation of a specific level of functional hsps

represses further hsp production. They have shown that
aberrant hsps, synthesized and accumulated during
incubation with the amino acid analog canavanine, do not
have any autoregulatory capabilities. Additionally, the
abnormal hsps do not selectively localize during hs, and
do not allow the development of thermal tolerance -- the
two strongly conserved characteristics of native hsps.

When the soybean seedlings are returned to 28°-30°
after a 2 to 4 hr incubation at 40°, there is a rapid
decay of hsp synthesis coupled with the appearance of
normal protein labeling patterns. By 4 hr at 30° hsp
synthesis is essentially undetectable (Key et al., 1981).
However, those hsps synthesized during the initial hs
treatment are generally stable for at least 21 hours,
regardless of whether the tissue is returned to 28° or
maintained at the hs temperature. The hsps can accumulate
to levels which permit their detection by Coomassie
staining, in some cases becoming the major stainable
proteins (unpublished data).

Although the molecular weights and isoelectric points
of hsps vary considerably from species to species (Key et
al., 1983A), all plants seem to synthesize a much more
complex pattern of low molecular weight hsps than other
organisms studied to date. In <u>Drosophila</u>, the four major
low MW hsps are 22, 23, 26 and 27 kD. In plants, there is
a very complex group of 15 to 27 kD proteins, comprised of
20-30 different polypeptides. Interestingly, this large
diversity of hsps in plants compared to other organisms is
restricted to the low molecular weight polypeptides. The
high molecular weight hsps (68-70, 84, 92 kD) are highly
conserved between plants, animals and bacteria. Because
of the unique nature and complexity of the 15 to 18 kD
hsps of plants, much of our work in soybean has
concentrated on these proteins and the mRNAs encoding
them.

Studies on hs mRNAs.

Following hs at 40°, a very abundant class of RNA
appears in seedlings which is either absent or present at
very low levels in control (30°) tissue (Schöffl and Key,
1982; Dawson and Grantham, 1981). In the case of soybean,
20 or more poly(A)RNA sequences of about 800 nucleotides
in length accumulate to as many as 20,000 copies per cell
within 2 hr at 40° based on cDNA/poly(A)RNA kinetic
hybridization studies (Schöffl and Key, 1982). These

mRNAs are undetectable in 30° tissue as analyzed by
Northern hybridization using cloned hs-specific cDNA
probes (Schöffl and Key, 1982; Czarnecka et al., 1984).
The hs mRNAs are easily detected during the initial 3 to
5 min at hs temperatures, and they accumulate during the
next 1 to 4 hr, depending upon the exact temperature of
hs. At 40° the level of hs mRNAs remains approximately
constant from 2 to 6 hr of hs, and then declines
gradually over the next few hr. The decline in hs mRNAs
during continuous 40° treatment is much slower than when
tissue is returned to 30° after a 2 to 4 hr hs. Under the
latter conditions the hs mRNA levels decline with a
half-life of about 1 hr. The relative contributions of
the rate of hs mRNA synthesis and the rate of decay to
these levels is not known at this time. In both cases,
however, the loss of hs mRNAs correlates with the loss of
hsp synthesis under the same conditions, and these data are
consistent with the view that autoregulation occurs at the
level of transcription of hs mRNAs coupled with decreased
translation of those mRNAs, as described above in
Drosophila. A large percentage of normal poly(A)RNAs
persist during hs at 40°, indicated by the minimal change
in complexity of the mRNA populations during hs. However,
some mRNAs decline significantly during hs (e.g. an
auxin-regulated sequence, pJCW1, Walker and Key, 1982;
ribulose bisphosphate carboxylase small subunit mRNA,
Vierling and Key, in preparation) while other sequences
do not decline at all (e.g. an actin mRNA, unpublished
data of Kimpel; ribulose bisphosphate carboxylase large
subunit mRNA, Vierling and Key, in preparation).

The Development of Thermotolerance and Localization of
the hs Proteins.

     Several lines of evidence support the view that in
soybean, as in other organisms, an initial hs treatment at
a tolerant temperature provides thermal protection to
otherwise lethal temperatures (Key et al., 1982; Key et al.,
1983B; Lin et al., 1984; Schlesinger et al., 1982A). In
soybean seedlings, several different pretreatments: (1) 2
hr at 40°, (2) 30 min at 40° followed by 2 to 4 hr at
30°, (3) 7 to 10 min at 45° followed by 2 hr at 30°, all
permit the soybean seedlings to tolerate a 2 hr 45° hs
treatment, an otherwise lethal treatment (Lin et al.,
1984). All of these treatments which lead to the
development of thermotolerance cause the accumulation of

high levels of hsps prior to the otherwise lethal
treatment. We have found one other treatment that also
induces synthesis of nearly the full spectrum of hs
proteins: incubation in 50-75 μM arsenite for 3 hours.
The treatment also confers the ability to tolerate a 2 hr
45° treatment. As described here later in more detail,
arsenite is unique among many other stresses that we have
studied in its ability to closely mimic the hs response.
A number of functions are clearly protected by these
pretreatments, including the capacity to synthesize hsps
and hs mRNAs at these otherwise inhibitory conditions and
the ability of the seedlings to survive and grow rather
normally upon return to the normal (e.g. 30°) growth
conditions. Thus, all vital functions must be protected
sufficiently during the otherwise lethal treatment to
permit those functions to persist and recover to
essentially a normal state within a few hr upon return to
the normal temperature. The recovery time depends to some
extent upon the temperature-time relationships of the hs
treatment. Slower recovery is observed if the hs
treatment exceeds the "breakpoint" compared to hs at or
below the breakpoint temperature.

The selective localization of hsps with cell
structures during hs likely provides the basis of
thermoprotection (Schlesinger et al., 1982A, 1982B; Lin
et al., 1984). It is known in Drosophila that some hsps
localize specifically within the nucleus (Velazquez et al.,
1980; Arrigo et al., 1980, Arrigo and Ahmad-Zadeh, 1981;
Velazquez et al., 1983), and there is suggestive evidence
that hsps also associate with other organelle fractions
during hs including the cytoskeleton in animal cells
(Schlesinger et al., 1982B), and mitochondria and
ribosomes in soybean (Lin et al., 1984). We do not know
the mechanism of association of hsps with the plant
organelle fractions, but the association is hs-dependent;
hs proteins made during arsenite treatment are found
throughout the cytoplasm, but they become localized
immediately when the temperature is shifted up. The hsps
become associated with organelles during hs, chase from
these fractions at 30° over a period of 3 to 4 hr, and
rapidly "reassociate" (within 15 min) during a subsequent
hs in a temperature-dependent manner (Lin et al., 1984).
The 15-18 kD hsps, the most abundant hs proteins in plants,
associate with all organelles examined in soybean --
ribosomes, nuclei, mitochondria, and plasma membrane
fractions. Two hsps, 20 and 24 kD, appear to specifically

localize in the mitochondrial fraction, but they do not
"chase" significantly during a subsequent 30° incubation
(Key et al., 1982; Lin et al., 1984). This may indicate
specific transport of these proteins into the mitochondria.
Recently, we have identified at least one hsp which is
post-translationally transported into isolated chloroplasts
(Vierling et al., 1984). Localization of hsps into
chloroplasts might also be expected, since thermoprotection
of some chloroplast functions seems to occur (Arntzen,
this volume). We currently are assessing the nature of
hsp localization in the various organelle fractions of the
cell using in vitro uptake analyses.

We are also taking advantage of the selective
localization of hsps with the ribosomal fraction to
prepare extracts highly enriched in heat-shock proteins.
Ribosomal pellets obtained immediately after a heat shock
treatment are fractionated on an ion exchange resin to
separate basic ribosomal proteins from the heat shock
proteins, which have pIs between 5 and 7. This
hsp-enriched fraction is now being used to develop
monoclonal antibodies which will enable us to more
precisely define and quantitate the localization of these
proteins during heat shock.

Heat (Stress) Shock as a Natural Phenomenon in Field
Grown Soybean Plants.

The studies summarized above were accomplished using
etiolated seedlings in shake culture under laboratory
conditions. Field conditions during the summer of 1983 in
Georgia permitted an assessment of hs mRNA accumulation at
ambient temperatures of 39° to 40° under both irrigated and
non-irrigated conditions (Kimpel and Key, in preparation).
Under dry conditions where leaf temperatures approach or
exceed ambient temperatures (Jung and Scott, 1980), hs
mRNAs accumulated to very high levels between 10:00 a.m.
and 5:00 p.m. Irrigated plants, where leaf temperatures
probably remain significantly below ambient, accumulated
low levels of hs mRNA at the peak ambient temperature of
39°-40°, levels comparable to 35° or so under laboratory
conditions in etiolated seedlings. Silver-stained
O'Farrell gels also showed the accumulation of hsps in
non-irrigated plants but not in irrigated plants
(unpublished data of Mansfield). These results have
recently been confirmed using growth chamber-grown green
plants (Kimpel and Key, in prepartion).

HS Multigene Families.

Several lines of evidence demonstrate that the 30 to
40 hsps in the 15 to 27 kD range in soybean consist of
families of hsps synthesized from mRNAs in gene families
ranging in number from 3 or so up to as many as 13 related
genes (Schöffl and Key, 1982; unpublished data). There
likely is over-lap in sequence homology even among these
seemingly different families of hs genes (unpublished data
of Nagao et al.). Southern hybridization analyses
(unpublished data) show that cDNAs cloned from hs-specific
mRNAs hybridize to multiple bands of total DNA digested by
a range of site-specific restriction endonucleases. In
vitro translations of cDNA/hybrid selected mRNAs also
demonstrate that individual cDNA clones hybrid select from
one up to 13 mRNAs when the translation products are
analyzed on 2D O'Farrell gels, as summarized below:

cDNA Clone	Number of Proteins	Size (kD)
pFS 2005*	13	15-18
pCE 53**	13	15-18
pFS 2019*	1	18
pCE 75**	9	15-16
pFS 2033*	3	20-24
pCE 54**	4-5	27

*Clones identified by Schöffl and Key, 1982. **Clones
identified by Czarnecka et al., 1984.

While pFS 2019 and other related clones hybrid
selected mRNA which translated into only one major
protein, appropriate gel analyses demonstrated that this
protein was most likely a predominant member of the group
of 13 selected by pFS 2005, pCE 53 and other related
clones (unpublished data of Mansfield). DNA sequence
analysis of genomic clones has confirmed that pFS 2019
represents primarily a 3' untranslated unique region of
one member of the 13 gene products showing high homology
to pFS 2005 and pCE 53 (unpublished data by Nagao,
Czarnecka and Gurley). Cross hybridization analyses among
λ genomic clones support this conclusion as well as
establish, in conjunction with the hybrid select/
translation analyses and DNA sequence analyses, some
clustering of the hs genes. However, chromosomal "walks"
have not yet been made to identify the level of clustering
of the hs genes. In Drosophila the four 22 to 27 kD genes

are clustered on a 12,000 base pair fragment of DNA
(Corces et al., 1980). Sequence analysis of the cDNA
clones provides additional evidence for "intra family"
sequence relationships among the cDNA groups noted above.
    To date only members of the 53/2005/2019 group of
isolated λ genomic clones have been sequenced, although we
are currently analyzing clones to the 2033, 75 and 54
groups noted above in addition to λ genomic clones
recently isolated which code for 70 and 84 kD hsps (Nagao,
Roberts and Czarnecka, unpublished data). Those genes
sequenced to date have no introns, have rather
characteristic TATA sequences positioned 5' to the
transcription initiation site(s) determined by S_1 nuclease
mapping (unpublished data of Nagao, Czarnecka, Gurley and
Schöffl). Additionally, all of the sequences contain a
consensus "hs promoter" (see Pelham, 1982) positioned 5'
to the TATA region; usually there are over-lapping
duplicated consensus "hs promoters" in these soybean
genes. Additionally, there are considerable regions of
homology around the sequences noted above in the 5'
untranscribed region of these genes. There is much less
homology in the 3' untranslated region of these genes.
One characteristic, however, is apparent multiple poly(A)
addition sites based on S_1 mapping and possible "consensus"
sequences for such. The coding region shows about 90%
amino acid homology among the four genes which are
completely sequenced (Nagao, Czarnecka, Schöffl and Gurley,
unpublished data); additionally there are numerous "silent"
nucleotide substitutions among the genes. While the amino
acid sequences are considerably different than those of
the 20 to 25 kD Drosophila hs genes (Ingolia and Craig,
1982; Southgate et al., 1983), they have similarly located
regions of high hydrophobicity and hydrophilicity
(Southgate et al., 1983) and possibly some highly related
structural features (see also Vollemy et al., 1983).
There is one sequence of hydrophobic amino acids in the 3'
region which is completely conserved between the four
soybean genes and the Drosophila genes, the relevance of
which is not understood. Based on sequence homology data
of the 70 kD genes of Drosophila (Ingolia et al., 1980)
and yeast (Craig et al., 1982) and the dnaK protein of E.
coli (Bardwell and Craig, 1984) and cross reactivity of
chicken 70 kD antibody across a broad spectrum of
organisms (see Schlesinger et al., 1982B) greater homology
of the high molecular weight protein genes of soybean to

other organisms is expected than is seen for the 15-18 kD
genes of soybean to the 20-25 kD genes of Drosophila.
In fact, Drosophila 70 and 84 kD cDNA clones (kindly
supplied by V. Corces) have been used at low stringency
to isolate corresponding λ genomic clones of soybean DNA
(unpublished data of Roberts and Nagao). (Shah in this
volume used this strategy to isolate a corn hsp 70 gene.)
A detailed analysis of the structure of these gene families
is essential to our understanding of the hs system of
soybean and its relationship to other related or very
phylogenetically distant groups of organisms.

Relationship of Heat Stress to Other Stresses in Soybean
Seedlings.

    Based on the observations that a wide range of stress
agents induce the "hs response" in Drosophila (see
Ashburner and Bonner, 1979), including recovery from
anoxia, coupled with the dramatic effects of anaerobiosis
on plant protein synthesis (Lin and Key, 1966; Sachs et
al., 1980; Sachs and Freeling, 1978) and general effects
of water stress on the protein synthetic apparatus (Hsiao,
1970; 1973; Moriella et al., 1973), we have made a rather
detailed analysis of the possible relationship of other
physical stress agents to high temperature (hs) stress
(Czarnecka et al. 1984; Edelman et al., submitted).
These studies generally utilized hs cDNA clone probes in
Northern hybridization analyses of poly(A)RNAs isolated
from "stressed" soybean seedlings. In the case of
arsenite, in vivo protein labeling was also achieved, but
with most of these stress agents amino acid uptake and
incorporation was so impaired that meaningful in vivo
studies were not feasible. The "stress" agents ranged
from PEG to impose up to -8 bars water stress, salt
stress, anaerobiosis, high levels of various plant
hormones, amino acid analogues, heavy metals, arsenite
and a range of respiratory and/or phosphorylation
inhibitors (Czarnecka et al., 1984). Generally very low
levels of "hs mRNAs" were detected using various hs cDNA
probes following treatment with this range of agents; the
levels which were detectable above untreated controls were
generally 100-fold (or more) lower than hs-induced mRNA
levels. However, arsenite resulted in induction and
accumulation of very significant levels of the hs mRNAs
homologous to all of the available hs cDNA probes, although
the induction was slower and often somewhat lower levels

accumulated than with hs. Also hs proteins accumulate in
response to arsenite treatment (Edelman et al., submitted).
These data are consistent with the observations on the
development of thermotolerance following a 28° to 30°
treatment with 50 to 75 µM arsenite mentioned earlier in
this chapter, yet there are some quantitative differences
among the various hsps between hs and arsenite treatment.
Additionally, cadmium induced the accumulation of
significant levels of several of the hs mRNAs (Czarnecka
et al., 1984; Edelman et al., submitted).

One set of mRNAs homologous to clone pCE54 that
accumulates from 3- to 10-fold above the normal control
level following hs was induced to near hs levels by
essentially all of the "stress" agents tested (Czarnecka
et al., 1984). These mRNAs translate into 4 to 6 27 kD
proteins and are the only "hs mRNAs" that we have detected
to date in non-hs or non-stressed soybean seedlings. The
physiological significance of these mRNAs/proteins is not
known, but they certainly seem to serve as a sensitive
"detector" or barometer for stress in soybean seedlings.
We do not yet know if there is a homologous counterpart to
these 27 kD proteins in other species, although
hybridization analyses of poly(A)RNAs among species to
pCE54 have failed to detect these sequences at our normal
hybridization stringency. These proteins in soybean do
not selectively localize during hs and remain as "soluble"
supernatant proteins.

Our stress treatments have generally been imposed
over only a 2 to 4 hr period as with hs. So while most
of these stress agents do not induce significant levels of
most of the "hs mRNAs", we have not exhausted treatment
conditions such that it can be concluded that none of
these agents, except arsenite and to a lesser extent
cadmium, are able to mimic the hs response. However, the
major hs proteins seem for the most part to be totally
distinct from the major anaerobic proteins (which generally
are catabolic enzymes for enhanced carbon glucose-
utilization for energy production with ethanol and/or
lactate accumulating as "terminal" electron acceptors) of
corn and soybean. Another point of interest is that hs
seems to over-ride other stress-induced proteins just as
it does most normal protein synthesis, including anaerobic
proteins of soybean (unpublished data), and pathogen or
elicitor induced proteins (e.g. Yarwood, 1967; Hadwiger
and Wagoner, 1983; Hahlbrock, in this volume).

Heat Shock: Potential Areas for Major Research Activity.

As noted in the introduction, minimal progress has
been made in gaining any insight into mechanisms operative
in the transcriptional and translational "switches" or
regulatory mechanisms which come into play following the
imposition of high temperature stress, whether by a
dramatic hs or a gradual rise to hs temperatures. While
the mechanisms will likely prove to be reasonably
conserved among the diverse groups of organisms that
exhibit a typical hs response, no information on
mechanisms of hs switching of transcription and
translation is available in plant systems.
What are the hsps and what is their function(s)?
Certainly, their synthesis and selective localization play
a role in thermotolerance. Do the hsps function as
structural entities as implicated by chromatin interband
localization of hsp 70 in Drosophila, or as a component
of the nuclear matrix, and/or as a component of the
cytoskeleton during hs of avian cells and/or do some
of the hsps play critical enzymatic functions? What is
the role, if any, of hs protein granules (e.g. Sanders et
al., 1982; Nover et al., 1983). What is the mechanism(s)
of selective localization of some of the hsps during hs?
Specifically in plants, does pollen not hs as indicated by
a few studies (e.g. Mascarenhas and Altschuler, 1983;
Cooper et al., 1984). Does pollen acquire some level of
thermotolerance (Mascarenhas and Altschuler, 1983; Cooper
et al., 1984) in the absence of hsp synthesis? Does loss
of pollen viability and impaired fertilization under high
temperatures relate to the inability of pollen (and
possibly even the ovary/stigma tissues) to hs and acquire
high levels of thermotolerance at high temperatures?
There is little question that a major consequence of high
temperature stress to crop production relates to failure
to achieve significant seed set at abnormally high
temperatures, especially when coupled with moisture stress
conditions. What is the relationship, if any, of
short-term hs and the acquisition of thermotolerance as
noted above to the long-term temperature adaptations that
have been studied (e.g. Berry and Bjorkman, 1980)? Is
there any commonality between high temperature stress
adaptation and cold stress acclimation (the latter covered
by Li in this volume)? The questions could go on, but the
foregoing may provide a basis for future studies and
hopefully will provide some recognition for the need for

enhanced efforts to understand temperature stress in plants, and possibly how to minimize the effects in the future. Certainly, additional information at the physiological, biochemical, molecular and genetic levels is needed if we are to make significant progress in gaining an understanding of these phenomenon and possibly how to manipulate them in a positive way.

REFERENCES

1. Arrigo AP, Ahmad-Zadeh C (1981). Immunofluorescence localization of the small heat shock proteins (hsp 23) in salivary gland cells of Drosophila melanogaster. Mol gen Genet 184:73.
2. Arrigo AP, Fakan S, Tissieres A (1980). Localization of the heat shock-induced proteins in Drosophila melanogaster tissue culture cells. Dev Biol 78:86.
3. Ashburner N, Bonner JF (1979). The induction of gene activity in Drosophila by heat shock. Cell 17:241.
4. Ballinger DG, Pardue ML (1983). The control of protein synthesis during heat shock in Drosophila cells involves altered polypeptide elongation rates. Cell 33:103.
5. Bardwell JCA, Craig E (1984). Major heat shock gene of Drosophila and the Escherichia coli heat-inducible dnaK gene are homologous. Proc Natl Acad Sci. 81:848.
6. Barnett T, Altschuler M, McDaniel CN, Mascarenhas JP (1980). Heat shock induced proteins in plant cells. Dev Gen 1:331.
7. Baszczynski CL, Walden DB, Atkinson BG (1982). Regulations of gene expression in corn (Zea mays L) by heat shock. Can J Biochem 60:569.
8. Berry J, Bjorkman O (1980). Photosynthetic response and adaptation to temperature in higher plants. In Briggs WR, Green PB, Jones RL (eds): "Annual Review Plant Physiology", Vol 31, California: Annual Reviews Inc., p 491.
9. Cooper P, Ho T-HD (1983). Heat shock proteins in maize. Plant Physiol. 71:215.
10. Cooper P, Ho T-HD, Hauptmann RM (1984). Tissue specificity of the heat-shock response in maize. Plant Physiol 75:431.
11. Corces V, Holmgren R, Freund R, Marimoto R, Meselson M (1980). Four heat shock proteins of Drosophila

melanogaster coded within a 12-kilobase region in chromosome subdivision 67B. Proc Natl Acad Sci 77:5390.

12. Craig E, Ingolia T, Slater M, Manseau L, Bardwell J (1982). Drosophila, yeast and E. coli genes related to the Drosophila heat-shock genes. In Schlesinger M, Ashburner M, Tissieres A (eds): "Heat Shock: From Bacteria to Man," New York: Cold Spring Harbor Laboratory, p 11.

13. Czarnecka E, Edelman L, Schöffl F, Key JL (1984). Comparative analysis of physical stress responses in soybean seedlings using cloned heat shock cDNAs. Plant Mol Biol 3:45.

14. Daniels CJ, McKee AHZ, Doolittle WF (1984). Archaebacterial heat-shock proteins. The EMBO J 4:745.

15   Dawson WO, Grantham GL (1981). Inhibition of stable RNA synthesis and production of a novel RNA in heat stressed plants. Biochem Biophys Res Commun 100:23.

16. DiDomenico BJ, Bugaisky GE, Lindquist S (1982A). The heat shock response is self-regulated at both the transcriptional and post-transcriptional levels. Cell 31:593.

17. DiDomenico BJ, Bugaisky GE, Lindquist S (1982B). Heat shock and recovery are mediated by different translational mechanisms. Proc Natl Acad Sci 79:6181.

18. Edelman L, Roberts JK, Czarnecka E, Key JL. Induction of heat shock-specific poly(A)+RNAs and proteins in soybeans at non-heat shock temperatures by arsenite and cadmium. Submitted to Plant Physiology.

19. Glover CVC, Vavra KJ, Guttman SD, Gorovsky MA (1981). Heat shock and deciliation induce phosphorylation of histone H1 in T. pyriformis. Cell 23:73.

20. Hadwiger LA, Wagoner W (1983). Effect of heat shock on the mRNA-directed disease resistance response of peas. Plant Physiol 72:553.

21. Hsiao TC (1970). Rapid changes in levels of polyribosomes in Zea mays in response to water stress. Plant Physiol 46:281.

22. Hsiao TC (1973). Plant responses to water stress. In Briggs WR (ed): "Annual Review Plant Physiology," Vol 24, California: Annual Reviews Inc, p 519.

23. Ingolia TD, Craig EA (1982). Four small Drosophila heat shock proteins are related to each other and to mammalian a-crystallin. Proc Natl Acad Sci 79:2360.

24. Ingolia TD, Craig EA, McCarthy BJ (1980). Sequence of three copies of the gene for the major Drosophila heat shock induced protein and their flanking regions. Cell 21:669.

25. Jung PK, Scott HD (1980). Leaf water potential, stomatal resistance, and temperature relations in field-grown soybeans. Agronomy J 72:986.

26. Kelley PM, Freeling M (1982). A preliminary comparison of maize anaerobic and heat shock proteins. In Schlesinger M, Ashburner M, Tissieres A (eds): "Heat Shock: From Bacteria to Man," New York: Cold Spring Harbor Laboratory, p 315.

27. Key JL, Czarnecka E, Lin CY, Kimpel J, Mothershed, C, Schöffl F (1983A). A comparative analysis of the heat shock response in crop plants. In Randall DD, Blevins DG, Larson RL, Rapp BJ (eds): "Current Topics in Plant Biochemistry and Physiology Symposium," Missouri: University of Missouri, p 107.

28. Key JL, Lin CY, Ceglarz E, Schöffl F (1982). The heat shock response in plants. In Schlesinger M, Ashburner M, Tissieres A (eds): "Heat Shock: From Bacteria to Man", New York: Cold Spring Harbor Laboratory, p 329.

29. Key JL, Lin CY, Ceglarz E, Schöffl F (1983B). The heat shock response in soybean seedlings. In Ciferri O, Dure L (eds): "Structure and Functions of Plant Genomes NATO ASI Series A, Life Sciences Vol 63" New York: Plenum Press, p 25.

30. Key JL, Lin CY, Chen YM (1981). Heat shock proteins of higher plants. Proc Natl Acad Sci 78:3526.

31. Li GC, Werb Z (1982). Correlation between synthesis of heat shock proteins and development of thermotolerance in chinese hamster fibroblasts. Proc Natl Acad Sci 79:3218.

32. Lin CY, Key JL (1966). Dissociation and reassembly of polyribosomes in relation to protein synthesis in the soybean root. J Mol Biol 26:237.

33. Lin CY, Roberts JK, Key JL (1984). Acquisition of thermotolerance in soybean seedlings: Synthesis and accumulation of heat shock proteins and their cellular localization. Plant Physiol 74:152.

34. Loomis WF, Wheeler S (1980). Heat Shock response of Dictyostelium. Dev Biol 79:399.

35. Loomis WF, Wheeler S (1982). The physiological role of heat-shock proteins in Dictyostelium. In Schlesinger M, Ashburner M, Tissieres A (eds): "Heat

Shock: From Bacteria to Man", New York: Cold Spring
Harbor Laboratory, p 353.

36. Mascarenhas JP, Altschuler M (1983). The response of
    pollen to high temperatures and its potential
    applications. In Mulcahy DL, Ottaviano E (eds):
    "Pollen: Biology and Implications for Plant Breeding",
    New York: Elsevier Science Publishing Co, p 3.

37. McAlister L, Finkelstein DB (1980). Heat shock
    proteins and thermal resistance in yeast. Biochem
    Biophys Res Commun 93:819.

38. McAlister L, Finkelstein DB (1980B). Alterations in
    translatable ribonucleic acid after heat shock of
    Saccharomyces cerevisiae. J Bacteriol 143:603.

39. Meyer Y, Chartier Y (1983). Long-lived and short-lived
    heat-shock proteins in tobacco mesophyll protoplasts.
    Plant Physiol 72:26.

40. Moriella CA, Boyer JS, Hageman RH (1973). Nitrate
    reductase activity and polyribosomal content of corn
    (Zea mays L.) having low leaf water potentials.
    Plant Physiol 51:817.

41. Neidhardt FC, Van Bogelen RA (1981). Positive
    regulatory gene for temperature-controlled proteins
    in Escherichia coli. Biochem Biophys Res Commun
    100:894.

42. Nover L, Scharf K-D (1984). Synthesis, modification
    and structural binding of heat-shock proteins in
    tomato cell cultures. Eur J Biochem 139:303.

43. Nover L, Scharf K-D, Neumann D (1983). Formation of
    cytoplasmic heat shock granules in tomato cell
    cultures and leaves. Molec and Cell Biol 9:1648.

44. Parker CS, Topol J (1984). A Drosophila RNA
    polymerase II transcription factor binds to the
    regulatary site of an hsp 70 gene. Cell 37:273.

45. Pelham HRB (1982). A regulatory upstream promoter
    element in the Drosophila hsp 70 heat-shock gene.
    Cell 30:517.

46. Reiter T, Penman S (1983). "Prompt" heat shock
    proteins: Translationally regulated synthesis of new
    proteins associated with the nuclear matrix-
    intermediate filaments as an early response to heat
    shock. Proc Natl Acad Sci 80:4737.

47. Sachs MM, Freeling M (1978). Selective synthesis of
    alcohol dehydrogenase during anaerobic treatment of
    maize. Molec gen Genet 161:111.

48. Sachs MM, Freeling M, Okimoto R (1980). The
    anaerobic proteins of maize. Cell 20:761.

49.   Sanders MM, Feeney-Triemer D, Olsen AS, Farrell-Towt
      J (1982). Changes in protein phosphorylation and
      histone H2b disposition in heat shock in Drosophila.
      In Schlesinger M, Ashburner M, Tissieres A (eds):
      "Heat Shock: From Bacteria to Man", New York: Cold
      Spring Harbor Laboratory, p 235.
50.   Scharf DK, Nover L (1982). Heat-shock-induced
      alternations of ribosomal protein phosphorylation in
      plant cell cultures. Cell 30:427.
51.   Schlesinger MJ, Aliperti G, Kelley PM (1982B). The
      response of cells to heat shock. Trends in Biochem
      Sci. 6:222.
52.   Schlesinger MJ, Ashburner M, Tissieres A (1982A). In
      "Heat Shock: From Bacteria to Man," New York: Cold
      Spring Harbor Laboratory.
53.   Schöffl F, Key JL (1982). An analysis of mRNAs for a
      group of heat shock proteins of soybean using cloned
      cDNAs. J Mol Appl Genet 1:301.
54.   Southgate R, Ayme A, Voellmy R (1983). Nucleotide
      sequence analysis of the Drosophila small heat shock
      gene cluster at locus 67B. J Mol Biol 165:35.
55.   Tilly K, McKittrick N, Zylicz M, Georgopoulos C
      (1983). The dnak protein modulates the heat-shock
      response of Escherichia coli. Cell 34:641.
56.   Velazquez J, DiDomenico B, Linquist S (1980).
      Intracellular localization of heat shock proteins in
      Drosophila. Cell 20:679.
57.   Velazquez J, Sonoda S, Bugaisky GE, Lindquist S
      (1983). Are HS proteins present in cells that have
      not been heat shocked. J Cell Biol 96:286.
58.   Vierling E, Mishkind ML, Schmidt GW, Key JL. A
      specific heat shock protein in plants is transported
      into chloroplasts. J Cell Biology (Abstract) in press.
59.   Voellmy R, Bromley P, Kocher HP (1983). Structural
      similarities between corresponding heat-shock
      proteins from different eucaryotic cells. J Biol
      Chem 258:3516.
60.   Walker JC, Key JL (1982). Isolation of cloned cDNAs
      to auxin-responsive poly(A)+RNAs of elongating
      soybean hypocotyl. Proc Natl Acad Sci 79:7185.
61.   Wang CH, Gomer RH, Lazarides E (1981). Heat shock
      proteins are methylated in avian and mammalian cells.
      Proc Natl Acad Sci 78:3531.
62.   Wu C (1984). Two protein-binding sites in chromatin
      implicated in the activation of heat-shock genes.
      Nature 309:229.

63. Yamamori T, Yura T (1982). Genetic control of heat-shock protein synthesis and its bearing on growth and thermal resistance in Escherichia coli K-12. Proc Natl Acad Sci 79:860.
64. Yarwood CE (1967). Adaptation of plants and plant pathogens to heat. In Prosser CL (ed): "Molecular Mechanisms of Temperature Adaptation", Washington, D.C.: Amer Assoc Sci Publ No 84, p 75.

This research supported by DOE contract DE-AS09-80ER10678 and a contract from Agrigenetics Research Associates. We thank Virginia Goekjian, Kenlock Westberry and Coco Whelchel for excellent technical assistance during these studies.

Cellular and Molecular Biology of Plant Stress, pages 181–200
© 1985 Alan R. Liss, Inc.

STRUCTURE AND EXPRESSION OF MAIZE HSP 70 GENE

Dilip M. Shah, Dean E. Rochester,
Gwen G. Krivi, Cathy M. Hironaka,
Thomas J. Mozer, Robert T. Fraley
and David C. Tiemeier

Division of Biological Sciences,
Corporate Research & Development,
Monsanto Chemical Company,
St. Louis, MO 63167

ABSTRACT   A set of at least eight heat shock proteins
(hsp's) is rapidly induced when the incubation
temperature of excised maize coleoptile tissue is raised
from 30°C to 42.5°C. *In vivo* translation of poly(A)
RNA isolated from nonheat-shocked (30°C) and heat-shocked
(42.5°C) tissue indicates that hsp's are translated from
newly synthesized mRNAs. We have used a cloned
Drosophila hsp 70 gene to select specific mRNA species
from a complex mixture of maize heat shock poly(A) RNA
and have shown that this mRNA translates *in vitro* into
70 kd polypeptide (hsp 70). We describe the isolation of
a recombinant λ phage containing a 9.6 kb EcoRI genomic
fragment spanning complete hsp 70 coding and 3'
noncoding sequences. The primary nucleotide sequence of
this clone encompassing first 354 amino acids has been
determined. The predicted amino acid sequence is 75%
homologous to the amino acid sequence of *Drosophila* hsp
70 polypeptide. The coding sequence is interrupted by
an intron in the codon specifying amino acid 71.
Although the inducible hsp 70 genes of *Drosophila* are
devoid of introns, a closely-related and constitutively
expressed *Drosophila* heat-shock-cognate gene (hsc 1)
contains an intron in exactly the same position. However,
unlike *Drosophila* cognate gene, transcription of our
cloned maize gene does appear by Northern analysis to be
enhanced by heat shock.

INTRODUCTION

The induction of heat shock proteins as a general response to thermal stress has been demonstrated in a wide spectrum of organisms including bacteria, fungi, animals and higher plants. Earlier studies examining this phenomenon in higher plants have shown that protein synthesis shifts rapidly from a complex pattern of normal proteins to a relatively simple pattern of heat shock proteins. When the plant tissue or cells in culture are exposed to elevated temperatures (1,2,4,10), a set of approximately 8 to 10 hsp's, whose mol.wt.'s range from 16 to 102 kd, is induced with a concomitant decrease in the synthesis of normal proteins. More recently it has been shown that the mRNAs encoding low mol.wt. hsp's accumulate rapidly after heat shock in soybean hypocotyl tissue suggesting de novo transcription of genes for this group of hsp's.

In Drosophila, the major polypeptide expressed after heat shock has a mol.wt. of 70 kd (hsp 70). The mRNA encoding hsp 70 is the most abundant species accumulated after heat treatment (12,14). A major heat shock-inducible protein of mol.wt. 70 kd is present in many eukaryotic organisms. A recent report by Kelley and Schlesinger (9) has shown that a polyclonal antibody made against chick embryo fibroblast hsp 70 cross-reacts with similar sized proteins from a wide spectrum of eukaryotic organisms indicating strong evolutionary conservation of this protein. This observation also argues in favor of hsp 70 serving a critical, but as yet unidentified, housekeeping role in eukaryotic cells exposed to thermal stress. We have undertaken the cloning of a maize hsp 70 gene with the aim of identifying DNA sequences responsible for thermo-inducible promotion of gene expression in higher plants.

RESULTS

Heat shock response in maize coleoptiles:

The induction of heat shock proteins (hsp's) was examined in excised coleoptile tissue incubated for 3 hr at various temperatures ranging from 30°C to 42.5°C.

The proteins, pulse-labeled with [^{3}H]-leucine, were
analyzed on SDS-polyacrylamide gel (Figure 1). At least,
six major hsp's with apparent mol.wt.'s of 104 kd, 88
kd, 86 kd, 72 kd, 24 kd and 16 kd are synthesized at
elevated levels in tissue incubated at 42.5°C. Two minor
hsp's with mol.wt. of roughly 20 kd are also induced.
These proteins are not synthesized in control tissue
incubated at 30°C, although polypeptides comigrating with
the 88 kd and 72 kd hsp's are detected at much reduced
levels. The initial coordinate induction of hsp's is
observed at 35°C and becomes more pronounced as the
temperature is raised from 35°C to 42.5°C. However,
synthesis of these proteins is strongly repressed at 45°C
(data not shown). A marked decline in the synthesis of
normal proteins is not observed even at the optimal heat
shock temperature of 42.5°C. Thus, in heat-shocked
coleoptile tissue, synthesis of hsp's takes place in
addition to normal protein synthesis. A study of the
time course of hsp synthesis in coleoptiles indicates
that heat shock response is initiated very rapidly. Hsp
synthesis is observed during the first hr of incubation
at 42.5°C and persists thereafter for at least 4 hr.
Maximal labeling of hsp's however takes place during the
second hr of incubation (data not shown).

In order to determine the control mechanism of this
heat shock response in maize, poly(A) mRNAs were isolated
from tissue incubated at 42.5°C and at 30°C and translated
in a reticulocyte cell-free system. A comparison of the
translated products of mRNA from heat-shocked (42.5°C)
tissue to the ones programmed by RNA from nonheat-shocked
(30°C) tissue is shown in Figure 2. At least, eight new
translational products are evident in the total translational products of 42.5°C mRNA. The mol.wt.'s of
these new polypeptides closely approximate the mol.wt.'s
of hsp's shown in Figure 1. The concentration of normal
mRNAs does not change dramatically in heat-shocked
tissue which supports the protein data discussed above.
This however does not imply that transcription of normal
mRNAs is continuing during heat shock, since the half
life of these mRNAs may be several hrs. It thus appears
that the synthesis of a new set of proteins results from
a *de novo* synthesis of a new set of mRNAs.

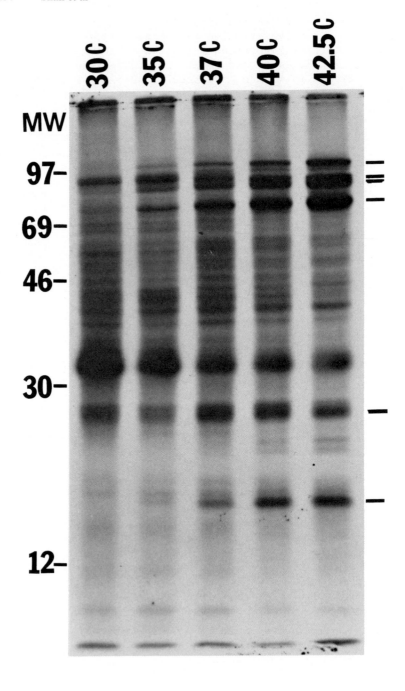

FIGURE 1.   Induction of heat shock proteins in
maize coleoptile tissue.   One-cm-long excised coleoptile
tissue of 4-day-old maize seedlings was incubated at
various temperature for 3 hr; 400 µCi of [^{3}H-leucine]
was added for the last 2 hr of incubation (10).   The
proteins were fractionated on a 12% SDS gel and
visualized by fluorography.   Molecular weights of
protein standards are shown in kd.   Arrows denote major
hsps.

Drosophila hsp 70 clone selects maize hsp 70 mRNAs
from a complex population:

Kelley and Schlesinger (9) have recently demonstrated
that hsp 70 proteins from diverse organisms are anti-
genically related thereby suggesting that the primary
amino acid sequence of this protein may be evolutionary
conserved.   Amino acid sequence conservation does not
necessarily imply strict DNA conservation, however, since
third base codon substitutions may occur.   Recent evidence
indicates that Drosophila and yeast hsp 70 genes share
significant nucleotide sequence homology (8).   We therefore
reasoned that the cloned Drosophila hsp 70 gene may
cross-hybridize to maize gene encoding hsp 72 and therefore
may serve as a hybridization probe to isolate the maize
gene encoding this protein.   For convenience, we will
refer to maize hsp 72 as hsp 70, since as we show below,
there is fairly extensive nucleotide sequence homology
between the maize gene encoding this protein and Drosophila
hsp 70 gene.   The specificity of the hybridization of
Drosophila hsp 70 sequences to maize nucleic acid was
demonstrated by Northern blot analysis of 42.5°C mRNA.
Poly(A) RNA was fractionated on a formaldehyde-agarose
gel, blotted onto a nitrocellulose filter, and hybridized
with ^{32}P-labeled hsp 70 clone (232.1) of Drosophila
melanogaster.   Figure 3 shows that, at low stringency of
hybridization (30% formamide, 37°C), Drosophila hsp 70
clone hybridizes to two mRNAs of 2.2 kb and 2.6 kb.   To
further determine if these mRNAs encode hsp 70 polypeptide,
clone 232.1 was bound to nitrocellulose and hybridized
with 42.5°C total poly(A) RNA.   Hybridized RNA was then
tested for its ability to direct the synthesis of hsp 70
polypeptide in a rabbit reticulocyte cell free system.

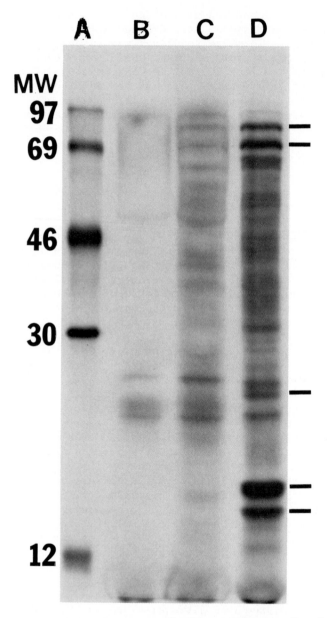

FIGURE 2.   *In vitro* translation products of poly(A)
RNA isolated from nonheat-shocked (30°C) and heat-shocked
maize coleoptile tissue.   Poly(A) RNAs were translated

in rabbit reticulocyte cell-free system.   The translation
products were analyzed on 12% SDS-polyacrylamide gel and
fluorographed.   A. Mol.wt. standards, B. no maize RNA,
C. poly(A) RNA from nonheat-shocked (30°C) and D. from
heat-shocked (42.5°C) coleoptile tissue.   Arrows denote
major hsp's translated from mRNA.

The mRNA selected by 232.1 translated into a product
which comigrated on a denaturing gel with the authentic
hsp 70 polypeptide (data not shown).

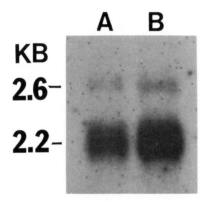

    FIGURE 3.   Hybridization of Drosophila hsp 70 clone
(232.1) to poly(A) RNA isolated from heat-shocked
(42.5°C) tissue.   A. 2.5 µg, B. 5.0 µg of poly(A) RNA.
The sizes of the transcripts are shown in kb.

Isolation of a recombinant λ phage encoding maize hsp 70
gene:

    Drosophila clone was used as a hybridization probe
to screen the genomic library of maize DNA constructed in
λgtWES·λB.   About 4.2 x 10⁴ phage recombinants were
screened.   Analysis of a total of eight positively
hybridizing plaques resulted in the isolation of a 9.6
kb EcoRI fragment of maize DNA containing sequences
homologous to Drosophila hsp 70 coding sequence.
Restriction map of this fragment, shown in Figure 4A,

was derived by analysis of the single and double digests
of phage DNAs with various restriction endonucleases. A
4.0 kb EcoRI-BamHI subfragment of this clone sharing
homology with the Drosophila clone was subcloned in pUC9
and named pMON9502. Further restriction mapping of this
clone was carried out (Figure 4B) and the coding
sequence was further defined to a segment of
approximately 2.2 kb by hybridization to yeast YG100
clone containing the entire hsp 70 coding sequence. The
direction of transcription was determined by hybridizing
the restriction fragments of pMON9502 with the
Drosophila 232.1 clone which contains only the 5' half of
the coding sequence.

Analysis of maize hsp 70 transcripts:

     Total RNAs from the nonheat-shocked (30°C) and heat-
shocked (42.5°C) tissue were electrophoresed on a
denaturing formaldehyde-agarose gel and transferred to
nitrocellulose. Poly(A) RNA from heat-shocked tissue
(42.5°C) was also run on the same gel. The 4.0 kb insert
purified from pMON9502 was labeled by nick translation
and hybridized to the filter. Two RNA species of 2.2 kb
and 2.6 kb were detected in the poly(A) RNA of heat-
shocked tissue as shown in Figure 5. However, when total
RNA from this same tissue was analyzed on the same gel,
the 2.6 kb transcript was found to migrate somewhat
anomalously with the apparent size of 2.4 kb. Reason for
the anomalous migration of this RNA is not clear. It is
likely that such apparent reduction in the size of 2.6
kb transcript results from the compression of the band
by 18S ribomomal RNA which also migrates in the same
region of the gel. The 2.2 kb transcript, on the other
hand, comigrated in both lanes. The 2.2 kb transcript
was also detected in total RNA from nonheat-shocked
(30°C) tissue. Heat shock at 42.5°C increases the steady
state level of this mRNA about 3-fold. A smaller
transcript of 1.0 kb was also detected in the poly(A)
RNA of a heat-shocked tissue. Further analysis has
revealed that this transcript shares homology with the
5' end of the coding sequence. pMON9502 was further
used in hybrid selected mRNA translation experiment to
show that it does indeed select RNAs which translate to

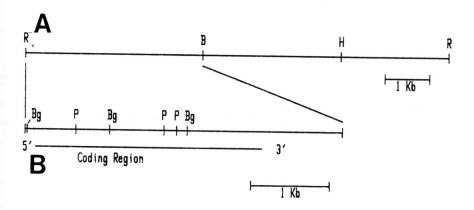

FIGURE 4. Restriction maps of the maize inserts in
A. λ M1, B. pMON9502 DNAs.

yield a polypeptide comigrating with hsp 70. The result
shown in Figure 6 is identical to one obtained using
Drosophila hsp 70 clone.

Primary sequence of maize hsp 70 gene:

        In Drosophila, hsp 70 is coded for by a small
multigene family consisting of five members. Nucleotide
sequences of the polypeptide coding regions of these
genes are about 97% conserved (7). The predicted amino
acid sequence of Drosophila hsp 70 protein consists of
641 amino acids. A partial nucleotide sequence of our
cloned maize hsp 70 gene in pMON9502 has been determined
using the chemical cleavage method of Maxam and Gilbert
(11). The nucleotide sequence shown in Figure 7
corresponds to the first 344 amino acids of the maize
hsp 70 polypeptide and about 80 nucleotides of the 5'
non-translated region. Some striking features pertaining

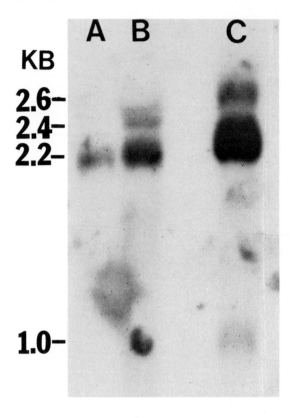

FIGURE 5. Northern analysis of RNAs from
nonheat-shocked (30°C) and heat-shocked (42.5°C)
coleoptile tissue. A. 10 µg of 30° total RNA, B. 10 µg
of 42.5°C total RNA, C. 5 µg of 42.5°C poly(A) RNA. The
4.0 kb EcoRI-BanHI insert of pMON9502 was used as
a hybridization probe. The sizes of transcripts are
indicated in kb.

to the structure of the gene have emerged from a comparison with the protein coding region of Drosophila hsp 70
gene. The coding sequence of the maize gene is interrupted by an intron which splits the codon specifying
aspartic acid at position 71. Ingolia and Craig (6) and
Craig et al. (5) have recently reported in Drosophila yet
another family of genes related to the hsp 70 genes that

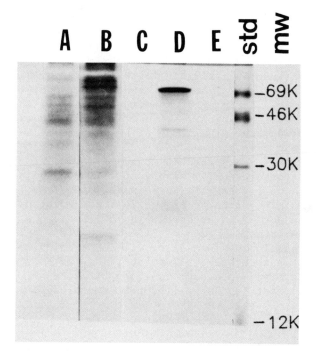

FIGURE 6.   Translation of mRNAs selected by
hybridization to cloned maize hsp 70 gene. Poly(A)
mRNAs from nonheat-shocked (30°C) and heat-shocked
(42.5°C) tissue homologous to pMON9502 were selected by
hybridization to pMON9502 immobilized on nitrocellulose
and translated in vitro. Translational products of A.
unselected 30°C poly(A) mRNA, B. unselected 42.5°C
poly(A) mRNA, C. selected 30°C poly(A) mRNA, D. selected
42.5°C poly(A) mRNA, E. selected 42.5°C poly(A) mRNA
using pBR322 as a control. Mol.wt.'s of protein
standards are indicated.

are normally expressed during development but not induced
by heat shock. These genes, termed heat-shock-cognate
genes (hsc 1, hsc 2, etc), have been partially sequenced.

One member of this family of genes (hsc 1) contains an
intron whose position corresponds precisely to the
position of an intron in the maize gene (Figure 8). The
sequences at the splice junctions are similar to the
eukaryotic consensus sequences (13). In contrast, the
major heat shock-induced genes of Drosophila contain no
introns. In the region analyzed, maize gene contains a
single insertion of two amino acids and a single deletion
of one amino acid. The predicted amino terminus of the
maize gene has five additional amino acids compared with
Drosophila hsp 70 gene. This type of heterogeneity in
amino termini has been observed among Drosophila heat-
shock-cognate genes and yeast hsp 70 genes (5,8).
Because of this heterogeneity, analysis of the amino
acid sequence homology began with amino acid proline at
position 7. As elucidated in Figure 9, the predicted
amino acid sequence of maize hsp 70 is 75% homologous
to both Drosophila hsp 70 (256 of 345 amino acids
compared) and hsc 1 proteins (55 of 210 amino acids
compared) (5). A sequence of 11 amino acids (aa 5 to 15)
is strictly conserved among hsp 70 genes of maize,
Drosophila and yeast (8) as well as heat-shock-cogante
genes of Drosophila (5).

```
TCTCTGCGAT TTCTCTAGAT CTCGACTACC CCCCACTAGT TTGGTTCCT
                                       M   A   K   S   E
CCTTTCGTTC GAGAGAGCGA TTCTGGTGGCA   ATG GCG AAG AGC GAG
  G   P   A   I   G   I   D   L   G   T   T   Y   S   C
GGT CCG GCG ATC GGG ATC GAC CTC GGC ACC ACC TAC TCG TGC
  V   G   V   W   Q   H   D   R   V   E   I   I   A   N
GTC GGC CTG TGG CAG CAC GAC CGG GTG GAG ATC ATC GCC AAC
  D   Q   G   N   R   T   T   P   S   Y   V   G   F   T
GAC CAA GGG AAC CGC ACC ACG CCG TCC TAT GTC GGC TTC ACC
  D   T   G   R   L   I   G   D   A   A   K   N   Q   V
GAC ACC GAG CGG CTC ATC GGC GAC GCT GCC AAG AAC CAA GTC
  A   M   N   P   T   N   T   V   F
GCC ATG AAC CCC ACC AAC ACC GTC TTC G gtacgcgctcac
ttcgccctct gcctttgtta ctgtcacgtt tctctagtgc tctcttgtgt
caggtggt .........................478 bp ..............
ttgtgctctc ctacctcctg atggtatctg atatctacga acgtacacta
tattg cttctcttac atacgtatct tgctcgatgc cttctcccag
tattgaccag tgtactcaca tagtcttgct cattcattgt aatgcag
  D   A   K   R   L   I   G   R   R   F   S   S   P   A
AT GCC AAG CGG TTG ATC GGC AGG AGG TTC TCT AGC CCT GCA
```

```
 V   Q   S   S   M   K   L   W   P   S   R   H   L   G
GTG CAG AGT AGC ATG AAG CTA TGG CCG TCA AGG CAC CTA GGG
 L   G   D   K   P   M   I   V   F   N   Y   K   G   E
CTT GGT GAC AAA CCC ATG ATT GTA TTC AAC TAC AAG GGC GAG
 E   K   Q   F   A   A   G   G   I   S   S   M   V   L
GAG AAG CAG TTT GCT GCT GAG GAG ATC TCC TCC ATG GTC CTC
 I   K   M   K   E   I   A   E   A   Y   L   G   S   T
ATC AAG ATG AAG GAG ATT GCT GAA GCC TAC CTT GGT TCC ACC
 I   K   N   A   V   V   T   V   P   A   Y   F   N   D
ATC AAG AAC GCA GTG GTG ACA GTG CCG GCC TAT TTC AAC GAC
 S   Q   R   Q   A   T   K   D   A   G   V   I   A   G
TCG CAG AGG CAG GCC ACC AAG GAC GCC GGT GTC ATT GCG GGC
 L   N   V   M   R   I   I   N   E   P   T   A   A   A
CTC AAT GTG ATG CGT ATC ATC AAC GAG CCC ACT GCT GCT GCT
 I   A   Y   G   L   D   K   K   A   T   S   S   G   E
ATT GCC TAC GGT CTT GAC AAG AAG GCC ACC AGC TCC GGC GAG
 K   N   V   L   I   F   D   L   G   G   G   T   F   D
AAG AAC GTG CTC ATC TTC GAC CTT GGT GGT GGC ACG TTT GAT
 V   S   L   L   T   I   E   E   G   I   F   E   V   K
GTG TCG CTC CTC ACC ATC GAG GAG GGC ATC TTC GAG GTG AAG
 A   T   A   G   D   T   H   L   G   G   E   D   F   D
GCC ACT GCG GGG GAC ACT CAC CTT GGG CGC GAG GAC TTC GAC
 N   R   M   V   N   H   F   V   Q   E   F   K   R   K
AAT CGC ATG GTG AAC CAC TTC GTC CAA GAG TTC AAG CGC AAG
 N   K   K   D   I   S   G   N   P   R   A   L   R   R
AAT AAG AAG GAC ATA AGC GGC AAC CCC CGT GCA CTG CGC CGG
 L   R   T   A   C   E   R   A   K   R   T   L   S   S
CTG CGC ACG GCG TGC GAG CGC GCC AAG CGC ACG CTG TCA TCG
 T   A   Q   T   T   I   E   I   D   S   L   F   E   G
ACT GCC CAG ACG ACC ATT GAG ATC GAC TCC CTG TTC GAG GGC
 I   D   F   T   P   R   S   S   R   A   R   F   E   E
ATC GAT TTC ACT CCA CGA TCA TCT AGG GCT CGC TTC GAG GAG
 L   N   M   D   L   F   R   K   C   M   E   P   V   E
CTC AAT ATG GAC TTG TTC CGG AAG TGC ATG GAA CCT GTG GAG
 K   C   L   R   D   A   K   M   D   K   S   S   V   H
AAG TGC TTC CGC GAC GCC AAG ATG GAC AAG AGC AGC GTG CAC
 D   V   V   L   V   G   G   S   T   R   I   P   K   V
GAC GTC GTG CTC GTC GGT GGC TCC ACC CGC ATC CCC AAG GTG
 Q   Q   L   Q
CAG CAG CTG CAG
```

FIGURE 7. A partial nucleotide sequence of the maize hsp 70 gne (pMON9502). The sequence spans the first 354 amino acids of maize hsp 70 polypeptide, 81 nucleotides

of the 5' non-translated region and part of the intron.
The deduced amino acid residues are denoted above each
codon in the standard one letter code.


```
Drosophila hsp 70        ATC TTC G                    AC GCC
                          |  ||  |                    |  |||
Drosophila hsc 70        GTG TTT G GTGAGT    TGCAG AT GCC
                         ||  ||  | | ||      ||||| ||  |||
Maize      hsp 70        GTC TTC G GTACGC    TGCAG AT GCC
                          |  |||            |  |||
Eukaryotic consensus         CA G GTAAGT    TNCAG G
                             A     G        C T
```

FIGURE 8.   Comparison of the exon-intron junctions.
The maize hsp 70 and Drosophila hsc 1 (6) exon-intron
junction sequences are aligned with the eukaryotic
consensus (13).   Drosophila hsp 70 sequence (6) is also
shown to illustrate the conservation of intron position
in maize hsp 70 and Drosophila hsc 1 genes.

Maize hsp 70 gene is induced by heat shock:

    In order to study the transcription of the maize
gene, a 21-nucleotide-long synthetic oligonucleotide
corresponding to the 5' transcribed but non-translated
region of the gene was used as a probe.   This oligonucleotide probe was shown not to hybridize to the
other members of the maize hsp 70 gene family (data not
shown) and was therefore deemed appropriate as a gene
specific probe.   This probe was hybridized to RNAs from
heat-shocked and nonheat-shocked tissue electrophoresed
on denaturing agarose gels and transferred to
nitrocellulose filter.   The pattern of hybridization,
shown in Figure 10, was almost identical to the pattern
obtained using the complete coding sequence of pMON9502
as a hybridization probe.   This result strongly
demonstrates that the transcription of the maize hsp 70
gene in pMON9502 is induced by heat shock.

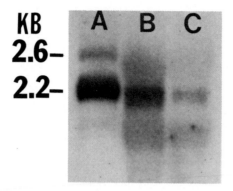

FIGURE 10. Northern analysis of mRNAs from nonheat-shocked (30°C) and heat-shocked (42.5°C) coleoptile tissue using a gene specific oligonucleotide probe. A. 5 µg of 42.5°C poly(A) mRNA, B. 10 µg of 42.5°C total RNA, C. 10 µg of 30°C total RNA. A 21-nucleotide-long oligonucleotide complementary to a sequence immediately upstream of the ATG initiation codon was chemically synthesized, labeled with ^{32}P using T4 polynucleotide kinase, and used as probe.

DISCUSSION

Heat shock response was examined in the excised coleoptile tissue of the maize seedlings. A set of heat shock proteins whose mol.wt.'s ranged from 16 to 104 kd was identified in this tissue. Heat shock response has been observed in all tissues examined, including both plumules and roots (3,4). The pattern of hsp synthesis in coleoptiles is not identical to that observed in roots. For example, hsp 76 is expressed in roots, but not in coleoptiles (4, unpublished results). In Drosophila, when the temperature is raised to 37°C, repression of preexisting translation takes place almost immediately. In contrast, maize tissue continues to synthesize most of the normal proteins at slightly reduced levels. The selective translation of hsp's does not appear to be highly pronounced in plant cells. Induction of hsp synthesis and recovery from heat shock is quite rapid in maize coleoptile tissue. Similar observations have been made in various other tissues of higher plants (1,4,10,

```
Maize       hsp 70   M A K S E G P A I G I D L G T T Y S C V G V W Q H D R V E I I A
Drosophila  hsp 70   M P K L * V * * * * G * * * * * * * * * * * Y * * G K * * * * N *
Drosophila  hsc 70   * * * * * * * * * * * * * * * * * * * * * * F * * G K * G * * * *

Maize       hsp 70   N D Q G N R T T P S Y V V G F T D T G R L I G D A A K N Q V A M N
Drosophila  hsp 70   Y * * * * * * * * * * * A * * * * * * * * * * * * S * N E P * * * *
Drosophila  hsc 70   * * * * * * * * * * * * A * * * G * * * * * * * * G S E * * * * * *

Maize       hsp 70   P T N T V F D A K R L I G R R F S S P A V Q S S M K L W P S R H
Drosophila  hsp 70   * * R * * * * * * * * * * * * * * * * Q S A E D * H * * F K V
Drosophila  hsc 70   * * N * * * * * * * * * K Y D D * D D A T * M D K H * * F E A

Maize       hsp 70   L G L G D K P M I V F N Y K G E E K Q F A A G G I S S M V L I K
Drosophila  hsp 70   V S D * G * * * * * K * G V E * * S * R * * P * * * * * * * * T *
Drosophila  hsc 70   F A G N G * R * * * * * * R * * * * * * * * * * * * * * * * * *

Maize       hsp 70   M K E I A E A Y L G S T I K N A V V T V P A Y F N D S Q R Q A T
Drosophila  hsp 70   * * T * * * * * * * * E S * T D * I * * * * * * * * * * * * * * *
Drosophila  hsc 70   * *

Maize       hsp 70   K D A G V I A G L N V M R I I N E P T A A A I A Y G L D K K A T
Drosophila  hsp 70   * * H * * * * * * * * L * * * * * * * * * * * L * * * * * N L K
Drosophila  hsc 70   * *

Maize       hsp 70   S S G E K N V L I F D L G G G T F D V S L L T I E E G A I F E V
Drosophila  hsp 70   A A * * R * * * * * * * * * * * * * * * * * * * I * * D * S L * *
Drosophila  hsc 70   * * * * * * * * * * * * * * * * * * * * * * V * * D * * Δ Δ * *
```

```
Maize       hsp 70    K A T A G D T H L G G E D F D N R M V N H F V Q E F K R K N K K
Drosophila  hsp 70    R S * * * * * * * * * * * * * * L * * L T * L A E * * * Y * * K
Drosophila  hsc 70    * * * * * * * * * * * * * * * * L * * Q * * * H G *

Maize       hsp 70    D I S G N P R A L R R L R T A C E R A K R T L S S T A Q T T I
Drosophila  hsp 70    * L R S * * * * * * * * * A * * * * * * * * S T E A * *
Drosophila  hsc 70    * L G Q * K * * * * * * * * * * * * * S T * A S *

Maize       hsp 70    E I D S L F E G I D F T P R S S R A R F E E L N M D L F R K C M
Drosophila  hsp 70    * * A * * * * Q * * Y T L V * * * * * * C N A * * N T L
Drosophila  hsc 70    * * * * * * * V * * Y T S V T * * * * * G * * V * * G T

Maize       hsp 70    E P V E K C L R D A K M D K S S V H D V V L V G G S T R I P K V
Drosophila  hsp 70    Q * * * * * A * N * * * * * * S Q I * * I * * * * * * * *
Drosophila  hsc 70

Maize       hsp 70    Q Q L Q
Drosophila  hsp 70    * S * L
Drosophila  hsc 70
```

FIGURE 9. Comparison of the amino acid sequences of maize hsp 70, Drosophila hsp 70 (7) and Drosophila hsc 1 (6) polypeptides: Amino acid sequences were deduced from the nucleotide sequences. Amino acids encoded by Drosophila hsp 70 and hsc 1 genes are shown where differences occur relative to maize polypeptide. Only 210 amino acids of hsc 1 polypeptide were available for comparison. Since amino termini of these polypeptides are heterogeneous, comparison began with proline residue at position 7.

unpublished results). We have shown that hsp's are translated from newly synthesized mRNAs. Transcriptional activation of a group of heat shock genes has been recently reported in soybean hypocotyl tissue (14).

A fragment of maize DNA encoding hsp 70 was isolated on the basis of homology to a Drosophila melanogaster hsp 70 gene. DNA sequence analysis of this gene has shown it to contain an intron the location of which is conserved in one member of a family of heat-shock-cognate genes (hsc 1) recently discovered in Drosophila (5). Such conservation of the position of an intron in the coding sequence of these two widely divergent species strongly suggests that these two genes have evolved from a common ancestor. Preliminary DNA sequence analysis of another maize hsp 70 gene indicates that it also contains an intron in the same position (unpublished results). At least, two Drosophila heat-shock-cognate genes are known to contain an intron and are not inducible by heat shock. In contrast, our cloned maize gene does appear by Northern analysis to be heat shock-inducible. In Drosophila, the heat inducible hsp 70 genes are devoid of introns.

The amino acid sequence of maize hsp 70 polypeptide is 75% homologous to Drosophila hsp 70 and hsc 70 polypeptides. The amino acid sequences of Drosophila and yeast hsp 70 polypeptides have been previously shown to be about 72% homologous (8). These observations strongly demonstrate remarkable sequence conservation of hsp 70 genes throughout eukaryotic evolution.

ACKNOWLEDGEMENTS

We would like to thank Drs. R. Morimoto and M. Meselson for sending us Drosophila hsp 70 clone 232.1. We are grateful to Dr. E. Craig for providing the yeast hsp 70 clone, YG100 used in this study. We also are grateful to Patsy Guenther for typing this manuscript on such short notice.

REFERENCES

1.   Altschuler M, Mascarenhas JP (1982). Heat shock proteins and effects of heat shock in plants.

Plant Mol Biol 1:103.
2. Barnett T, Altschuler M, McDaniel CN, Mascarenhas
    JP (1980). Heat shock induced proteins in plant
    cells. Dev Gen 1:331.
3. Baszczynski CL, Walden DB, Atkinson BD (1982).
    Regulation of gene expression in corn (Zea mays L.)
    by heat shock. II. In vitro analysis of RNAs from
    heat shock seedlings. Can J Biochem Cell Biol
    61:395.
4. Cooper P, Ho TD (1983). Heat shock proteins in
    maize. Plant Physiol 71:215.
5. Craig E, Ingolia T, Slater M, Manseau L (1982).
    Drosophila and yeast multigene families related to
    the Drosophila heat shock genes. In Schlesinger
    MJ, Ashburner M, Tissieres A (eds): "Heat Shock,
    From Bacteria to Man". Cold Spring Harbor
    Laboratory: p 11.
6. Ingolia TD, Craig EA (1982). Drosophila gene
    related to the major heat shock-induced gene is
    transcribed at normal temperatures and not induced
    by heat shock. Proc Natl Acad Sci USA 79:525.
7. Ingolia TD, Craig EA, McCarthy BJ (1980). Sequence
    of three copies of the gene for major Drosophila
    heat shock induced protein and their flanking
    regions. Cell 21:669.
8. Ingolia TD, Slater MR, Craig EA (1982).
    Saccharomyces cerevisiae contains a complex
    multigene family related to the major heat shock-
    inducible gene of Drosophila. Mol Cell Biol 2:1388.
9. Kelly PM, Schlesinger MJ (1982). Antibodies of two
    majro chicken heat shock proteins cross-react with
    similar proteins in widely divergent species. Mol
    Cell Biol 2:267.
10. Key JL, Lin CY, Chen YM (1981). Heat shock proteins
    of higher plants. Proc Natl Acad Sci USA 78:3526.
11. Maxam A, Gilbert W (1977). A new method for
    sequencing DNA. Pro Natl Acad Sci USA 74:560.
12. McKenzie SL, Henikoff S, Meselson M (1975). Locali-
    zation of RNA from heat induced polysomes at puff
    sites in Drosophila melanogaster. Proc Natl Acad
    Sci USA 72:1117.
13. Mount SM (1982). A catalogue of splice junction
    sequences. Nucleic Acids Res 10:459.
14. Schöffl F, Key JL (1982). An analysis of mRNAs
    for a group of heat shock proteins of soybean using
    cloned cDNAs. J. Mol Appl Genet 1:301.

15. Spradling A, Penman S, Pardue ML (1975). Analysis of Drosophila mRNAs by in situ hybridization: sequences transcribed in normal and heat shock cultured cells. Cell 4:395.

Cellular and Molecular Biology of Plant Stress, pages 201–216
© 1985 Alan R. Liss, Inc.

POTATO COLD HARDINESS AND FREEZING STRESS

Paul H. Li

Laboratory of Plant Hardiness, Department of
Horticultural Science and Landscape Architecture,
University of Minnesota, St. Paul, MN 55108

INTRODUCTION

Two approaches have been used to study plant cold
hardiness. One compares studies between cold-accli-
mated and non-acclimated plants of the same species,
and the other compares studies among species within a
genus with different inherent cold hardiness. Both
approaches have been used with the potato. The
following summarizes some of the knowledge learned from
this crop.
    Definition. The potato here refers to the tuber-
bearing Solanum species in addition to the commonly
cultivated S. tuberosum. Cold hardiness (freezing
tolerance) is a term used to describe the ability of
the tissue that can survive ice formation in the cell
when subjected to subzero temperatures. Cold acclimation (cold hardening) is a term used to describe the
transition of the plant from a freezing sensitive state
to a tolerant state when exposed to acclimating
conditions.

FREEZING STRESS

Under rate-controlled cooling conditions, two types
of freezing, intracellular and extracellular, can be
observed. When cells are cooled rapidly, e.g. 4°C/min.
(2) or faster, cells undergo intracellular freezing.
It is safe to say that plant cells can not survive when

intracellular freezing occurs. In nature, fortunately, plants cool slowly and ice always forms in the intercellular spaces or on the surface of the tissue resulting in extracellular freezing. Intracellular freezing can be avoided if extracellular freezing is allowed to occur (16). In the laboratory, extra-cellular freezing can be introduced by inoculating tissues with ice at just below 0°C. In nature, it depends on the ice-nucleating ability of plants and/or external sources of ice nucleation.

During extracellular freezing, water moves from the inside of the cell, mainly from the vacuole, to inter-cellular spaces, forming ice; this process results in cell dehydration. As freezing proceeds, ice growth causes cell contraction. Thus, during extracellular freezing, a cell can experience three types of stress: freeze-induced dehydration stress, osmotic stress due to removal of water from the vacuole, and mechanical stress by ice growth and cell contraction. Tolerance to extracellular freezing is probably the major means in nature by which plants survive freezing stress (the presence of ice in tissues). However, extracellular freezing can fatally injure the cell when cooled beyond a tolerable limit. Levitt (23) concluded that toler-ance to freeze-induced dehydration and avoidance of intracellular freezing are the survival mechanisms of a plant cell.

Cellular water seldom freezes at the freezing point of water. It supercools to varying degrees before forming ice, even in tender plant species (3). The extent of supercooling in nature is primarily dependent upon the ice-nucleating ability of the plant and/or its surroundings. The ice-nucleating ability is a charac-teristic of a given species and may vary considerably among different parts of the plant (20). Asahina (2) has demonstrated that tuber sprouts of S. tubersoum could remain supercooled for 18 hrs at -5.5°C and 4 hrs at -7.5°C. Rajashekar et al. (30) reported that leaf cells of other tuber-bearing Solanum species could remain supercooled above -6.9°C, and intact plants from -5°C to -7°C when cooled at a rate of 2°C/hr. General-ly, the extent of supercooling decreases with slower cooling rates. Reducing the population of ice nucle-ators (26), or the high concentration of cell sap may induce a few more degrees of supercooling; thus, the cell could survive a greater degree of subzero temperature stress by avoiding ice formation.

In vivo Freezing Tolerance.

Freezing tolerance of 60 tuber-bearing Solanum species has been assessed by an excised leaf tissue test (24). Tolerance is expressed by killing temperatures (°C), and the commonly cultivated S. tuberosum potato, a tender species, can survive only above -3°C (25).

Using the pulsed nuclear magnetic resonance spectrometer, Chen et al. (10) studied six potato species including one hybrid with cold hardiness ranged from -3 to -6°C. They found that highly significant correlations exist between cold hardiness of leaf cells and the % unfrozen water at killing temperatures. The hardier the species the less percentage of unfrozen water at killing temperatures was observed. The differences in cold hardiness among these species also do not result from the avoidance of ice formation (supercooling), because no relationship was observed between cold hardiness and melting point depression of cell sap. These findings confirm a previous report that potato cold hardiness is a true tolerance of extracellular freezing (34).

It is generally believed that the thermal stability of proteins plays a significant role in freezing tolerance. Levitt (22) in 1962 proposed a sulfhydryl-disulfide hypothesis to interpret how the stability of proteins works under freezing stress.

Potato leaves have abundant ribulose bisphosphate carboxylase-oxygenase (RuBPCase) located within the chloroplasts. It is probably the most abundant and important protein in the leaf tissue. A comparison of its structure and the function has thus been made between S. commersonii (Sc), a cold-hardy potato, and S. tuberesum (St), a non-hardy potato, in terms of SH group distributions on the enzyme and enzyme stability to freeze-thaw cycles (17).

The titration of native Sc and St RuBPCase with 5,5'-dithiobis(2-nitrobenzoic acid) (DTNB) indicates the presence of at least two populations of SH groups.

The first-order rate constant for slow titrating sulf-
hydryl groups for the native Sc RuBPCase was only 22%
of the first-order rate constant for St RuRPCase. The
difference in the kinetics of SH titration indicates a
difference in native structure. Upon complete denaturation of the enzyme from both sources in the presence
of 1% sodium dodecyl sulfate (SDS), the same number of
SH groups were titratable with DTNB. However, the
first-order rate constant was drastically different
under these conditions indicating that the structure of
the denatured state of the two enzymes was also
different.

The enzyme in the potato has a molecular weight of
about 540,000 daltons and is made up of eight large
(54,000 daltons) and eight small (14,000 daltons)
subunits (17). Gel scans of St RuBPCase, prepared and
electrophoresed in the presence of SDS without β-mercaptoethanol, indicate that the relative proportion of
the large subunit to the small subunit decreased more
significantly than that of Sc RuBPCase. Similar
results were also observed after several cycles of
freeze-thaw of the enzyme from both sources. Under the
conditions of the absence of β-mercaptoethanol and
after freeze-thaw cycles, a polypeptide of about
108,000 daltons, which has a correct molecular weight
for a dimer of the large subunit, was observed from the
denatured St RuBPCase. This was not, however, found
from Sc RuBPCase (Fig. 1 and 2). Results suggest that
the large subunit of RuBPCase from cold hardy Sc
species is structurally more stable to freeze-thaw
cycles than the large subunit from St species.

Differences in either the primary or secondary
structure of the same protein from genetically related
species, with decreased degree of exposure of surface
SH groups, would represent a novel mechanism for re-
sisting the denaturation of the protein by the sulf-
hydrly to disulfide interchange (22). Sc RuBPCase and
St RuBPCase are examples clearly supporting this suggestion that the native structure of the former has
fewer exposed SH groups than that of the latter even
though the total number of SH groups titratable with
DTNB is the same from both species. Therefore, species
within a genus, which are better able to withstand
freezing stress, may have proteins which are more
stable to freezing as a result of less exposure of SH

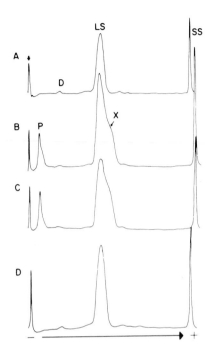

FIGURE 1. Spectrophotometric scans of SDS-polyacrylamide gels of Sc RuBPCase (A) Purfied sample prepared and electrophoresed in the presence of 2% (v/v)
β-mercaptoethanol. (B) Purified sample prepared and
electrophoresed in the absence of β-mercaptoethanal.
(C) Sample used in B subjected to five successive
freeze-thaw cycles. The sample was frozen at -20°C for
40 min and thawed at room temperature. (D) β-mercaptoethanol added to a final concentration of 2% to Sc
RuBPCase which had been subjected to five successive
freeze-thaw cycles. Arrow at the left of scan A indicates the peak which represents the interface between
the stacking and separating gels. In all cases, 15 to
20μg of protein was applied to each gel. D, dimer of
the large subunit; P, high relative mass polymeric
form; X, 45,000 to 50,000 polypeptide; LS, large
subunit; and SS, small subunit (17).

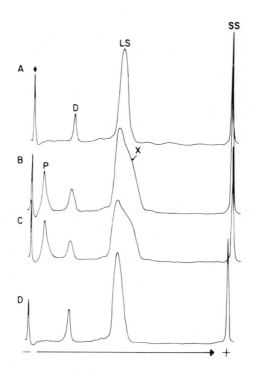

FIGURE 2.   Spectrophlotometric scans of SDS-polyacrylamide gels of St RuBPCase.   Experimental conditions
were exactly the same as those described for Fig. 1A,
1B, 1C and 1D (17).

groups than those species, which are less able to
withstand freezing stress.

In vitro Freezing Tolerance.

In nature, the cell wall is always involved during
a   freeze-thaw   cycle.   However,   its   influence   on
freezing tolerance has been considered to be neglig-
ible.   This is in accord with the generally accepted

view that endurance to freeze-induced dehydration of
the protoplast is the mechanism of freezing tolerance
(23). Accumulated evidence in recent years, however,
indicates that the freezing behavior differs between in
vivo and in vitro (15); that the pattern of freezing
injury is more complex in vivo than in vitro (21); that
the level of freezing tolerance from the same plant
source is much greater in vitro than in vivo (35).

In a study with S. tuberosum cultured cells, the
results clearly indicate that, when exposed to the same
degree of protoplasmic dehydration, the released protoplasts, plasmolyzed cells and cells showing cytorrhysis
(during dehydration the cell wall remains in contact
with the dehydrated protoplast) exhibit significant
differences in freezing tolerance (35). It is known
that freezing tolerance increases at the moment when
cellular water is removed from the cell. However,
cells showing cytorrhysis do not show any increase in
freezing tolerance (Fig. 3). They are killed at -3°C,
similar to that of cells in vivo, whereas the plasmolyzed cells and released protoplasts increases in
freezing tolerance (Fig. 4). These differences in
freezing tolerance are in contrast to the explanation
that the increase in freezing tolerance caused by dehydration is solely due to the removal of cellular
water. Tao et al. (35) suggested that the difference
in freezing tolerance is also related to the presence
of the cell wall and its degree of contact with the
plasma membrane during a freeze-thaw cycle. In a report on a drought tolerant moss, Malek and Bewley (27)
suggested that the presence of the cell wall may be
involved in causing the irreversible damage that occurs
during slow freezing, in addition to the dehydration
induced by extracellular ice formation. The evidence
provided by Malek and Bewley (27) and Tao et al. (35)
supports Iljin's cell-wall-mechanical-injury theory
(19) proposed some 50 years ago. The observations obtained with isolated potato protoplasts reflect maximal
capacity of the organelles in terms of freezing tolerance. This clearly cannot be extended to the actual
tolerance of the protoplasts in their cellular context.

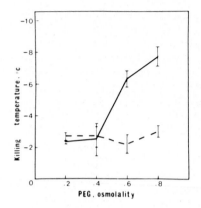

FIGURE 3. Killing temperature of S. tuberosum callus cells frozen in PEG 1000 (——) and PEG 6000 (----) solutions at varied osmolality. LT_{50}, mean lethal temperature; and PEG, polyethylene glycol (35).

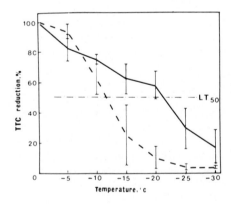

FIGURE 4. TTC reduction of S. tuberosum callus cells (----) and released protoplasts (——) frozen in 0.5 M sucrose solutions. TTC, triphenyltetrazolium chloride; and 0.5 M sucrose yields 0.6 osmolality (35).

COLD ACCLIMATION

An adapted plant species has internal mechanisms that permit it to survive in its ecological niche. These mechanisms are extremely complex and involve adaptive responses to temperatures, photoperiod, light intensity and quality, water availabilty, nutrients, and others. In most cases, low temperatures combined with short days provide the ideal condition for cold acclimation (9, 12, 13, 31).

Potato cold acclimation can be achieved either by directly exposing plants to constant day/night low temperatures (4) or by a stepwise lowering of temperatures (9). The hardiness level that is achieved is dependent upon the degrees of low temperatures used, with lower temperatures resulting in greater cold hardiness. The optimum temperature is about 2°C with an upper limit of 12-13°C (5). Under constant, low temperatures, the potato attains maximum cold hardiness in about 2 weeks (4) and the increase in cold hardiness initiates on the 4th day of cold acclimation (8).

The potato can be cold acclimated in vitro (4,8). The hardiness induced in vitro is similar to that of intact plant systems. Results encourage the use of tissue culture as an alternative system for studying the mechamisms of cold acclimation.

Classification.

On the basis of the cold hardiness of leaves and their ability to acclimate to cold, five groups of tuber-bearing Solanum species have been classified (5). They are: (a) freezing tolerant and able to cold-acclimate; (b) freezing tolerant but unable to cold-acclimate; (c) freezing sensitive but able to cold-acclimate; (d) freezing sensitive and unable to cold-acclimate; and (e) chilling sensitive. The commonly cultivated S. tuberosum belongs to group d, which is freezing sensitive and unable to cold-acclimate.

Acclimation Mechanism

It is generally believed that alterations in the plasma membrane occur during cold acclimation and that these allow cells to survive lower extracellular freezing. Upon exposure to cold-acclimating conditions, intramembrane particles (IMP) aggregated in the plasma membranes of S. acaule callus cells, reaching a peak at about 5 to 10 days; after 15 days the IMP redistributed to the control level (36). S. acaule is a freezing tolerant and able to cold acclimate potatoes. No such aggregation and redistribution of IMP occurred in S. tuberosum callus under similar experimental conditions. The callus tissue of S. acaule, after 15 days, can cold acclimate from -6° to -9°C, whereas S. tuberosum remains at -3°C. The aggregation of IMP when the membrane is cooled towards its phase-transition is well documented in some organisms (1). No observations of aggregation during acclimation in plants, other than the potato, have been reported. The redistribution of IMP aggregates in S. acule callus membrane after 15 days could be interpreted as a decrease in phase-transition temperature of these membranes. Such a partial lipid crystallization of the S. acaule membrane (IMP aggregation) during the early stage of acclimation may be a necessary step for acclimation (36) because it may trigger the operation of self-regulating mechanisms to restore membrane fluidity (29).

The protective effect of sugars such as sucrose against freezing injury has been repeatedly demonstrated in vitro (33). It is, however, unlikely that such a protective role occurs in vivo when tissue sugars increase during cold acclimation. In potatoes, increases in sugars during cold acclimation have been observed both in cold-acclimating S. acaule, S. commersonii, and in S. tuberosum, which does not cold acclimate (5). The sugar level of S. tubersoum grown at different low temperatures varies, but no differences in cold hardiness develop. In addition, sugar increases in both S. acaule and S. commersonii are not proportional to the increases in cold hardiness of the corresponding species. The increase in sugar usually occurs prior to any measurable increase in cold hardiness (7). It appears that in potatoes an increase in sugar is a universal response to low temperature and not directly related to the cold acclimation.

In $\underline{S}$. $\underline{acaule}$ and $\underline{S}$. $\underline{commersonii}$ potatoes, the net increases in cold hardiness were correlated with increases in soluble proteins (6). It suggests that protein metabolism during cold acclimation plays an important role in the development of cold hardiness. Cox and Levitt (11) suggested that only those species that are able to conduct active protein synthesis at low temperature have the capability to cold-acclimate. Their suggestion is supported by studies in the potato (8), $\underline{Chlorella}$ (14) and winter wheat (37) in which cycloheximide, an inhibitor of cytoplasmic ribosomal protein synthesis, was found to inhibit the development of cold hardiness. Paldi and Devay (28) reported that in frost resistant wheat cultivars, cold acclimation induces an increase in rRNA synthesis, which is primarily related to a higher rRNA cistron number in the cultivars, and that the activity of rRNA cistrons varies as a function of temperature. The inability of $\underline{S}$. $\underline{tuberosum}$ to synthesize protein during cold acclimation may explain in part why this species fails to cold-acclimate (6). The effect of cold-acclimating temperatures on the structure and function of soluble proteins such as RuBPCase has been studied extensively in "Puma" rye by Huner $\underline{et}$ $\underline{al}$. (18). Their results indicate that (a) the large subunits from the acclimated sources are less susceptible to sulfhydryl-disulfide aggregation under freezing stress than are those from non-acclimated materials and (b) RuBPCase from acclimated sources has a higher affinity for CO_2 at low temperature compared with the emzyme from non-acclimated sources, thus retaining their functional efficiency at low temperatres and enabling to survive freezing stress.

Chen et al. (8) reported that cold acclimated $\underline{S}$. $\underline{commersonii}$ plants had free ABA content which increased 3-fold on the 4th day during cold acclimation and declined to its initial level thereafter. This, however, was not observed in $\underline{S}$. $\underline{tuberosum}$ when acclimated under indentical low temperature. Applied ABA could induce cold hardiness in $\underline{S}$. $\underline{commersonii}$ leaves whether plants were grown under a warmer (25°C) or cold (2°C) temperature regime. The development of cold hardiness could be completely inhibited when

TABLE 1
EFFECTS OF LOW TEMPERATURE (LT), ABA (20 MG/L),
AND CYCLOHEXIMIDE (10^{-5}M) ON CHANGES IN COLD
HARDINESS IN S. commersonii AGAR-CULTURED
PLANTS (8). COLD HARDINESS WAS DETERMINED
AFTER 15 DAYS OF TREATMENT.

Treatments	Killing temperature (°C)
Control (20/15°C, day/night)	-4
LT (2°C, day/night)	-10
LT + cycloheximide (added at 0 day)	-4
Lt + cycloheximide (added at 5th day)	-8
ABA (20/15°C, day/night)	-10
ABA + cycloheximide (added at 0 day)	-4

cycloheximide was added to a culture system at the beginning of cold acclimation or at the beginning of ABA treatment in a warm regime. When inhibitor was added at the 5th day during cold acclimation (After ABA peak), the induction of cold acclimation was not inhibited (Table 1). Results of the cycloheximide-ABA study (8) suggest that the low temperature triggers an elevation of endogenous ABA, which induces the synthesis of proteins, which in turn are responsible for the development of cold hardiness. ABA probably is not directly involved in the development of cold hardiness but is essential in the process of cold acclimation because of its effect on protein systhesis.

One might hypothesize that sugar accumulation during cold acclimation may play a key, but indirect, role in the development of cold hardiness in the potato by triggering the endogenous increase in ABA through the elevation of osmotic concentration (7). It has been shown that the osmotic concentration increase to a critical level that can raise the endogenous ABA content in the plant (32). Perhaps, S. tuberosum, a species that does not cold-acclimate, can not accumulate enough sugars during acclimation to elevate the osmotic concentration to a critical level needed to result in an ABA buildup. The increase in sugar content and the elevation of cellular osmotic concentration in S. tuberosum during cold acclimation were considerably less than in S. commersonii, a species that can cold-acclimate (7). No ABA increase was observed in S. tuberosum during cold acclimation (8). This line of thinking is further supported by an experiment in which applied ABA did indeed increase the cold hardiness in S. tuberosum (4). Nonetheless, the question concerning the specific protein synthesized during cold acclimation that is directly involved in the development of cold hardiness is still unknown.

ACKNOWLEDGEMENT

My thanks to former research colleagues whose ideas and work have contributed directly to my views of the potato cold hardiness. Special thanks to Dr. Norman P.A. Huner, Dr. Jiwan P. Palta, Dr. Tory H.H. Chen, Dr. Maria A. Toivio-Kinnucan, Dr. Dali Tao, and Dr. Paul M. Chen. Scientific Series No. 13903 of the University of Minnesota Agriculture Experiment Station.

REFERENCES

1. Armond PA, Staehelin LA (1979). Lateral and vertical displacement of integal membrane proteins during lipid phase transition in Anacystis nidulans. Proc Natl Acad Sci USA 76:1901.
2. Asahina E (1954). A process of injury induced by the formation of spring frost on potato sprout. Low Temp Sci, Ser B 11:13
3. Cary JW, Mayland HF (1970). Factors influencing freezing of super-cooled water in tender plants. Agron J 62:715.
4. Chen HH, Gavinlertvatana P, Li PH (1979). Cold acclimation of stem-cultured plants and leaf callus of Solanum species. Bot Gaz 140:142.
5. Chen HH, Li PH (1980). Characteristics of cold acclimation and deacclimation in tuber-bearing Solanum species. Plant Physiol 65:1146.

6.   Chen HH, Li PH (1980). Biochemical changes in tuber-bearing Solanum species in relation to frost hardiness during cold acclimation. Plant Physiol 66:414.

7.   Chen HH, Li PH (1982). Potato cold acclimation. In Li PH, Sakai A (eds):"Plant Cold Hardines and Freezing Stress". Vol 2. New York: Academic Press, p 5.

8.   Chen HH, Li PH, Brenner ML (1983). Involvement of abscisic acid in potato cold acclimation. Plant Physiol 71:362.

9.   Chen P, Li PH (1976). Effect of photoperiod, temperature and certain growth regulators on frost hardiness of Solanum species. Bot Gaz 137:105.

10.   Chen PM, Burke MJ, Li PH (1976). The frost hardiness of several Solanum species in relation to the freezing of water, melting point depression, and tissue water content. Bot Gaz 137: 313.

11.   Cox W, Levitt J (1976). Interrelations between environmental factors and freezing resistance of cabbage leaves. Plant Physiol 57:553.

12.   Gusta LV, Fowler DB, Tyler NJ (1982). Factors influencing hardening and survival in winter wheat. In: Li PH, Sakai A (eds): "Plant Cold Hardiness and Freezing Stress". Vol. 2, New York: Academic Press, p 23.

13.   Hatano S, Sadakane H, Tutumi M, Watanabe T (1976). Studies on frost hardiness in Chlorella ellipsoidea. I. Development of frost hardiness of Chlorella ellipsoidea in synchronous culture. Plant & Cell Physiol 17:451.

14.   Hatano S (1978). Studies on frost hardiness in Chlorella ellipsodea: effects of antimetabolites, surfactants, hormone and sugars on the hardeining process in light and dark. In Li PH, Sakai A (eds): "Plant Cold Hardiness and Freezing Stress". New York: Academic Press, p 175.

15.   Heber U, Schmitt JM, Krause GH, Klosson RJ, Santarius KA (1981). Freezing damage to thylakoid membranes in vitro and in vivo. In Morris, GJ and Clarke A (eds): "Effects of Low Temperatures on Biological Membranes", London: Academic Press, p 263.

16. Hudson MA, Brustkern P (1956). Resistance of young and mature leaves of Mnium undulatum to frost. Planta 66:135.

17. Huner NPA, Paulta JP, Li PH, Carter JV (1981). Comparison of the structure and function of ribulose bisphosphate carboxylase-oxygenase from a cold-hardy and nonhardy potato species. Can J Bio Chem 59:280.

18. Huner NPA, Hopkins WG, Elfman B, Hayden DB, Griffith M (1982). Influence of growth at cold-hardening temperature on protein structure and function. In Li PH, Sakai A (eds): "Plant Cold Hardiness and Freezing Stress," Vol. 2, New York: Academic Press, p 129.

19. Iljin WS (1934). Veber den Kaeltetod der Pflanzen und seine Ursachen. Protoplasma 20:105.

20. Kaku S (1973). High ice nucleating ability in plant leaves. Plant and Cell Physiol. 14:1035.

21. Krause GH, Klosson RJ, Tröster U (1982). On the mechanism of freezing injury and cold acclimation of spinach leaves. In Li PH, Sakai A (eds): "Plant Cold Hardiness and Freezing Stress". Vol. 2, New York: Academic Press, p 55.

22. Levitt J (1962). A sulfhydryl-disulfide hypothesis of frost injury and resistance in plants. J Theoret Biol 3:355.

23. Levitt J (1980). Responses of plants to environmental stress. Vol. 1 2nd ed, New York: Academic Press, p 497.

24. Li PH (1977). Frost killing temperatures of 60 tuber-bearing Solanum species. Amer Potato J 54: 452.

25. Li PH, Palta JP (1978). Frost hardening and freezing stress in tuber-bearing Solanum species. In Li PH, Sakai A (eds): "Plant cold Hardiness and Freezing Stress", New York: Academic Press, p 49.

26. Lindow SE, Arny DC, Upper CD, Barchet WR (1978). The role of bacterial ice nuclei in frost injury to sensitive plants. In Li PH, Sakai A (eds): "Plant Cold Hardiness and Freezing Stress", New York: Academic Press, p 249.

27. Malek L, Bewley JD (1978). Effects of various rates of freezing on the metabolism of a drought-tolerant plant, the moss Tortula ruralis. Plant Physiol 61:334.
28. Palki E, Devay M (1983). Relationship between the cold-induced rRNA synthesis and the rRNA cistron number in wheat cultivars with varying degrees of frost hardiness. Plant Sci letters 30:61.
29. Pugh EL, Kates M (1979). Membrane-bound, phospholipid desaturases. Lipids 14:159.
30. Rajashekar C, Li PH, Carter JV. (1983). Frost injury and heterogenous ice nucleation in leaves of tuber-bearing Solanum species. Plant Physiol 71:749.
31. Rajashekar C, Tao DL, Li PH (1983). Freezing resistance and cold acclimation in turfgrasses. HortScience 18:91.
32. Rikin A, Waldman M, Richmond AE, Dovrat A (1975). Hormonal regulation of morphogenesis and cold-resistance. Modification by abscisic acid and gibberellic acid in alfalfa (Medicago sativa L.) seedlings. J Exp Bot 26:175.
33. Santarius KA (1982). The mechanism of cryoprotection of biomembrane systems by carbohydrates. In Li PH, Sakai A (eds): "Plant Cold Hardiness and Freezing Stress". Vol 2, New York: Academic Press, p 475.
34. Sukumaran NP, Weiser CJ (1972). Freezing injury in potato leaves. Plant Physiol 50:564.
35. Tao DL, Li PH, Carter JV (1983). Role of cell wall in freezing tolerance of cultured potato cells and their protoplasts. Physiol Plant 58:527.
36. Toivio-Kinnucan MA, Chen HH, Li PH, Stushnoff C (1981). Plasma membrane alterations in callus tissues of tuber-bearing Solanum species during cold acclimation. Plant Physiol 67:478.
37. Trunova TI (1982). Mechanism of winter wheat hardening at low temperature. In Li PH, Sakai A (eds): "Plant Cold Hardiness and Freezing Stress". Vol. 2, New York: Academic Press, p 41.

Cellular and Molecular Biology of Plant Stress, pages 217–226
© 1985 Alan R. Liss, Inc.

Adh1 AND Adh2[1]: TWO GENES INVOLVED IN THE MAIZE ANAEROBIC
RESPONSE

M.M. Sachs[2], E.S. Dennis, J. Ellis, E.J. Finnegan,
W.L. Gerlach, D. Llewellyn, and W.J. Peacock

CSIRO, Division of Plant Industry, GPO Box 1600,
Canberra, ACT 2601, AUSTRALIA

INTRODUCTION

Anaerobic treatment results in a drastic alteration in
the pattern of protein synthesis in maize seedlings (1).
Pre-existing (aerobic) protein synthesis is repressed while
synthesis of approximately 20 novel "anaerobic"
polypeptides is induced (1). This is most likely a plant's
natural response to flooding.

Studies on the maize anaerobic response stemmed from
the extensive analysis of the maize alcohol dehydrogenase
(ADH) system by Schwartz and coworkers (reviewed by
Freeling and Birchler: 2). Hageman and Flesher (3) were
the first to show that ADH activity increases as a result
of flooding maize seedlings. Freeling (4) later showed
that ADH activity increased at a zero order rate between 5
hr and 72 hr of anaerobic treatment reflecting a
simultaneous expression of two unlinked genes, Adh1 and
Adh2. Schwartz (5) showed that ADH activity is required to
allow maize seeds and seedlings to survive anaerobic
treatment and presumably flooding under natural conditions.

COMPARISON OF THE ANAEROBIC AND HEAT-SHOCK STRESS RESPONSES

As a stress response, the effect of anaerobiosis on
maize seedlings appears to be analogous to the heat-shock

---

[1]Alcohol dehydrogenase: gene = Adh; protein, enzyme
= ADH.
[2]Present Address: Department of Biology, Washington
University, St. Louis, MO 63130, U.S.A.

response which has been studied in animals, plants, fungi
and bacteria and characterized most extensively in
Drosophila melanogaster. This treatment mediates a repression of pre-existing (non-stress) protein synthesis while
inducing transcription and translation of a small number of
heat-shock proteins (hsps, reviewed by Ashburner and
Bonner, and Schlesinger et al.: 6,7). In Drosophila and
some other animals, other forms of stress such as azide
poisoning, amino acid analogue treatment, and recovery from
anoxia induce the same stress response. Interestingly, in
organisms as diverse as E. coli, maize, mice, Dictyostelium
and yeast, at least some of the proteins induced by heat-
shock are closely related to those of Drosophila (7). This
is especially true for a class of heat-shock induced proteins with an approximate molecular weight of 70 kd. Moreover, a control signal involved in the induction of transcription of Drosophila heat-shock genes (CTgGAAtnTTCtAGa)
is recognized by heterologous transcription systems such as
monkey COS cells and Xenopus oocytes, but only when these
systems are subjected to heat-stress (8). A homologous
sequence is found upstream to heat-shock genes of soybeans
and maize (J.L. Key and D.M. Shah, pers. comm.).

    The maize anaerobic response is analogous to the heat-
shock response, but except for one possible overlap
involves a different set of proteins (1,9,10). In an early
study (11), anaerobic treatment was shown to cause a near
complete dissociation of polysomes and a rapid repression
of protein synthesis in soybean roots. Maize seedlings
respond to anaerobic treatment by redirecting protein
synthesis. As with the heat-shock systems described above,
there is an immediate repression of pre-existing protein
synthesis and an induction of synthesis of a new set of
proteins (1). During the first five hours of anaerobic
treatment there is a transition period, during which there
is a rapid increase in the synthesis of a class of polypeptides with an approximate molecular weight of 33 kd.
These have been referred to as the transition polypeptides
(TPs). After approximately 90 minutes, the synthesis of an
additional group of ~20 polypeptides is induced. This
group of 20 anaerobic polypeptides (ANPs) represents more
than 70 percent of the total label incorporation after five
hours of anaerobic treatment. By this time synthesis of
the TPs is at a minimal level; however, pulse-chase experiments have shown that the TPs are very stable. The
synthesis of the ANPs continues in a quantitatively stable

ratio for up to 72 hours of anaerobic treatment, at which
time protein-synthesis decreases as the seedlings begin to
die (1).

The function of some of the ANPs is known. ADH1 and
ADH2 have been identified as ANPs through the use of
genetic variants (1,12). More recently glucose phosphate
isomerase and fructose-1, 6-diphosphate aldolase have been
identified as ANPs (Kelley and Freeling, pers. comm.).
Pyruvate decarboxylose activity has also been shown to be
induced by anaerobiosis (13,14) and therefore may also be
an ANP. The functions of the remaining ANPs or of the TPs
are as yet unknown. However, the five ANPs that have been
identified are all glycolytic enzymes. So it appears that
at least one function of the anaerobic response is to
enable the plant to produce as much ATP as possible during
short term flooding.

In the presence of air, each maize organ examined,
including the roots, coleoptyle, mesocotyl, endosperm,
scutellum and anther wall synthesizes a tissue-specific
spectrum of polypeptides. Under anaerobic conditions all
of the above organs synthesize only the ANPs. Moreover,
except for a few characteristic qualitative and
quantitative differences, the patterns of anaerobic protein
synthesis in these diverse organs are remarkably similar
(15). On the other hand, maize leaves, which have emerged
from the coleoptyle, do not incorporate label under
anaerobic conditions, and do not survive even very short
periods of anaerobiosis (15).

The rapid repression of pre-existing protein
synthesis, as in soybeans (11), is correlated with a near
complete dissociation of polysomes in anaerobically treated
maize tissue (Dennis and Pryor, unpublished). This is not
due to degradation of "aerobic" mRNAs, since the mRNAs
encoding the "aerobic" proteins remain translatable in an
in vitro system at least five hours after anaerobic
treatment was initiated (1). This is in agreement with the
observation that the polysomes dissociated by anaerobiosis
rapidly reform, up to 80-90 percent their pretreatment
levels, even in the absence of new RNA synthesis, when
soybean seedlings are returned to air (11).

The induction of anaerobic polypeptide synthesis
occurs in two phases; the immediate synthesis of the TPs

and then the appearance after a 90 minute lag of the 20
ANPs. In heat shock, the appearance of the hsps is immediate with kinetics very similar to the rapid appearance of
the TPs. Both the hsps and the TPs exhibit a decreased
rate of synthesis after 5 hours of stress treatment (1,9).
It is interesting that the only possible overlap in protein
synthesis occuring in response to heat-shock or anaerobiosis is that of a TP and a 33 kd hsp (10).

ANAEROBIC SPECIFIC cDNA CLONES

The molecular analysis of the maize anaerobic response
was initiated by synthesizing cDNA clones from high
molecular weight poly(A) RNA of anaerobically treated maize
seedling roots (16). Anaerobic specific cloned cDNAs were
identified by colony hybridization analysis, which included
using probes of labelled cDNAs of mRNA from anaerobic and
aerobic roots. Colonies were selected that hybridized
specifically with the "anaerobic" cDNAs. The anaerobic
specific cDNA clones were grouped into families on the
basis of cross hybridization to each other, and several of
these families were analyzed by hybrid selected translation
and by RNA blot (Northern) hybridization.

The Adh1 and Adh2 cloned cDNA families were identified
from this anaerobic-specific cDNA clone library and these
cDNA clone families were extensively analyzed.

MOLECULAR ANALYSIS OF THE MAIZE Adh GENES

One of the genes induced by anaerobic treatment,
Alcohol dehydrogenase 1, is among the best characterized
gene systems in higher organisms (reviewed by Freeling and
Birchler: 2). Much is also known about Adh2 (17).
Genetically, the Adh1 and Adh2 loci are unlinked. Adh1 is
located at position 127 on the long arm of maize chromosome
1 (18), while Adh2 is located at approximately position 46
on the short arm of chromosome 4 (17). The gene products
of Adh1 and Adh2 are very similar to each other, and
although the ADH1 subunits migrate faster on an SDS gel
than the ADH2 subunits, on the basis of DNA sequence
information (see below) it is most probable that the sizes
of the polypeptide subunits are identical with a molecular
weight of 38 kd. The ADH enzymes are active as dimers and

can be composed of either homodimers or ADH1:ADH2
heterodimers. ADH1 and ADH2 subunits share some antigenic
determinants.

In seedlings, Adh1 and Adh2 are expressed only at a
very low level under normal conditions. Expression of both
genes is induced by anaerobic treatment (e.g., flooding) or
by treatment with the synthetic auxin, 2,4-D (4). Adh1 is
constitutively expressed in the scutellum and pollen,
whereas Adh2 expression is very low in the scutellum and
undetectable in pollen. On the other hand, Adh2 expression
is predominant in organs such as the prop root (Schwartz,
pers. comm.) and tassel nodes (Sachs, unpublished).
Alleles of Adh1 also exhibit differential organ
specific expression. When the Adh1-1S allele is compared
to the Adh1-1F allele, in heterozygotes, the "slow"
electrophoretic variant is expressed to a greater extent in
the scutellum and to a lesser extent in the induced primary
root, relative to the "fast" variant (18,19).

Two of the anaerobic-specific clones described above
were shown to contain ADH1 or ADH2 sequence information
(16,20). The identification of these clones was confirmed
in many ways, but perhaps most cogently, by use of genetic
variants. In the case of the Adh1 cDNA clone, use of the
mutant Adh1-U725 (12) whose product shows altered migration
on SDS PAGE was used in hybrid selected translation
experiments (16). In addition, the naturally occurring
electrophoretic variants, Adh1-1S and Adh1-1F exhibit
altered mRNA patterns on RNA blot (Northern) hybridization
analysis. The "slow" allele exhibits one mRNA class with
approximate size of 1,650 bases, while the "fast" allele
exhibits an additional mRNA class of 1,750 bases. A gamma
irradiation induced crm⁻ Adh1 null (Adh1-FkFγ25: 2)
exhibits no detectable mRNA product (21). The Adh2 cDNA
clone was first identified by its cross hybridization to an
Adh1 probe in colony hybridization analysis. Its identity
as an Adh2 cDNA was confirmed by using a naturally
occurring null allele, Adh2-33 (17) as maize lines
homozygous for this allele exhibit no detectable mRNA which
hybridizes to the presumptive Adh2 cDNA probe (20).

The Adh cDNA clones were used as probes to isolate the
Adh2 gene from the Berkeley Fast inbred line (20) and
several interesting alleles of Adh1, which include two
naturally occuring variant alleles, Adh1-1S and Adh1-1F,

and transposable element induced mutants and revertants
(21-26).

A comparison of the nucleotide sequences of the cloned
cDNAs and the genomic clones suggests that the two Adh
genes arose from a single ancestral gene which was
duplicated. The coding regions of Adhl and Adh2 appear to
be identical in length and are more than 80% homologous at
both the amino acid and nucleotide sequence level. Both
genes are interrupted by nine introns in identical
positions within the gene, but for any intron neither the
nucleotide sequence nor the length is conserved (20).
Except for an 11 bp region which includes the TATA box and
a few other regions of 8 bp or less, the 5' and 3' flanking
sequences of these two genes show little or no homology
(20). It appears that while the structure and protein
coding sequences of Adhl and Adh2 are highly conserved, the
nontranslated sequences have almost completely diverged
since the duplication event occurred.

Both genes exhibit many of the transcriptional and
post transcriptional signal sequences previously described
for animal genes (20,21). These include presumptive TATA
sequences (27) occurring approximately 35 bp upstream from
the experimentally determined transcription start points.
In addition the sequences at the intron-exon splice
junctions fit the previously determined consensus for
animal genes (28). However, no poly(A) addition signal
(AATAAA: 29) has been found at appropriate positions in
either gene. It has become clear that although this signal
seems necessary for poly(A) addition to animal mRNAs, a
number of other plant genes (e.g., soybean leghemoglobin
and maize zein) also do not have this consensus AATAAA in
the standard animal location: 10-30 bp upstream of the
poly(A) tail (30,31). Adh2 was found to have a single
Poly(A) site 15 bp downstream from the sequence AATAAT. On
the other hand, experimental determination of the poly(A)
addition site for Adhl (25) reveals that there are in fact
multiple sites. In the Adhl-1S allele there appear to be
four closely spaced poly(A) sites and in the Adhl-1F allele
there are seven, including an additional major poly(A) site
approximately 120 bases downstream from the most distal in
Adhl-1S, thus accounting for the longer mRNA size class
observed in RNA blot (Northern) analysis of Adhl-1F RNA.
None of the poly(A) sites of Adhl are close to a AATAAA or
related sequence or any other obvious consensus sequences.

A recent report by Dhease et al. (32) on poly(A) addition in T-DNA genes indicates this phenomenon may be common in plants. Multiple poly(A) addition sites and the lack of a functional AATAAA signal is also often observed in yeast (33,34).

The anaerobic-specific cDNA clones were also used as probes to measure gene expression. The levels of mRNA hybridizable to these cDNA clones increases during anaerobic treatment. This has been measured and quantified rigorously in the case of Adhl and Adh2. The kinetics of mRNA increase are the same for Adhl and Adh2. In both cases the mRNA level first appears to increase at 90 minutes of anaerobic treatment. The level then continues to increase until it plateaus at fifty times above the aerobic level at five hours of anaerobiosis. This level is maintained until after 48 hours in the case of Adhl but starts declining after 10 hours in the case of Adh2 (20). This pattern of mRNA level increase and decrease is reflected in the previously described rates of in vivo anaerobic protein synthesis (1).

A comparison of the 5' regions of the Adhl and Adh2 genes reveals an 11 bp homologous region which includes the TATA sequence and three additional 8 bp regions of homology. One or more of these sequences might account for the anaerobic control of these genes (20). In vitro muta-genesis coupled with expression studies and the analysis of the 5' region of the other anaerobic genes will be neces-sary to determine which, if any, of these regions is important for the induced expression of the anaerobic response genes.

REFERENCES

1. Sachs MM, Freeling M, Okimoto R (1980). The anaerobic proteins of maize. Cell 20:761.
2. Freeling, M, Birchler JA (1981). Mutants and variants of the alcohol dehydrogenase - 1 gene in maize. In Setlow JK and Hollaender A (eds): "Genetic Engineering Principles and Methods." Vol. 3 Plenum Press, p 223.
3. Hageman RH, Flesher D (1960). The effect of anaerobic environment on the activity of alcohol dehydrogenase and other enzymes of corn seedlings. Arch Biochem Biophys 87: 203.

4. Freeling M (1973). simultaneous induction by anaerobiosis or 2,4-D of multiple enzymes specified by two unlinked genes: Differential Adhl-Adh2 expression in maize. Mol Gen Genet 127:215.

5. Schwartz D (1969). An example of gene fixation resulting from selective advantage in suboptimal conditions. Am Nat 103:479.

6. Ashburner M, Bonner JJ (1979). The induction of gene activity in Drosophila by heat shock. Cell 17:241.

7. Schlesinger M, Ashburner M, Tissieres A (1982). Heat Shock From Bacteria to Man. Cold Spring Harbor Laboratory Press.

8. Pelham HRB, Bienz M (1982). A synthetic heat-shock promoter element confers heat-inducibility on the herpes simplex virus thymidine kinase gene. EMBO J 1:1473.

9. Cooper P, Ho T-HD (1983). Heat shock proteins in maize. Plant Physiol 71:215.

10. Kelley PM, Freeling M (1982). A preliminary comparison of maize anaerobic and heat-shock proteins. In Schlesinger MJ, Ashburner M, Tissieres A (eds): "Heat Shock From Bacteria to Man," Cold Spring Harbor Laboratory Press, p 315.

11. Lin CY, Key JL (1967). Dissociation and reassembly of polyribosomes in relation to protein synthesis in the soybean root. J Mol Biol 26:237.

12. Ferl RJ, Dlouhy SR, Schwartz D (1979). Analysis of maize alcohol dehydrogenase by native-SDS two dimensional electrophoresis and autoradiography. Mol Gen Genet 169:7.

13. Wignarajah K, Greenway H (1976). Effect of anaerbiosis on activities of alcohol dehydrogenase and pyruvate decarboxylase in roots in Zea mays. New Phytol 77:575.

14. Laszlo A (1981). Maize pyruvate decarboxylase: An inducible enzyme. Ph.D. dissertation. University of California, Berkeley.

15. Okimoto R, Sachs MM, Porter EK, Freeling M (1980). Patterns of polypeptide synthesis in various maize organs under anaerobiosis. Planta 150:89.

16. Gerlach WL, Pryor AJ, Dennis ES, Ferl RJ, Sachs MM, Peacock WJ (1982). cDNA cloning and induction of the alcohol dehydrogenase gene (Adhl) of maize. Proc Natl Acad Sci USA 79:2981.

17. Dlouhy SR (1980). Genetic, biochemical and physiological analysis involving the Adh2 locus of Zea mays. Ph.D. dissertation. Indiana University.

18. Schwartz D (1971). Genetic control of alcohol dehydrogenase – A Competitive model for regulation of gene action. Genetics 64:411.

19. Woodman JC, Freeling M (1981). Identification of a genetic element that controls the organ-specific expression of Adhl in maize. Genetics 98:354.

20. Dennis ES, Sachs MM, Gerlach WL, Finnegan EJ., Peacock WJ. A comparison of the Adhl and Adh2 genes of maize. In preparation.

21. Dennis ES, Gerlach WL, Pryor AJ, Bennetzen JL, Inglis A, Llewellyn D, Sachs MM, Ferl RJ, Peacock WJ. The Adhl gene of maize. Submitted.

22. Peacock, WJ, Dennis ES, Gerlach WL, Llewellyn D, Lorz H, Pryor AJ, Sachs MM, Schwartz D, Sutton WD (1983). Gene transfer in maize: Controlling elements and the alcohol dehydrogenase genes. In: Proc of the 15th Miami Winter Symp. Academic Press.

23. Sutton WD, Gerlach WL, Schwartz D, Peacock WJ (1984). Molecular analysis of Ds controlling element mutations at the Adhl locus of maize. Science 223:1265.

24. Sachs MM, Peacock WJ, Dennis ES, Gerlach WL (1983). Maize Ac/Ds controlling elements – A molecular viewpoint. Maydica 28:289.

25. Sachs MM, Dennis ES, Gerlach WL, Peacock WJ. Complex nucleotide sequence and poly(A) addition site polymorphisms exist in two alleles of maize Alcohol dehydrogenase 1. In preparation.

26. Bennetzen JL, Swanson J, Taylor WC, Freeling M. DNA insertion in the first intron of maize Adhl affects message levels: Cloning of progenitor and mutant Adhl alleles. Submitted.

27. Goldberg ML (1979). Ph.D. dissertation. Stanford University.

28. Breathnach R, Chambon P (1981). Organization and expression of eucaryotic split genes coding for proteins. Ann Rev Biochem 50:349.

29. Proudfoot NJ, Brownlee GG (1976). 3' non-coding region sequences in eukaryotic messenger RNA. Nature 263:211.

30. Janson EO, Paludan K, Hyidig-Neilson JJ, Jorgensen P, Marcker KA (1981). The structure of a chromosomal leghaemoglobin gene from soybean. Nature 291:677.

31. Geraghty D, Peifer MA, Rubenstein I, Messing J (1981).
     The primary structure of a plant storage protein:
     Zein Nucleic Acids Res 9:5163.
32. Dhaese P, DeGreve H, Gielen J, Seurinck J, Van Montagu
     M, Schell J (1983). Identification of sequences
     involved in the polyadenylation of higher plant
     nuclear transcripts using Agrobacterium T-DNA genes as
     models. EMBO J 2:419.
33. Zaret KS, Sherman F (1982). DNA sequence required for
     efficient transcription termination in yeast. Cell
     28:563.
34. Henikoff S, Kelly JD, Cohen EH (1983). Transcription
     terminates in yeast distal to a control sequence.
     Cell 33:607.

Cellular and Molecular Biology of Plant Stress, pages 227–245

ENERGY METABOLISM AND SYNTHESIS OF NUCLEIC ACIDS
AND PROTEINS UNDER ANOXIC STRESS

Alain Pradet,[1] Bernard Mocquot,[1] Philippe Raymond,[1]
Christiane Morisset,[2] Lorette Aspart,[3] and Michel Delseny[3]

[1]Station de Physiologie Végétale, I.N.R.A.
Centre de Recherches de Bordeaux
33140 Pont de la Maye, France

[2]Laboratoire de Cytologie Végétale Expérimentale
Université Pierre et Marie Curie
12, rue Cuvier, 75230 Paris Cedex 05, France

[3]Laboratoire de Physiologie Végétale
E.R.A. 226 du C.N.R.S., Université de Perpignan
66025 Perpignan Cedex, France

ABSTRACT   Recently, it was shown that plants under
anoxia can synthesize DNA, RNA and proteins. This
ability is variable and depends on the plant, as well
as on the fermentative metabolism, the ATP-production
and the adenylate energy charge (AEC) level. The
correlation between these phenomena is discussed.

INTRODUCTION

In the soils, roots and seeds are often submitted to
hypoxic and anoxic stresses. The effect of such
environmental stresses varies greatly, depending on the
cells : most root tips are killed after a few hours of
anoxic treatment, whereas certain imbibed seeds are
capable of surviving for several weeks.
The main phenomenon characterizing plant anoxic
stresses is a reduction of the synthesis of ATP (1). In
some tissues, the rate of ATP-regeneration is extremely
low and does not allow a significant biosynthetic
activity. In other tissues, ATP-production is high and
DNA, RNA and protein syntheses can be observed.

The estimation of the ATP-production under anoxia is usually carried out by measuring either the sugar consumption, or the accumulation of end-products. However, these estimations are difficult when all the substrates and end-products are not known, as is the case for most plant tissues.

The tissue ATP/ADP and AEC ratios (2) are sometimes misused in estimating the ATP-regeneration rate in plants submitted to various stresses. Nevertheless, it is now established that these ratios are correlated with the metabolic fluxes in tissues subjected to hypoxic and anoxic conditions and they provide an indirect means for estimating the effect of anoxic stress on ATP-regeneration (3), as discussed in this paper.

Using the AEC as indicator, it was possible to show that starchy seeds (e.g. rice seeds) exhibit a high metabolic activity under anoxia (4,5). In fatty seeds (e.g. lettuce seeds),on the other hand, the metabolic activity is extremely low (5,6). Preliminary studies of the biosynthetic capacities and of the gene expression in these seeds under anoxia are also presented.

EFFECT OF STRESSES ON ADENINE NUCLEOTIDE RATIOS

The metabolic activity of plant tissues may be limited by numerous environmental or nutritional factors, such as temperature, lack of oxygen, and sugar or ammonium starvation. All of these factors may be felt by the cells as stresses. Many authors have tried to study the effect of numerous stresses on the adenine nucleotide status of the cells because it was believed that the ATP/ADP, ATP/AMP or AEC ratios could be useful indicators of metabolic activity. For a long time, it was believed that the value of these ratios was correlated to the activity of the metabolic fluxes. However, some years ago, D. ATKINSON postulated that it was impossible to predict any correlation between adenine nucleotide ratios and metabolic fluxes (2). The problem now seems to be on the way to being clarified. In order to understand or predict the effect on the adenine nucleotide ratios of factors affecting the metabolic activity, it is necessary to know whether the metabolism is affected primarily at the level of ATP utilization or at the level of ATP-regenerating sequences (3).

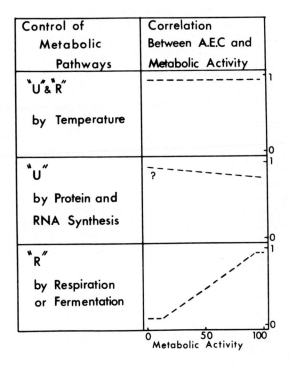

Control of Metabolic Pathways	Correlation Between A.E.C and Metabolic Activity
"U" & "R" by Temperature	
"U" by Protein and RNA Synthesis	
"R" by Respiration or Fermentation	

FIGURE 1 . Relationships between AEC and metabolic activity according to the type of control.

Temperature.

A variation of temperature simultaneously affects the enzymes of the "U" and of the "R" pathways (fig. 1). When tissues growing at temperatures between 20 and 30°C are transferred to low (0 to 5°C) or high (40 to 45°C) temperatures, the respiratory rate and, consequently, the ATP production are lowered or increased, respectively. However, the adenine nucleotide ratios do not vary (7).

Nitrogen Starvation

A limitation of protein synthesis by ammonium starvation, or by inhibitors such as cycloheximide, produces a lowering of the ATP demand, thus inducing a reduction of the ATP regeneration. The reduction of the metabolic activity is primarily produced at the level of ATP utilization sequences. In this case, an increase of the AEC could be expected in the range of high value where it is experimentally difficult to detect. No variation or sometimes very small increase were observed (8).

Oxygen Limitation.

A reduction of oxygen availability below the critical oxygen pressure reduces ATP regeneration by limiting oxidative phosphorylation. In this case, a correlation between the metabolic activity and the adenine nucleotide ratios is observed.
    Under anoxia, ATP regeneration in plants originates mainly from glycolysis and ethanol production. The metabolic flux through these pathways can be controlled by sugar availability, or by inhibitors such as fluoride. In this case, the metabolic activity is primarily controlled at the level of ATP-regenerating sequences and, here too, a correlation between adenine nucleotide ratios and metabolic activity is observed (9,10).
    Consequently, these ratios can be used as indicators of metabolic activity under hypoxia and anoxia.

Variation of the AEC in Lettuce and Rice Seeds.

When dry seeds are imbibed under aerated conditions, the level of the AEC increases rapidly from the low values usually observed in dry seeds, to values ranging between 0.8 and 0.9, the values observed in all tissues under normal metabolic conditions.
    During imbibition of lettuce seeds under anoxia, the AEC increases slightly from 0.25 to 0.35 during the first 6 hrs. Afterwards, it decreases to 0.25 and remains at this value for several days. The seeds are not killed by this treatment and recover high AEC values within a few minutes when transferred to air (fig. 2A).
    During the imbibition of rice seeds under anoxia, the AEC value increases rapidly to about 0.6 and then increases slowly before stabilizing at a high value (about 0.8) during the following days. The seeds are still alive after 30 days under anoxia (fig. 2B).
    From these observations, it may be predicted that metabolic activity and biosynthetic processes are active in rice seeds and much less so in lettuce seeds.

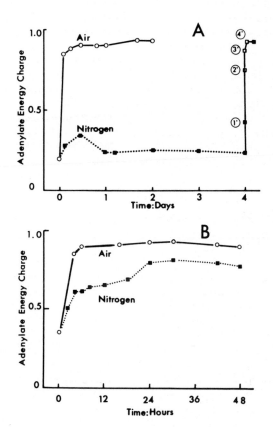

FIGURE 2 . Changes in AEC in lettuce (A) and rice (B)
seeds soaked under air or nitrogen.

BIOSYNTHETIC ACTIVITIES UNDER ANOXIA

Synthesis of Adenine Nucleotides in Rice and Lettuce
Embryos.

    In rice seedlings germinated in air for 48 hrs, the
transfer to anoxia induces a 70% drop of the adenine
nucleotide (AdN) pool, (ATP+ADP+AMP), in the coleoptile.
This pool increases 10-fold during the following 48 hrs
(11). The AdN level also increases in the rice coleoptiles

when the seeds are soaked under anoxia (4).

In lettuce seeds imbibed under anoxia, the AdN pool increases two-fold in 48 hrs. However, in seeds pre-imbibed in aerated water and then transferred to anoxia, the AdN pool remains constant (9).

An increase of the AdN pool size may result either from a de novo synthesis, or from the degradation of preformed RNA : as the energy cost of synthesis is much higher than that of RNA degradation, the synthesis is likely to be blocked under anoxia, particularly in lettuce. However, the incorporation of ^{14}C-adenine or ^{32}P-phosphate into AMP shows that the synthesis of adenine nucleotides occurs under anoxia (table 1).

TABLE 1
LABELING OF AMP IN LETTUCE EMBRYOS UNDER ANOXIA[a]

Labeling time (hrs)	specific radioactivity of AMP (cpm/pmol)	
	^{14}C-adenine	^{32}P-phosphate
N$_2$ 0.5	–	0.3
1	12	0.8
2	18	2
20	65	13
Air 0.2	–	0.6
1	30	4

[a]Germinating lettuce embryos in 1 ml water were made anaerobic with an O_2-free N_2 gas flow. The labeled precursor was then added. The N_2 flow was maintained during incubation for the time indicated. AMP was purified by paper electrophoresis and TLC. Radioactivity was counted and AMP was determined by the firefly method (9).

DNA Synthesis in Rice Embryos germinated under Anoxia.

It has been shown that, during the growth of the rice coleoptile under anoxic conditions, there is incorporation of ^{3}H-thymidine into DNA (12). The labeled nucleic acids were extracted, and it was checked that they were sensitive to DNase. Moreover, we have analysed this DNA on a CsCl gradient in order to characterize this product and to compare its buoyant density with that of known markers. Figure 3 shows that DNA banded in a single region with a buoyant density of 1.705 g/cm³.

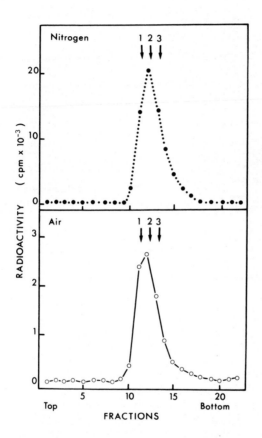

FIGURE 3 . Analysis by CsCl gradient of DNA fro rice coleoptiles grown in air or under nitrogen. ³H-labeled DNAs were centrifuged in MSE superspeed 65 ultracentrifuged (SW 6x5 ml rotor) for 64 hrs at 40,000 rpm at 20°C. DNA markers were used : 1-Calf thymus (1.699 g/cm³; 2. E. Coli (1.710 g/cm³); 3-Micrococcus lysodeikticus (1.731 g/cm³).

DNA synthesis (duplication and/or repair) has also

been demonstrated by autoradiography in rice radicle even without growth and mitosis (13 ; fig. 4).

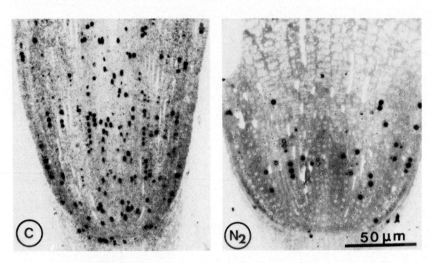

FIGURE 4. Thymidine incorporation under anoxia. Rice seedlings were germinated for 48 hrs in aerated (C) or oxygen deprived (N_2) conditions. 5 hrs incubation in 10 µCi/ml ^{3}H-thymidine took place before the end of each experiment.

All the results are in agreement with the cytological observation of Kordan (14), who has shown the initiation of the primordia of adventitious roots and mitotic figures in the coleoptile.

We have also shown that the level of deoxyribonucleoside triphosphates is drastically reduced under anoxia and that DNA polymerase activity as well as thymidine phosphorylating activities are present (12 and unpublished results).

RNA Metabolism in Rice Embryos during Adaptation to Anoxia.

When aerobic rice seedlings are transferred to an anaerobic environment, RNA is synthesized (15). The accumulation of the newly made molecules is modulated during the adaptation period. We have shown, that during this period the overall rate of RNA synthesis (which is

the resultant of the transcription and degradation rates)
parallels changes in the AEC. As there is no change in
the total amount of nucleic acids during the first 24
hours of anoxia, the degradation rate is close to the
synthesis rate. The analysis of RNA has shown that rRNA
and mRNA are synthesized and it appears that the
processing of ribosomal RNA precursor is altered. At the
beginning of the anoxic treatment, the processing activity
is very limited and during the adaptation there is a
partial recovery of the processing rate. This result
agrees with autoradiographic studies showing that labeling
occurs in all nuclei under anoxia with some delay for
cytoplasmic labeling (fig. 5).

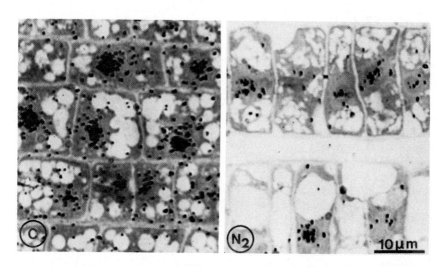

FIGURE 5 . Uridine incorporation under anoxia. Rice
seedlings were germinated for 48 hrs in aerated
conditions. Then air (C) was replaced by a nitrogen flow
(N_2) for increasing lengh of time, from 1 hr to 24 hrs
(7 hrs in this figure). They were incubated in 120 μCi/ml
^{3}H-uridine, for 1 hr, before the end of each experiment.

The analysis of the poly(A) content indicates that
about 40% of the poly(A) RNA is degraded very rapidly and
no recovery is observed during adaptation, indicating that
newly made mRNA equilibrates the degradation.
The analysis of polyribosome profiles indicates that

polyribosomes are present throughout the anoxic treatment. Nevertheless, there is an initial decrease in their number during the first 30 min, followed by a recovery. This result agrees with the observed capacity of rice embryos to synthesize numerous polypeptides under anoxia (11). We have also shown that the newly made RNAs (labeled with ^{3}H-uridine) are incorporated into monosomes and polyribosomes under anoxia and preferentially into the latter.

The correlation between the time-course of AEC and RNA metabolism changes suggests a role of the nucleotide pools in the control of the response to the anaerobic treatment. This response may correspond to an adaptation to these conditions.

Protein Synthesis under Anoxia.

Protein synthesis in rice embryos during adaptation to anoxia. We have previously reported (11) that an important synthesis of polypeptides occurs during the transfer of rice embryos to anoxia. The protein synthesis is immediately reduced at the beginning of the anoxic treatment when the AEC is lowered to 0.60. The protein synthesis then increases to reach the highest level under anoxia when the AEC has recovered a high value (0.80). By two-dimensional polyacrylamide gel electrophoresis, we have shown that a small number of polypeptides are intensely labeled with a ^{14}C-aminoacid mixture during the first hours under anoxia. After 24 or 48 hrs of anaerobic treatment, the number of synthesized polypeptides increases. These patterns indicate a change in the expression of the genome during the anoxic treatment. A more important modification in the anaerobic protein pattern has also been found in maize roots by Sachs et al. (16) and in rice roots by Bertani et al. (17). The lower protein synthesis observed in maize and rice roots may be correlated with lower values of the AEC.

Effect of NaF on AEC and protein synthesis in rice embryos. We have studied the effect of sodium fluoride, an inhibitor of glycolysis, applied at different concentrations to aerobic rice embryos (2 day-old) under anoxia. After 1 hr of anaerobic treatment, the proteins were labeled for 2 hrs with a ^{14}C-aminoacid mixture, as previously described (11), and the AEC was determined in a parallel experiment. The results are presented in Table 2.

TABLE 2
EFFECT OF NaF ON ENERGY CHARGE AND PROTEIN SYNTHESIS
UNDER ANOXIA

NaF conc. mM	AEC	uptake	incorporation	incorporation/ uptake
			$cpm \times 10^{-3}/embryo$	
0	0.70	1,050	92	0.088
1	0.60	800	60	0.075
3	0.40	540	23	0.042
10	0.15	300	0.4	0.001

When the NaF concentration increases, there is a decrease of the AEC values and a parallel decrease in the uptake of labeled aminoacids and in their incorporation into an acid-insoluble product. The apparent rate of protein synthesis, estimated by the ratio between incorporation and uptake, is also lowered. It is noticeable that, in the presence of 10 mM NaF, the AEC is very low (0.15) and protein synthesis is nearly blocked. The analysis by two-dimensionnal polyacrylamide gel electrophoresis of the protein patterns, after NaF treatment, shows that they are similar. There is a decrease of the synthesis of all of the polypeptides synthesized under anoxic conditions. Thus, when glycolysis is limited by NaF under anoxic conditions, ATP production is lowered and the capacity of protein synthesis is limited.

Protein synthesis in rice embryos germinated under anoxia. In rice seeds imbibed under anoxic conditions the AEC reaches 0.80 after 24 hrs and remains stable. After 3 days under anoxia, we have labeled the proteins and a typical anaerobic pattern was obtained by two-dimensional polyacrylamide gel electrophoresis (fig. 6), indicating that a great number of polypeptides are synthesized.

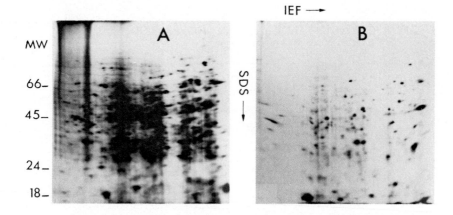

FIGURE 6 . Fluorogram of two-dimensional gels of aerobic (A) and anaerobic (B) proteins of rice embryos. Rice seeds were germinated in air for two days and under nitrogen for three days and then labeled with ^{14}C-aminoacids for 2 hrs as described by Mocquot et al. (11). Molecular weight standards, 66, 45, 24, 18 kD were used.

Effect of anoxia on protein synthesis in lettuce embryo. Germinating lettuce seeds are able to survive under anoxic conditions for long periods. The AEC is low (0.3) and we can expect a low metabolic activity (9). We have studied the uptake of ^{14}C-leucine and its incorporation into an acid-insoluble product by isolated lettuce embryos. The results presented in figure 7 show that the ^{14}C-leucine uptake is reduced under anoxia and tends to decrease as time goes on. The incorporation is very low and close to the background.

Some indications concerning protein turn-over under anoxia. We have shown that, in rice embryos, during anoxic treatment, there is a stabilization of the protein level (11). Under these conditions, the labeling of many polypeptides indicates an important turnover of proteins. A preliminary experiment was undertaken to study protein degradation in rice embryos under anoxia.

Two day-old aerobic rice seedlings were labeled with a ^{14}C-aminoacid mixture for 6 hrs and then washed with water to remove external aminoacids. The seedlings were put in water under a nitrogen gas flow.

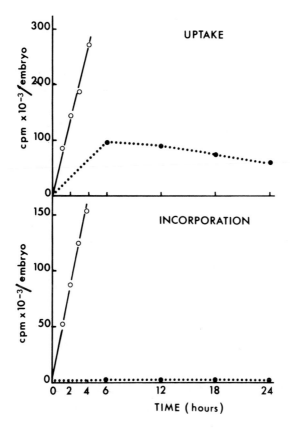

FIGURE 7.   Leucine uptake and incorporation into acid-insoluble fraction by lettuce embryos imbibed for 6 hrs in air and labeled for 6 hrs under aerobic ($\cdots\bullet\cdots$) and anaerobic (—o—) conditions with ^{14}C-leucine (3 μci/ml).

The residual acid-insoluble radioactivity was determined over the following days. Even after 2 days, there is no change in the total radioactivity and there is only a small decrease of the radioactivity incorporated into the proteins. As we know that protein synthesis occurs under anoxia in these embryos, this result shows that the counts released by the degradation of proteins nearly equals the counts incorporated into new proteins.

The small loss of radioactivity may be due to some catabolism of labeled aminoacids (table 3).

TABLE 3
PROTEIN TURNOVER IN RICE EMBRYOS UNDER ANOXIA

Treatment (hrs under N2)	total radioactivity	acid-insoluble radioactivity
	cpm x 10^{-6}/embryo	
0	3.55	2.33
6	3.35	2.27
24	3.45	2.22
48	3.25	1.96

The same study was performed with lettuce seeds. After 6 hrs of imbibition in air, lettuce embryos were labeled with ^{14}C-leucine for 6 hrs, washed with water and put under a nitrogen gas flow. The acid-insoluble radioactivity was measured after different times of incubation under anoxia at 20°C. The results presented in table 4 show that there is almost no loss of radioactivity from the labeled proteins. As we have shown that practically no synthesis occurs under anoxia, we can deduce that there is an inhibition of protein degradation in this plant material under anoxic conditions.

TABLE 4
PROTEIN TURNOVER IN LETTUCE EMBRYOS UNDER ANOXIA

Treatment (hrs under N2)	total radioactivity	acid-insoluble radioactivity
	cpm x 10^{-3}/embryo	
0	660	400
6	595	440
36	590	435

These results show that, under anoxia, the macromolecular machinery is operating in rice embryos and may play a role in their adaptation to anaerobic stress. On the other hand, it seems that, in lettuce embryos, most of the metabolic processes are strongly reduced, suggesting another mechanism of adaptation. The degradation of polyribosomes and the arrest of protein synthesis found by Lin and Key in soybean roots (18) can be explained by the low ATP-production and low energy charge probably occuring in these roots (10).

Comparison of anaerobic and heat shock proteins in rice. Two types of stress are now well studied in plants at the molecular level. The set of proteins induced by anaerobiosis was characterized in maize roots (16). Heat shock also induces the synthesis of proteins in soybean seedlings (19). Recently, it was shown that the two sets of proteins are very distinct in maize roots (20). We have compared the proteins synthesized in rice embryos at the beginning of anoxic treatment at 26 °C (between 1 and 5 hrs) with the proteins synthesized after a heat shock at 44°C (1 hr of pre-treatment and 2 hrs of labeling with a ^{14}C-aminoacid mixture. The patterns obtained after two-dimensional polyacrylamide gel electrophoresis indicate that the intensely labeled proteins induced by the two stresses are different (fig. 8).

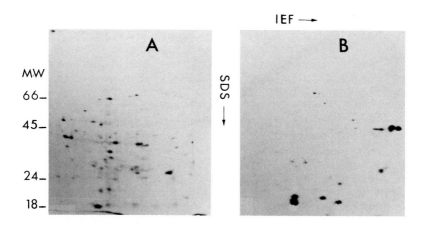

FIGURE 8.   Fluorogram of two-dimensional gels of anaerobic (A) and heat shock (B) proteins of rice embryos. Molecular weight standards, 66, 45, 24, 18 kD were used.

Nevertheless, some minor proteins may be common, as mentioned by Kelley and Freeling (20).

The heat shock treatment at 44 °C has no effect on the ATP level nor on the AEC value as compared with those of rice seedlings grown at normal temperature (26 °C).

Energy Cost of Rice Growth under N_2.

The ability to grow under strictly anaerobic conditions seems to be restricted to a few seeds; root growth of established plants, under anoxia, has not been reported.

Do these seeds possess some peculiar energy yielding pathway or regulatory mechanisms, or is the ability to grow under such conditions related to the large amount of stored fermentable reserves?

Similarly, is the amount of energy (in the form of ATP) provided by fermentation high enough to account for the growth of the rice coleoptile and the correlated synthesis of macromolecules?

The relationship of growth with energy production has been studied by microbiologists over the past 25 years (21) and has been expressed as YATP or ATP yield (the mass (d.w.) in grams, of microorganisms, which is accumulated per mole of ATP produced). The YATP is particularly easy to determine under anaerobic conditions, when the fermentative pathways are known.

The YATP is influenced by growth conditions, being higher in complex media than in minimal media.

Theoretical calculations of the ATP requirement for the formation of cell material are based on the macromolecular composition of the cells.

In rice embryos, polysaccharides and proteins are the major macromolecular components. Then nucleic acids and lipids contribute little to the energy cost of growth. Alpi and Beevers reported that, despite its greater length, the anaerobic coleoptile contains less proteins and carbohydrates than the aerobic coleoptile and its dry weight is lower (22). We have shown that protein synthesis results from the turn over of preexisting proteins and cellulose is likely to be formed from reserve polysaccharides.

The growth yield YATP for rice embryos, calculated from their macromolecular composition, (80% polysaccharides, 15% proteins) is close to 40 g/mol ATP.

This estimation must be corrected for the cost of maintenance and osmotic work. Hence, a value of 20 to 30 g/mol ATP seems a plausible estimation. It is similar to that found for microorganisms growing on complete media.

In agreement with the view that ethanol is the main fermentation product in plant tissues (23), we have found that other potential fermentation products, such as lactate, malate, alanine and succinate are formed very slowly as compared to ethanol. The rate of ATP regeneration, calculated from the rate of ethanol accumulation in rice embryos, is close to 250 nmol/hr. Thus, the increase of the embryo dry weight would be 0.15 to 0.20 mg/day.

The actual increase of embryo d.w. under anoxia was 0.15 mg/day. This experimental value is close to the theoretical one. It may be concluded that the rate of ethanol production is sufficient to account for the growth of rice embryos under anoxia.

SELECTIVE CONTROL OF ATP-CONSUMING PATHWAYS

It has been suggested (2) that, under adverse conditions, the scarce energy resources are conserved for use in essential maintenance processes : the ATP-consuming pathways related to non-essential processes would be more strongly inhibited than the more essential ones.

The synthesis of ADH, and other anoxia-related proteins in various organisms may be examples of a selective control of pathways under stress conditions. In the case of ADH, the control is at the transcriptional level (24).

However, when fermentation under anoxia is inhibited by fluoride, in rice embryos, both the AEC charge and the protein synthesis decrease; there is no evidence for a specific inhibition or activation of the synthesis of any protein. Similarly, the synthesis of adenine nucleotides, which does not seem to be necessary for survival under limiting conditions and which ATP cost is high, is not blocked under anoxia. In these cases, there is no specific control of pathways according to energy availability.

It is clear that during hypoxic and anoxic stresses, the rate of ATP regeneration limits quantitatively the ATP-utilizing pathways. Although the hypothesis cannot be discarded, we have found no evidence that a limitation of energy availability contributes to the selective control of gene expression associated with hypoxic and anoxic stresses.

REFERENCES

1. Pradet A, Bomsel JL (1978). Energy metabolism in plants under hypoxia and anoxia. In Hook DD, Crawford RMM (eds): "Plant Life in Anaerobic Environment", Ann Arbor: Ann Arbor Science, p. 89.
2. Atkinson DE (1977). "Cellular Energy Metabolism and its Regulation". New York: Academic, 293 pp.
3. Pradet A, Raymond P (1983). Adenine nucleotide ratios and adenylate energy charge in energy metabolism. Annu Rev Plant Physiol 34:199.
4. Pradet A, Prat C (1976). Métabolisme énergétique au cours de la germination du riz en anoxie. In Jacques R (éd): "Etudes de Biologie Végétale", Gif sur Yvette, France: Phytotron CNRS, p. 561.
5. Al-Ani A, Leblanc JM, Raymond P, Pradet A (1982). Effet de la pression partielle d'oxygène sur la vitesse de germination des semences à réserves lipidiques et amylacées: rôle du métabolisme fermentaire. C R Acad Sc Paris 295:271.
6. Raymond P, Al-Ani A, Pradet A (1983). Low contribution of non-respiratory pathways in ATP regeneration during early germination of lettuce seeds. Physiol Vég 21:677.
7. Pradet A (1969). Etude des adénosine 5'-mono-, di- et tri-phosphates dans les tissus végétaux. V. Effet in vivo sur le niveau de la charge énergétique d'un déséquilibre induit entre fourniture et utilisation de l'énergie dans les semences de laitue. Physiol Vég 7:261.
8. Cocucci MC, Marre E (1973). The effect of cycloheximide on respiration, protein synthesis and adenosine nucleotide levels in Rhodotorula gracilis. Plant Sci Lett 1:293.
9. Raymond P, Pradet A (1980). Stabilization of adenine nucleotide ratios at various values by an oxygen limitation of respiration in germinating lettuce (Lactuca sativa L.) seeds. Biochem J 190:39.
10. Saglio PHM, Raymond P, Pradet A (1980). Metabolic activity and energy charge of excised maize root tips under anoxia. Plant Physiol 66:1053.

11. Mocquot B, Prat C, Mouches C, Pradet A (1981). Effect of anoxia on energy charge and protein synthesis in rice embryo. Plant Physiol 68:636.
12. Mocquot B, Pradet A, Litvak S (1977). DNA synthesis and anoxia in rice coleoptiles. Plant Sci Lett 9:365.
13. Morisset C (1980). Biosynthèse de DNA pendant l'anoxie, décelée par histo-autoradiographie dans les radicules de riz (Oriza sativa L.). Biol Cell 38:22a.
14. Kordan HA (1976). Mitotic activity in rice seedlings germinating under oxygen deficiency. J Cell Sci 20:57.
15. Aspart L, Got A, Delseny M, Mocquot B, Pradet A (1983). Adaptation of ribonucleic acid metabolism to anoxia in rice embryos. Plant Physiol 72:115.
16. Sachs MM, Freeling M, Okimoto R (1980). The anaerobic proteins of maize. Cell 20:761.
17. Bertani A, Menegus F, Bollini R (1981). Some effects of anaerobiosis on protein metabolism in rice roots. Z Pflanzenphysiol 103:37.
18. Lin CY, Key JL (1967). Dissociation and reassembly of polyrisobomes in relation to protein synthesis in soybean root. J Mol Biol 26:237.
19. Key JL, Lin CY, Chen YM (1981). Heat shock proteins of higher plants. Proc Natl Acad Sci USA 78:3526.
20. Kelley PM, Freeling M (1982). A preliminary comparison of maize anaerobic and heat shock proteins. In Schlesinger ML, Ashburner M, Tissieres A (eds):"Heat Shock from Bacteria to Man", Cold Spring Harbor, New York : Cold Spring Harbor Laboratory, p. 315.
21. Stouthamer AH (1977). Energetic aspects of the growth of micro-organisms. In Haddock BA, Hamilton WA (eds): "Microbiol Energetics", Cambridge: Cambridge University, p. 285.
22. Alpi A, Beevers H (1983). Effects of O_2 concentration on rice seedlings. Plant Physiol 71:30.
23. Smith AM, ap Rees T (1979). Effect of anaerobiosis on carbohydrate oxidation by roots of Pisum sativum. Phytochem 18:1453.
24. Ferl RJ, Brenman MD, Schwartz D (1980). In vitro translation of maize ADH : evidence for the anaerobic induction of mRNA. Biochem Genet 18:681.

Cellular and Molecular Biology of Plant Stress, pages 247–262
© 1985 Alan R. Liss, Inc.

INDUCTION OF HYDROLASES AS A DEFENSE REACTION
AGAINST PATHOGENS [1]

Thomas Boller

Botanisches Institut, Universität Basel
CH-4056 Basel, Switzerland

ABSTRACT   Plants possess hydrolases that can attack
the cell walls of pathogens. Among them are chitinase,
lysozyme, and β-1,3-glucanase. Plant chitinases and
"lysozymes" both act primarily as endochitinases and
are basic proteins of ca. 30 kD. Most β-1,3-glucanases
are endoglucanases; many are also basic proteins of ca.
33 kD. Both chitinase and β-1,3-glucanase can be induced
by exogenous ethylene or in response to infection or
treatment with fungal cell wall components. Experiments
with inhibitors of ethylene biosynthesis indicate that
ethylene and infection are separate, independent stimuli
for the induction of these enzymes. *In vitro*, chitinase
and β-1,3-glucanase partially degrade isolated cell
walls of pathogens. Thus, the two enzymes have the potential to inflict damage on pathogens. However, it is
unknown, at present, if this potential is actually employed against invading pathogens, and to what extent
the induction of chitinase and β-1,3-glucanase contributes to an enhanced disease resistance *in vivo*.

INTRODUCTION

Plants possess a large arsenal of constitutive and inducible biochemical defenses against potential pathogens (1).
Among them, one might expect hydrolytic enzymes that attack
the cell walls of invading pathogens (1,2). The two hydrolases most often mentioned in this context are chitinase and
β-1,3-glucanase. This is based on an observation unrelated

[1]Supported by Swiss National Science Foundation,
Grant 3.400-0.83

to plant pathology: Abeles *et al.* (3) found that these two enzymes were coordinately induced by exogenous ethylene. Since chitin and β-1,3-glucan are important components of the cell walls of many fungi (4), Abeles *et al.* (3) postulated that chitinase and β-1,3-glucanase may function in defense against fungal pathogens. They further hypothesized that stress ethylene, formed endogenously in response to an attack by pathogens, may also induce the two enzymes and thereby enhance the defense potential of the plant.

Several research groups, including ourselves, have been attracted by this hypothesis and have examined the induction of chitinase and β-1,3-glucanase and its significance for the defense against pathogens in various plants. Here I want to summarize and discuss three aspects of these studies: the biochemical characterization of chitinase and β-1,3-glucanase; the mode of induction of these enzymes in stressed and diseased plants; and the contribution of lytic enzymes to the defense potential of a plant.

CHARACTERIZATION OF PLANT CHITINASES AND β1,3-GLUCANASES

Chitinases

The substrate of chitinases, chitin (a β-1,4-linked polymer of N-acetylglucosamine), occurs in animals, in fungi and in certain algae but has never been found in higher plants. Therefore, the widespread occurrence of chitinase in higher plants (5) remains unexplained in terms of the plants' own metabolism. The substrate for chitinase must come from another organism, from outside. Thus, chitinase has been suggested to act as a defense against chitin-containing pathogens (1-3). Lysozyme, which has been found in higher plants as well, may similarly represent a defense against bacteria. Below, it will be shown that "chitinase" and "lysozyme" describe two activities of one and the same enzyme.

Properties of purified plant chitinases. Chitinase can be purified easily by affinity chromatography on its insoluble substrate, chitin (6,7). The plant chitinases purified so far have all similar properties (Table 1): Their molecular weight, determined by sodium dodecyl sulfate gel electrophoresis, is around 30 kD. (Pea pods contain two chitinases with somewhat differing molecular weights.) The native enzymes have similar molecular weights, indicating that they exist as monomers in solution. Their isoelectric points are in the basic range.

TABLE 1
PHYSICAL PROPERTIES OF PLANT CHITINASES

Plant, tissue	Mol.wt.[a]	Isoel.pt.	References
Wheat germ	30,000	7.5 - 9.2	(6)
Tomato stem	31,000	8.5	(8)
Bean Leaf	30,000	9.0[b]	(7)
Pea Pod	32,000	9.1	-[b]
	36,000	9.0	

[a]Determined by SDS electrophoresis
[b]F. Mauch, unpublished results

The reaction products formed by the enzymatic degradation
of regenerated chitin have been studied for the chitinases
from wheat germ and bean leaf. Both produced primarily chito-
oligosaccharides; very little free N-acetylglucosamine was
formed (6,7). Therefore, in measuring plant chitinase activity, it is important to employ an assay for endochitinase
(6,7); the assay for plant chitinase that has been used most
often (3,8), which measures the free N-acetylglucosamine
only, often yields erratic results and may underestimate the
actual chitinase activity by orders of magnitude. It is also
important to be aware of the nonlinear relationship between
the amount of enzyme and the amount of reaction product
formed (7).

The purified chitinase from wheat germ was found to be
specific with regard to its polysaccharide substrate: It did
not hydrolyse cellulose (β-1,4-glucan), β-1,3-glucan or
β-1,6-glucan; it did not attack the carbohydrate chain of
ovalbumin or ovomucoid (6). Bean chitinase had a similarly
restricted specificity.(7).

Properties of purified plant lysozyme. Soon after his
discovery of lysozyme, Fleming detected a similar bacteriolytic activity in plants (9). Certain plant latices were
found to have particularly high amounts of lysozyme. In
papaya latex, which is generally known for its high content
of proteinases, lysozyme accounted for up to 30 % of the
soluble proteins (10). The proteins responsible for lysozyme
activity have been purified from several latices and from
turnip roots (Table 2). Their physical properties are strik-
ingly similar to those of plant chitinase (Table 1).

TABLE 2
PHYSICAL PROPERTIES OF PLANT LYSOZYMES

Plant, tissue	Mol.wt.	Isoel.pt.	References
Papaya latex	28,000[a]	10.5	(11,12)
Fig latex	29,000[a]	9	(13)
Rubber tree latex	26,000[b]	9.0	(14)
Turnip root	25,000[c]	10.0	(15)

[a]Analysis of sedimentation velocity
[b]SDS gel electrophoresis
[c]Gel filtration

When incubated with bacterial cell walls, plant lysozyme had a much lower specific activity than egg white lysozyme, exhibited a much narrower pH-optimum (pH 4 - 5), and
showed different enzyme kinetics (11-15). When incubated
with chitin, the purified plant lysozymes exhibited very
high endochitinase activities, much more so than egg white
lysozyme (12,13,15). These data indicate that "lysozyme" and
"chitinase" activities are due to the same enzyme, and that
the preferential substrate for the plant enzyme is chitin,
while for the animal enzyme, it is the bacterial cell wall
peptidoglycan.

In support of this conclusion, we found in our own work
on bean chitinase that lysozyme activity was induced in parallel to chitinase activity after an ethylene treatment, and
that it was co-purified with chitinase (7). The specific
activity of purified chitinase on chitin was ca. 20 times as
high as commercial *Streptomyces* chitinase; on the other hand,
its specific activity against bacterial cell walls was only
about 10 % of that of egg white lysozyme. While the pH-optimum of bean chitinase acting on chitin was broad, its lysozyme activity had a sharp pH-optimum at pH 5, with very little activity left at pH 6.2, the pH where animal lysozyme is
conventionally measured (Figure 1).

Localization of chitinase. We recently studied the localization of chitinase in ethylene-treated bean leaves (16).
Mesophyll protoplasts prepared from the leaves had essentially
the same specific activity of chitinase as the leaf homogenates, indicating that most of the activity was intra-

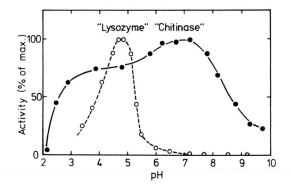

FIGURE 1. pH Activity profile for the lysozyme activity
and for the chitinase activity of purified bean chitinase.
Enzyme assays were performed as described (7)

cellular. Vacuoles were isolated from the protoplasts and
were found to contain most of the intracellular chitinase
activity. Thus, the ethylene-induced chitinase is deposited
in the central vacuole of the leaf cells. It will be interesting to examine how the enzyme, which appears to be formed
de novo (see below), is transferred into the vacuole.

The "lysozyme" of rubber tree latex was found to be
localized in the lutoid fraction, a fraction of small membrane-bound organelles of the latex which are considered
homologous with the vacuole (14).

β-1,3-Glucanases

In contrast to chitin, β-1,3-glucan (callose) is present
and is turned over in higher plants. Therefore, it must be
assumed that some of the described β-1,3-glucanase activities
from higher plants function in the plants' own metabolism.
Hence, the widespread occurrence of β-1,3-glucanases (2,17)
in higher plants is not surprising. However, since chitinase
and β-1,3-glucanase were induced in parallel in ethylene-
treated leaves (3), and since both chitin and β-1,3-glucan
are important fungal wall components (4), β-1,3-glucanase has
also been hypothesized to function in defense against pathogenic fungi (3). An indirect argument in favor of this is the
finding that a bean pathogen, *Colletotrichum lindemuthianum*,
produces an inhibitor that blocks the enzymatic activity of
bean β-1,3-glucanase (18).

TABLE 3
PHYSICAL PROPERTIES OF PLANT β-1,3-GLUCANASES

Plant, tissue	Mol.wt.	Isoel.pt.	References
Bean leaf	34,000[a]	11	(3)
Tobacco cells	33,000[c]	9.9	(19)
Soybean cotyledon	33,000[c]	10.5;8.7	(20)
Pea pod	32,000[c]	9.5	_[d]
Rye seedling	24,100[b]	10.6;9.2	(21)
Barley seedling	21,400[b]	9.8	(21)
Tobacco leaf	45,000[b]	4.9	(22, 23)
Pea seedling	22,000[c]	5.4	(24)
	37,000[c]	6.8	

[a]Analysis of sedimentation velocity
[b]Gel filtration
[c]SDS gel electrophoresis
[d]F. Mauch, unpublished results

Properties of purified plant β-1,3-glucanases. β-1,3-Glucanases have been purified from several plants (Table 3). Many are very similar to plant chitinase with regard to the molecular weight and to the basic isoelectric point. Others, like the β-1,3-glucanases purified from tobacco leaves (22, 23) and from pea seedlings (24) , have different molecular weights and acidic isoelectric points. The β-1,3-glucanases partially purified from tomato (25) are also acidic proteins. All enzymes listed in Table 3 act as endoglucanases.

An entirely different β-1,3-glucanase has been isolated from cultured soybean cells. This enzyme, which was found to be bound to the cell wall, attacks β-1,3-glucan as an exoglucanase; it also has β-glucosidase activity and can act as a glucosyl transferase (26).

INDUCTION OF CHITINASE AND β-1,3-GLUCANASE
BY ETHYLENE AND BY PATHOGEN STRESS

In specific tissues, e.g. in laticifers, chitinase
("lysozyme") may be accumulated constitutively in large
quantities. However, in general, chitinase (5,7) and
β-1,3-glucanase (2,17) are present at a small basal level in
plant tissues and become elevated only upon treatments with
exogenous ethylene or as a consequence of plant disease.

Induction by Ethylene

Abeles *et al.* (3) first reported the induction of chitinase and β-1,3-glucanase by ethylene in bean leaves. We
recently investigated this effect more closely (7). We found
that chitinase activity began to increase less than 6 h after
the onset of ethylene treatment, and that the enzyme was induced thirtyfold within 24 h (Fig. 2). *In vitro* translation
of RNA and immunoprecipitation of the translation products
by a specific antiserum against bean chitinase indicated that
the induction by ethylene was due to a dramatic increase in
the level of translatable mRNA for chitinase (U. Vögeli, unpublished results).

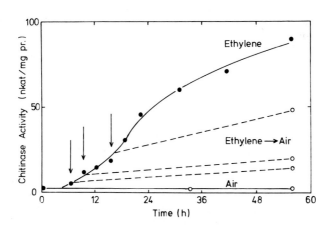

FIGURE 2. Time course of chitinase induction by ethylene
in bean seedlings. Some plants were put back from ethylene
into ethylene-free air at the times indicated by arrows.
Reproduced from (7)

Chitinase induction appeared to be closely linked to the presence of ethylene since it could be halted again, even after it had started, by withdrawal of ethylene (Fig. 2). About 1 ppm exogenous ethylene was required for half-maximal induction. Endogenously produced ethylene appeared to act as an inducer as well: Treatment of leaves with 1-aminocyclo-propane-1-carboxylic acid (ACC), the biosynthetic precursor of ethylene, strongly stimulated the endogenous ethylene pro-duction and induced chitinase (7). However, various stresses that caused only transient bursts of ethylene formation, e.g. mechanical wounding, heat (40 °C) and cold (2 °C) shocks, did not induce chitinase (U. Vögeli, unpublished).

Ethylene also induced chitinase in melon seedlings (27) and in many other plants (7). In bean leaves, among a number of hydrolases, only chitinase and β-1,3-glucanase in-creased in response to ethylene (7).

Induction by Pathogen Stress

Induction in response to viruses. Plants often respond to pathogen stress by the so-called hypersensitive reaction. The local lesions formed upon infection with certain viruses are an example of this. Interestingly, β-1,3-glucanase activ-ity increased fortyfold in the course of local-lesion in-fections of tobacco leaves with tobacco mosaic virus (TMV) (28). Chitinase was found to be similarly induced in tobacco leaves infected with TMV (A. Gehri, unpublished). Induction of chitinase and β-1,3-glucanase may belong to a syndrome of defense reactions displayed during the hypersensitive re-sponse; it is unlikely that they have a role in the pre-vention of virus spread (28).

Induction in response to fungi. Induction of chitinase and β-1,3-glucanase might be of more direct significance as a response to fungal infection. Therefore, several studies have examined the two enzymes in plants infected with fungi. In general, the activities of these enzymes were found to increase strongly in the course of infection (2,8,25,27,29-31).

The interpretation of these results is complicated by the fact that fungal enzymes may contribute to the observed increase of activity. In some studies, in order to distin-guish between fungal and host enzymes, the enzymes were sub-jected to electrophoresis or other separation techniques be-fore assay. Using such methods, it was concluded that the increased β-1,3-glucanase activity in melons infected with *Colletotrichum lagenarium* was due to an enzyme secreted by

the pathogen (29). In contrast, the increased activities of
β-1,3-glucanase (25) and chitinase (8) in tomato plants infected with *Verticillium albo-atrum* appeared to be due to
plant enzymes.

    Induction in response to elicitors. One way to avoid a
possible pathogen contribution to the measured enzyme level is to apply preparations of heat-killed pathogens, pathogen cell walls or related carbohydrates. Such preparations
are frequently used as elicitors of phytoalexin accumulation
(1). They can also induce hydrolase activities; for example,
autoclaved spores of *Verticillium* induced chitinase and
β-1,3-glucanase in tomato stems (30). Laminarin induced
β-1,3-glucanase in melon seedlings (31).

    We recently tested the effect of an elicitor from cell
walls of *Phytophthora megasperma* on chitinase activity in
cultured parsley cells (32). This elicitor caused a tenfold
increase of chitinase activity within 12 h. It also induced
phenylalanine ammonia lyase and had been shown previously
to act as an elicitor of phytoalexin production. The induction of both chitinase and phenylalanine ammonia-lyase was
preceded by a very rapid increase in the key enzyme of
ethylene biosynthesis, ACC synthase (Fig. 3).

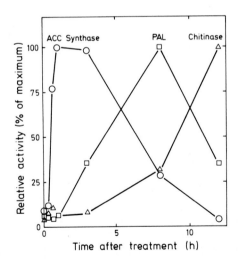

    FIGURE 3. Induction of chitinase, phenylalanine ammonia-
lyase and ACC synthase in parsley cell cultures by an elicitor
from *Phytophthora megasperma* (40 µg ml⁻¹). Reproduced from
(32)

This finding raises the question whether the induction of chitinase by elicitor is mediated by ethylene. A mediating role for stress ethylene in the induction of biochemical defenses has already been postulated in other cases (33).

Is chitinase and β-1,3-glucanase induction by fungal infection or elicitors mediated by ethylene? We choose immature pea pods to study this question (34). Infection of this tissue with *Fusarium solani* resulted in a rapid and reproducible increase of chitinase and β-1,3-glucanase. After a lag of 4 - 8 h, the activities increased up to tenfold within 24 h. Treatments with autoclaved spores or with chitosan, an elicitor of phytoalexin formation in this tissue, also induced the two enzymes. Fungal infection or elicitors increased stress ethylene production within 3 h. Chitinase and β-1,3-glucanase activities in uninfected pods were increased by a treatment with exogenous ethylene. However, aminoethoxyvinylglycine (AVG) or cobalt ions, known inhibitors of ethylene biosynthesis, did not reduce chitinase and β-1,3-glucanase induction in infected pods although they strongly inhibited ethylene production. Hence, stress ethylene does not appear to be a necessary signal for the induction of the two hydrolases; rather, ethylene and fungal elicitors or infection seem to be separate, independent stimuli for the induction of the enzymes (34).

CONTRIBUTION OF CHITINASE AND β-1,3-GLUCANASE TO THE PLANT'S DEFENSE AGAINST PATHOGENS

In certain plant-pathogen interactions, the invading fungi are known to be lysed within the plant (2,30,35). This may be due to the action of plant glycosidases, as has been suggested (2,30, 35). However, it might also merely represent a case of fungal autolysis. There is presently no direct evidence for an antifungal activity of plant hydrolases *in vivo*.

On the other hand, as summarized below, *in vitro* studies have established that chitinase and β-1,3-glucanase readily attack and degrade isolated fungal cell walls; furthermore, they may release soluble products with strong elicitor activities in the course of this process. It remains to be shown that either of these activities plays a role in the outcome of plant-pathogen interactions *in vivo*.

Degradation of Pathogen Cell Walls

Bacteria. Plant "lysozyme" has generally been thought
to act as an antibiotic (9-14). Its activity has been measured with *Micrococcus lysodeikticus* (11-14) or *Sarcina lutea*
(10); the walls of these bacteria are lysed to a considerable extent. However, to our knowledge, it is unknown whether
plant lysozyme has any lytic activity against cell walls of
potentially plant pathogenic bacteria.

Fungi. Enzymes present in crude plant extracts have
been shown to partially degrade isolated fungal cell walls
(2,7,35,36). Purified chitinase from tomato released soluble
fragments from cell walls of the tomato pathogen, *Verticillium albo-atrum*.(36); similarly, purified bean chitinase
readily attacked isolated cell walls of *Fusarium solani* f.sp.
phaseoli, a bean pathogen (7).

The fact that lysis was only partial (7,35,36) does not
preclude an important role in defense: Interestingly, plant
chitinase degrades "nascent" chitin formed by chitin synthetase *in vitro* much more effectively than preformed chitin
(6). Rapid dissolution of the newly formed chitin in hyphal
tips might be sufficient to stop a growing fungus.

Release of Elicitors

Cell wall fragments isolated by biochemical techniques
from various fungi have been found to be highly active elicitors of phytoalexin formation (1). Crude plant extracts have
been shown to release enzymatically fragments from fungal
cell walls with strong elicitor activities (37). Recently,
an endo-β-1,3-glucanase from soybean was purified (20). This
enzyme released a highly active elicitor from *Phytophthora
megasperma* cell walls (38).

Plant enzymes may also inactivate elicitors. The cell
wall-bound β-glucosyl hydrolase from suspension-cultured
soybean cells (26) inactivated the glucan elicitor isolated
from *Phytophthora megasperma* (39).

Induction of Chitinase and β-1,3-Glucanase in Relation to
Susceptibility and Resistance

In several studies, the induction of chitinase and
β-1,3-glucanase was compared in susceptible and resistant
plants in the hope to obtain correlative indications for a

role of these enzymes in defense. Netzer *et al.* (31) compared
the induction of β-1,3-glucanase in a susceptible and a resistant variety of muskmelon infected with *Fusarium oxysporum*. They found that the enzyme was induced in both cultivars, but that it increased more in the resistant than in
the susceptible cultivar. Thus, increased induction of
β-1,3-glucanase and increased resistance to the pathogen
were correlated. In contrast, Pegg and Young (30) found a
negative correlation between resistance and hydrolase induction in their work on tomato lines infected with *Verticillium albo-atrum*: Here, the increase in chitinase and
β-1,3-glucanase was greater in the susceptible than in the
resistant tomato variety, even when autoclaved spores were
used as elicitors.

In pea pods, chitinase and β-1,3-glucanase activities
were followed after infection with a compatible and an incompatible strain of *Fusarium solani*. There was no difference
in the time course and level of induction of either enzyme
(34).

CONCLUSIONS

In a recent theoretical paper, the author (who obviously
was unaware of the work summarized here) suggested to transfer a bacterial gene for chitinase into the genome of higher
plants (40). He argued that the expression of this gene
should confer increased resistance against fungi. Here,
plants seem to be one step ahead of the genetic engineers:
They have acquired chitinase genes already through evolution
and seem to regulate them in function of the stress exerted
by pathogens.

Plants possess endochitinase - which can also act as a
lysozyme - and endo-β-1,3-glucanase. While these two enzymes
are normally present at a low level, they are often coordinately induced by exogenous ethylene, by pathogen attack, or
by elicitors.

At present, there is little information about the mechanism of induction. We found that ethylene induces chitinase
by increasing the level of translatable mRNA for chitinase,
and that ethylene-induced chitinase is transferred into
the vacuole. It remains to be investigated whether or not
elicitors induce chitinase in a similar way. Furthermore,
it is an intriguing question whether induction of chitinase
and β-1,3-glucanase in infected tissue is part of a defense
syndrome which also includes other processes like the rapid

increase in stress ethylene production and the accumulation
of phytoalexins, or whether each of the different responses
to infection is elicited separately by specific induction
mechanisms.

In bean leaves, chitinase (7) and β-1,3-glucanase (3)
accumulate in considerable amounts in the course of an ethylene treatment. They represent 5 % and 10 %, respectively, of
the protein soluble at pH 5 (3,7) or 1 % and 2 % of the total
cellular protein. Because of their quantitative importance,
one might expect them to figure prominently among the "ethylene-induced", "disease-related" or "elicitor-induced" proteins newly appearing in the protein patterns of appropriately
treated plants. (Note, however, that they may escape detection in conventional two-dimensional gel electrophoresis because of their high isoelectric point.)

Because of their abundance, chitinase and β-1,3-glucan-
ase represent attractive models for the study of the molecular biology of responses to ethylene or pathogens. However,
it should be kept in mind that it is not their quantity that
renders them particularly important to the student of biological stresses; rather, it is the straightforward hypothesis that they represent an important defense against fungi.

Unfortunately, this hypothesis still rests on indirect
evidence only: While it has been clearly demonstrated that
the two enzymes have a potential to attack and partially
degrade fungal cell walls *in vitro*, it remains to be shown
in what manner and to what extent this potential contributes
to the resistance of a plant against an attacking pathogen
in vivo.

REFERENCES

1. Bell AA (1981). Biochemical mechanisms of disease re-
    sistance. Ann Rev Plant Physiol 32: 21.
2. Pegg GF (1977). Glucanohydrolases of higher plants: a
    possible defence mechanism against parasitic fungi. In
    Solheim B, Raa J (eds): "Cell Wall Biochemistry Related
    to Specificity in Host-Plant Pathogen Interactions,"
    Tromsø: Universitetsforlaget, p 305.
3. Abeles FB, Bosshart RP, Forrence LE, Habig WH (1970).
    Preparation and purification of glucanase and chitinase
    from bean leaves. Plant Physiol 47: 129.
4. Wessels JGH, Sietsma JH (1981). Fungal cell walls: a sur-
    vey. In Tanner W, Loewus FA: "Plant Carbohydrates II,"
    Encycl Plant Physiol NS, Vol 13B, Berlin: Springer, p 352.

5. Powning RF, Irzykiewicz H (1965). Studies on the chitinase system in bean and other seeds. Comp Biochem Physiol 14: 127.
6. Molano J, Polacheck I, Duran A, Cabib E (1979). An endochitinase from wheat germ. J Biol Chem 254: 4901.
7. Boller T, Gehri A, Mauch F, Vögeli U (1983). Chitinase in bean leaves: induction by ethylene, purification, properties, and possible function. Planta 157: 22.
8. Pegg GF, Young DH (1982). Purification and characterization of chitinase enzymes from healthy and *Verticillium albo-atrum*-infected tomato plants, and from *V. albo-atrum*. Physiol Plant Pathol 21: 389
9. Fleming A (1922). On a remarkable bacteriolytic element found in tissues and secretions. Proc Roy Soc 93: 306.
10. Smith EL, Kimmel JR, Brown DM, Thompson EOP (1955). Isolation and properties of a crystalline mercury derivative of a lysozyme from papaya latex. J Biol Chem 215: 67.
11. Howard JB, Glazer AN (1967). Studies of the physicochemical and enzymatic properties of papaya lysozyme. J Biol Chem 242: 5715.
12. Howard JB, Glazer AN (1969). Papaya lysozyme. Terminal sequences and enz matic properties. J Biol Chem 244: 1399.
13. Glazer AN, Barel AO, Howard JB, Brown DM (1969). Isol. and characterization of fig lysozyme. J Biol Chem 244: 3583.
14. Tata SJ, Beintema JJ, Balabaskaran S (1983). The lysozyme of *Hevea brasiliensis* latex: Isolation, purification, enzyme kinetics and a partial amino acid sequence. J Rubb Res Inst Malaysia 31: 35.
15. Bernier I, Van Leemputten E, Horisberger M, Bush DA, Jollès P (1971). The turnip lysozyme. FEBS Lett 14: 100.
16. Boller T, Vögeli U (1984). Vacuolar localization of ethylene-induced chitinase in bean leaves. Plant Physiol 74: 442.
17. Clarke AE, Stone BA (1962). β-1,3-Glucan hydrolases from the grape vine (*Vitis vinifera*) and other plants. Phytochemistry 1: 175.
18. Albersheim P, Valent BS (1974). Host-pathogen interactions VII. Plant pathogens secrete proteins which inhibit enzymes of the host capable of attacking the pathogen. Plant Physiol 53: 684.
19. Shinshi H, Kato K (1983). Physical and chemical properties of β-1,3-glucanase from cultured tobacco cells. Agric Biol Chem 47: 1455.
20. Keen NT, Yoshikawa M (1983). β-1,3-Endoglucanase from soybean releases elicitor-active carbohydrates from fungus cell walls. Plant Physiol 71: 460.

21. Ballance GM, Manners DJ (1978). Partial purification and properties of an endo-1,3-β-D-glucanase from germinated rye. Phytochemistry 18: 1539.

22. Moore AE, Stone BA (1972). A β-1,3-glucan hydrolase from *Nicotiana glutinosa*.I. Extraction, purification and physical properties. Biochim Biophys Acta 258: 238.

23. Moore AE, Stone BA (1972). A β-1,3-glucan hydrolase from *Nicotiana glutinosa* II. Specifity, action pattern and inhibitor studies. Biochim Biophys Acta 258: 248.

24. Wong YS, Maclachlan GA (1979): 1,3-β-D-Glucanases from *Pisum sativum* seedlings. Biochim Biophys Acta 571: 244.

25. Young DH, Pegg GF (1981). Purification and characterization of 1,3-β-glucan hydrolases from healthy and *Verticillium albo-atrum* infected tomato plants. Physiol Plant Pathol 19: 391.

26. Cline K, Albersheim P (1981). Host-pathogen interactions XVI. Purification and characterization of a β-glucosyl hydrolase/transferase present in the walls of soybean cells. Plant Physiol 68: 207.

27. Toppan A, Roby D (1982). Activité chitinasique de plantes de melon infectées par *Colletotrichum lagenarium* ou traitées par l'éthylène. Agronomie 2: 829.

28. Moore AE, Stone BA (1972). Effect of infection with TMV and other viruses on the level of a β-1,3-glucan hydrolase in leaves of *Nicotiana glutinosa*. Virology 50: 791.

29. Rabenantoandro Y, Auriol P, Touzé A (1976). Implication of β-(1-3)glucanase in melon anthracnose. Physiol Plant Pathol 8: 313.

30. Pegg GF, Young DH (1981). Changes in glycosidase activity and their relationship to fungal colonization during infection of tomato by *Verticillium albo-atrum*. Physiol Plant Pathol 19: 371.

31. Netzer D, Kritzman G, Chet I (1979). β-(1,3)Glucanase activity and quantity of fungus in relation to *Fusarium* wilt in resistant and susceptible near-isogenic lines of muskmelon. Physiol Plant Pathol 14: 47.

32. Chappell J, Hahlbrock K, Boller T (1984). Rapid induction of ethylene biosynthesis in cultured parsley cells by fungal elicitor and relation to the induction of phenylalanine ammonia-lyase. Planta, in press.

33. Boller T (1982). Ethylene-induced biochemical defenses against pathogens. In Wareing PF (ed): "Plant Growth Substances 1982," London: Academic Press, p 303.

34. Mauch F, Hadwiger LA, Boller T (1984). Ethylene: Symptom, not signal for the induction of chitinase and β-1,3-glucanase in pea pods by pathogens and elicitors. Submitted.

35. Wargo PM (1975). Lysis of the cell wall of *Armillaria mellea* by enzymes from forest trees. Physiol Plant Pathol 5: 99.
36. Young DH, Pegg GF (1982). The action of tomato and *Verticillium albo-atrum* glycosidases on the hyphal wall of *V. albo-atrum*. Physiol Plant Pathol 21: 411.
37. Yoshikawa M, Matama M, Masago H (1981). Release of a soluble phytoalexin elicitor from mycelial walls of *Phytophthora megasperma* var. *sojae* by soybean tissues. Plant Physiol 67: 1032.
38. Keen NT, Yoshikawa M, Wang MC (1983). Phytoalexin elicitor activity of carbohydrates from *Phytophthora megasperma* f.sp. *glycinea* and other sources. Plant Physiol. 71: 466.
39. Cline K, Albersheim P (1981). Host–pathogen interactions XVII. Hydrolysis of biologically active fungal glucans by enzymes isolated from soybean cells. Plant Physiol 68: 221.
40. Nitzsche W (1983). Chitinase as a possible resistance factor for higher plants. Theor Appl Genet 65: 171.

Cellular and Molecular Biology of Plant Stress, pages 263–273
© 1985 Alan R. Liss, Inc.

INDUCTION OF ETHYLENE BIOSYNTHESIS BY CELL WALL
DIGESTING ENZYMES[1]

James D. Anderson, Edo Chalutz[2] and A. K. Mattoo

Plant Hormone Laboratory, BARC, ARS, USDA,
Beltsville, Md. 20705
Department of Horticulture and Department of Botany
University of Maryland, College Park, Md. 20742

ABSTRACT Ethylene biosynthesis is induced in
tobacco leaf discs by a variety of cell wall
digesting enzyme mixtures (e.g., Cellulysin)
routinely used for protoplast isolation. This
induction appears to directly involve the rate
limiting enzyme, 1-aminocyclopropane-1-carboxylate
(ACC) synthase, which converts S-adenosylmethionine
to ACC. Preincubation of leaves for 6 to 18 h in
60 µl/l ethylene increased their subsequent response
to Cellulysin by increasing ethylene production up
to 10-fold over control leaves preincubated in air.
Sensitization of leaf tissue by ethylene was
inhibited by cycloheximide. A highly purified
active fraction from Cellulysin that induces
ethylene production was isolated by a combination of
membrane ultrafiltration, Sephacryl gel
fractionation and preparative isoelectric focusing.
The activity of the purified fraction was stable to
a 1 hr treatment with proteinase K, trypsin or
chymotrypsin but was totally abolished in 3% SDS or
by incubation at 70C.

[1]This work was carried out under the Cooperative
Agreements No. 58-32U4-1-216 and 58-32U4-2-394 of the
U.S. Dept. of Ag., ARS and the Univ. Md. Scientific
Article No. A3826 of The MD Agric. Expt. Station.
[2]On leave from Division of Fruit and Vegetable
Storage, ARO, The Volcani Center, Bet Dagan, Israel

INTRODUCTION

The synthesis and evolution of ethylene during plant growth and development is well established. Besides the "normal" production of ethylene during plant development, various stresses (e.g., wounding, application of herbicides or growth regulators, and host-pathogen interactions) are reported to induce ethylene biosynthesis (1). Recently, we reported induction of ethylene biosynthesis by the cell wall digesting enzyme mixture, Cellulysin (2). This finding that an enzyme mixture of fungal origin induces a rapid induction of ethylene biosynthesis in a higher plant leaf is new and might have implications in understanding the mechanism of ethylene production during the hypersensitive response in host-pathogen interactions. Perhaps some of the problems encountered with protoplast isolation and stability might be concerned with ethylene induction leading to premature senescence of the protoplasts.

In this report we summarize some of our results concerning the induction of ethylene biosynthesis by Cellulysin, stimulation of this phenomenon by ethylene, and properties of the partially purified, active ethylene-inducing factor from Cellulysin.

RESULTS AND DISCUSSION

Cellulysin-Induced Ethylene Production in Tobacco Leaf Discs and its Stimulation by Ethylene

The induction of ethylene by the cell wall digesting enzyme mixture, Cellulysin, is now well documented (2-5). This induction in tobacco leaf discs is relatively faster than the one by indoleacetic acid (2). In fact, induction of ethylene biosynthesis by IAA reaches maximal rates 24 hours after treatment (6). Pretreating tobacco leaf tissue with 1 μl/l (or higher) ethylene for 16 hours prior to incubation with Cellulysin stimulates ethylene biosynthesis up to 10-fold over that in freshly cut leaf discs (3,4) (Fig. 1).

The mechanism of stimulation of Cellulysin-induction by ethylene is not fully understood, although ethylene pretreatment was shown to stimulate the conversion of ACC to ethylene in tobacco leaf discs (4,5). Possibly,

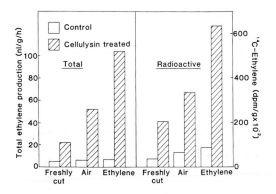

Figure 1. Comparison between rates of total and labeled ethylene production from [3,4-^{14}C]methionine by tobacco (cv Xanthi) leaf discs. Leaves were pretreated in air or ethylene, then discs were cut and treated with Cellulysin. The values represent average rates of ethylene production during the first 2 h of incubation after the addition of Cellulysin (3).

ethylene pretreatment stimulates ethylene formation via some or all of the following steps: 1) increases the amount of ethylene forming enzyme; 2) affects the vacuole where the ethylene forming enzyme is believed localized (7); 3) makes ethylene forming enzyme more accessible to ACC. But these are mere speculations at this stage. However, treating leaves with 10 μg/ml cycloheximide (CHI) prior to ethylene pretreatment, abolishes the ethylene-effect, possibly indicating a requirement for de novo protein synthesis.

CHI when applied with Cellulysin inhibits ethylene production by about 50% in the ethylene-pretreated tissue (Fig. 2). This protein synthesis inhibitor also inhibits ethylene stimulation in aged leaves. It now appears that the effect of CHI on Cellulysin-induced ethylene biosynthesis is not as clear-cut as was first envisioned (2). The specificity of CHI in inhibiting only protein synthesis in plants is suspect. It also affects membrane function. For instance CHI inhibit IAA-induced hydrogen-ion secretion (8) as well as affecting respiration and ion uptake (9). We know that several membrane perturbants inhibit ethylene

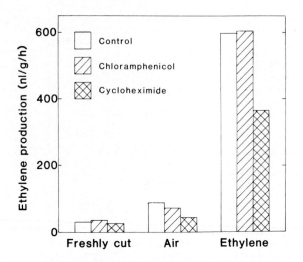

Figure 2. Influence of the protein synthesis inhibitors, chloramphenicol (100 µg/ml) and cycloheximide (10 µg/ml), on Cellulysin-induced ethylene production in freshly cut, aged and ethylene-pretreated leaves.

biosynthesis as well as its induction (10). Possibly, CHI might be acting as a membrane perturbant as well as a protein synthesis inhibitor in plants. Therefore further studies are needed to delineate the actual mechanism of CHI inhibition of ethylene production.

Effect of Other Cell Wall Digesting Enzymes on Ethylene Induction

Other enzyme preparations used for cell wall digestion and protoplast isolation, besides Cellulysin, are also capable of inducing ethylene biosynthesis (Table 1). Cellulases obtained from Worthington induced little or no ethylene biosynthesis in tobacco leaves, at least when assayed at 100 µg of protein/ml. Cellulase-RS was very active in inducing ethylene biosynthesis, especially in discs from ethylene-pretreated leaves. On the basis that Cellulases P, B or PB (Worthington) were ineffective in inducing ethylene biosynthesis, we conclude that this induction is not caused by injury or

TABLE 1

INDUCTION OF ETHYLENE IN TOBACCO (CV. BURLEY MAMMOTH) LEAF
DISCS BY CELL-WALL DIGESTING ENZYME PREPARATIONS DURING
THE 5TH HR OF INCUBATION WITH 100 μg PROTEIN/ML OF
DESALTED ENZYME

Enzyme Preparation*	Source	Ethylene Production	
		Freshly cut	Ethylene-treated
		nl/g.hr	
Control		13	14
Cellulysin	Calbiochem, USA	120	625
Cellulase (RS)	Yakult Pharm. Ind. Co.,Ltd	42	860
Cellulase	Sigma, USA	16	66
Pectinase	Sigma, USA	23	117
Rhozyme	Rohm and Hass, USA	16	53
Driselase	Kyowa Hakko Kogyo Co., Ltd	13	44
Pectolyase	Seishin Pharm Co., Ltd	61	131

*Other enzymes tested that gave little or no
ethylene-inducing response at 100 μg/ml protein were
Cellulases P,B, and PB (Worthington); Macerase
(Calbiochem) and Hemicellulase (Sigma) (5).

by major components of the enzyme mixture, and probably is
independent of protoplast formation. In view of this,
attempts were made to relate a component of several cell
wall digesting enzyme preparations to the ethylene
inducing factor. Single-dimension sodium dodecyl sulfate
polyacrylamide gel electrophoresis (SDS-PAGE) of these
proteins revealed very complex mixtures (5). Because of
this complexity it was impossible to assign any protein or
group of proteins to the ethylene inducing factor (EIF).

Of the five tobacco cultivars tested, Burley Mammoth
gave the best response to Cellulysin in stimulating
ethylene biosynthesis (4). The degree of response varied,
in that all plantings of a given cultivar did not respond
to the same degree. The reason for such different
responsiveness is not clear. However, the conversion of
ACC to ethylene was stimulated to the same degree in
ethylene pretreated leaves of both the non-responsive and
responsive plants (5). Thus the non-responsive plants do
respond to ethylene but seem to lack the ability to
react to Cellulysin.

Purification of the Active Ethylene-Inducing Factor (EIF)
from Cellulysin

We have partially purified an active, EIF from
Cellulysin by combining membrane-ultrafiltration,
Sephacryl S-200 chromatography and isoelectric focusing
(5). The active fraction was retained by an Amicon PM-10
membrane which retains molecules greater than 10,000-d
and interacts with Sephacryl S-200 (5). Thus EIF was
effectively separated from the bulk of the protein. This
purification step enabled a 50-fold (and greater) increase
in the specific activity. The active preparation was
further fractionated on a preparative isoelectric focusing
column. The activity was spread over a pH range of 7 to
10. Possibly, the wide pH range over which the activity
spread was caused by the presence of several components or
a single protein variously modified. Further purification
was accomplished using a Sephadex G100 column.

The active component is very sensitive to treatment
with SDS (Fig. 3). Seventy-five percent of the control
activity was lost upon incubation with 0.5% SDS.

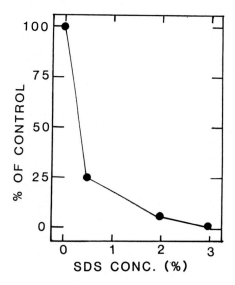

Figure 3.   Inhibition by sodium dodecyl sulfate (SDS)
of the Cellulysin-induced ethylene production.

The partially purified EIF is quite stable to
protease treatment (Table 2). Only a maximum of 25% of
the activity was lost upon treatment with various
proteases. Also, combination of four proteases had little
effect. However, EIF was very sensitive to heat treatment
(Fig. 4). About 60% of the activity was lost at 60C and
98% at 70C. The fact that the active component is a
charged molecule with an apparent molecular weight of
about 18K by gel filtration and is very sensitive to SDS
and temperature, suggests the proteinaceous nature of the
component. However, inability of several proteases to
inactivate EIF may suggest it to be a protein with various
modifications that could make it resistant to proteases.
This contention is supported by the fact that Cellulysin
is a mixture of enzymes and is known to contain
proteolytic activity (11).

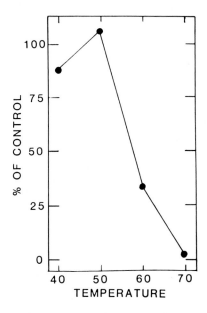

Figure 4. Sensitivity of EIF to temperature.
Aliquots of the partially purified component were
incubated for 10 min at indicated temperatures before
being bioassayed.

TABLE 2
EFFECT OF PROTEASES ON ETHYLENE INDUCING
ACTIVITY OF A PARTIALLY PURIFIED COMPONENT OF
CELLULYSIN

Protease	Activity*
	% of control
Proteinase K	74
Trypsin	93
Chymotrypsin	79
Staph. aureus V8	77
All four together	85

*The partially purified ethylene inducing factor from
Cellulysin was incubated for 1 hr at 30 C with 10 μg/ml
each protease before being taken to induce ethylene
biosynthesis in ethylene-pretreated leaf discs.

Possible Mechanism for Cellulysin-Induced Ethylene Production

We have proposed a model showing possible control
sites of the ethylene biosynthetic pathway (10). In a
modification of this model (Fig. 5) the EIF of Cellulysin
interacts at the plasma membrane producing a signal that
activates the ACC synthase system. The activation could
involve either conversion of an inactive to an active ACC
synthase or de novo synthesis of the enzyme. The evidence
supporting the activation mechanism is the observation
that protese inhibitors block the formation of ACC and
rate limit the production of ethylene (see ref. 10).
However, the increase in ACC synthase activity by de novo
synthesis of the protein cannot be ruled out since CHI, to
some extent, does inhibit the stimulation due to ethylene
pretreatment and the Cellulysin-effect in ethylene-
pretreated leaves (Fig. 2). Additional effect of CHI
could be at the step of conversion of ACC to ethylene or
to some other membrane response. An ethylene-effect also
seems likely at the conversion of ACC to ethylene since
the ethylene-treated leaves converted more ACC to ethylene
than the controls. This invokes a mechanism of auto-
stimulation of ethylene biosynthesis. There seems to be a

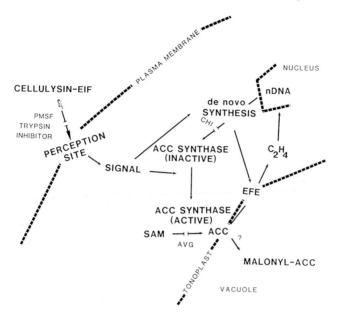

Figure 5.   Proposed model on the mechanism of
induction of ethylene biosynthesis by the ethylene
inducing factor (EIF) from Cellulysin.

close association of ACC synthase and the ethylene forming
enzyme because more ethylene is produced upon Cellulysin
induction than by feeding exogenous ACC (5).
     Implications of the above control mechanism are
manifold.   One of these is in the general observation that
ethylene production is an early event during host-pathogen
interactions.   Auto-stimulation of ethylene biosynthesis
could be part of a hypersensitive reaction that provides a
signal from one cell or one organ to the next to senesce
or abscise the part of the tissue under attack by a
foreign object, here a pathogen.

ACKNOWLEDGEMENTS

We thank Tommie Johnson, Marcia Sloger, and Annette
Thomas for assistance. Mention of a company name or
trademark does not constitute endorsement by the U. S.
Department of Agriculture over others of a similar nature
not mentioned.

REFERENCES

1. Lieberman M (1979). Biosynthesis and action of
    ethylene. Ann Rev Plant Physiol 30:533-591.
2. Anderson JD, Mattoo AK, Lieberman M (1982). Induction
    of ethylene biosynthesis in tobacco leaf discs by cell
    wall digesting enzymes. Biochem Biophys Res Commun
    107:588-596.
3. Chalutz E, Mattoo AK, Solomos T, Anderson JD (1984).
    Enhancement by ethylene of cellulysin induced ethylene
    production by tobacco leaf discs. Plant Physiol
    74:99-103.
4. Chalutz E, Mattoo AK, Anderson JD (1983). Cellulysin-
    induced ethylene production by tobacco leaf discs in
    relation to ethylene produced during host-pathogen
    interactions. Proceedings Plant Growth Regulator Soc
    Amer 18-24.
5. Anderson JD, Chalutz E, Mattoo AK (1984). Purification
    and properties of ethylene-inducing factor from the
    cell wall digesting mixture, Cellulysin. In: "Ethylene
    - Biochemical, Physiological and Applied Aspects."
    (eds. Fuchs, Y and Chalutz, E), Martinus Nijhoff/Dr. W
    Junk Publishers, in press.
6. Aharoni N, Anderson JD, Lieberman M (1979). Production
    and action of ethylene in senescing leaf discs.
    Effect of indoleacetic acid, kinetin, silver ion, and
    carbon dioxide. Plant Physiol 64:805-809.
7. Guy M, Kende H (1983). Ethylene formation in pea
    protoplasts. Plant Physiol Supp 72:209.
8. Rayle, DL (1973). Auxin-induced hydrogen-ion secretion
    in Avena coleoptiles and its implications. Planta
    114:63-73.
9. Ellis, RJ and MacDonald, IR (1970). Specificity of
    cycloheximide in higher plant systems. Plant Physiol
    46:227-232.

10. Mattoo AK, Anderson JD (1984). Wound-induced increase in 1-aminocyclopropane-1-carboxylate synthase activity: Regulatory aspects and membrane association of the enzyme. In: "Ethylene - Biochemical, Physiological and Applied Aspects." (eds. Fuchs, Y and Chalutz, E), Martinus Nijhoff/Dr. W. Junk Publishers, in press.

11. Boller T, Kende H. (1979). Hydrolytic enzymes in the central vacuole of plant cells. Plant Physiol 63:1123-1323.

Cellular and Molecular Biology of Plant Stress, pages 275–290
© 1985 Alan R. Liss, Inc.

INDUCTION OF GLYCEOLLIN BIOSYNTHESIS[1]
IN SOYBEANS

Hans Grisebach, Helga Börner, Marie-
Luise Hagmann, Michael G. Hahn[2],
Joachim Leube, and Peter Moesta[3]

Institut für Biologie II, Universität
Freiburg, D-7800 Freiburg, FRG

ABSTRACT   Accumulation of the phytoalexin
glyceollin in soybean infected by Phytophthora megasperma f.sp. glycinea or treated
with a glucan elicitor from this fungus is
preceded by increases in the activity of a
number of enzymes including those involved
in the biosynthesis of glyceollin. Infection leads to de novo enzyme synthesis which
is controlled at the level of gene transcription. A new enzyme involved in glyceollin biosynthesis is 3,9-dihydroxypterocar-
pan 6a-hydroxylase. Cyclic AMP is not involved as a second messenger in the plant's
response to infection. With a specific
radioimmunoassay for glyceollin I and an
indirect immunofluorescent stain for fungal hyphae an analysis of glyceollin accumulation and extent of penetration of fungal hyphae in compatible and incompatible

[1]This work was supported by Deutsche For-
schungsgemeinschaft (SFB 206) and by Fonds
der Chemischen Industrie
[2]Present address: Department of Plant Biology
The Salk Institute, San Diego CA 92138-9216
[3]Present address: Department of Chemistry
and Biochemistry, University of California,
Los Angeles CA 90024

interaction in soybean roots was carried
out. The results support the important
role of glyceollin in the plant's defense reactions.

INTRODUCTION

In the course of evolution plants have developed a variety of defense mechanisms against potential pathogens. Preformed defense mechanisms
are contrasted with defenses triggered by the in-
vader. Included with the latter is the accumulation of antimicrobial substances at the infection
site, the so called phytoalexins [1].
   In this article we describe briefly experiments on the biochemical mechanisms underlying
the accumulation of phytoalexins in soybean (Glycine max (L) Merill) infected by Phytophthora
megasperma f.sp. glycinea (Pmg), the causal agent
of Phytophthora rot. Furthermore a detailed analysis of the accumulation of phytoalexins in infected soybean roots is presented and the significance of this accumulation upon the defense reaction is discussed.

ACCUMULATION OF PHYTOALEXINS IN SOYBEAN
ENZYMES OF THE GLYCEOLLIN PATHWAY

Upon infection of soybean seedlings with Pmg
or, after treatment with a glucan elicitor from
this fungus, the glyceollin isomers I-III (Fig.1)
accumulate as major phytoalexins at the infection
site. The biosynthetic pathway of the glyceollins
and the known enzymes are shown in Fig.2. Pulse
and pulse chase experiments demonstrated that the
level of glyceollins is determined predominantly
by their rate of synthesis [2]. This conclusion
was further supported by observations that large
increases and subsequent decreases of a number of
enzymes involved in glyceollin biosynthesis, occur
after infection [3,4]. In Table 1 the enzymes are
listed for which activity changes after infection
with Pmg or after challenge with Pmg-elicitor,

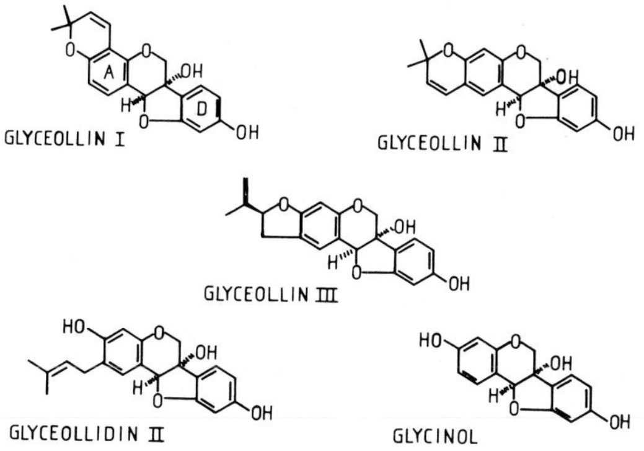

FIGURE 1. Structures of glyceollins I-III and of two precursors (glyceollidin II and glycinol) of the glyceollins.

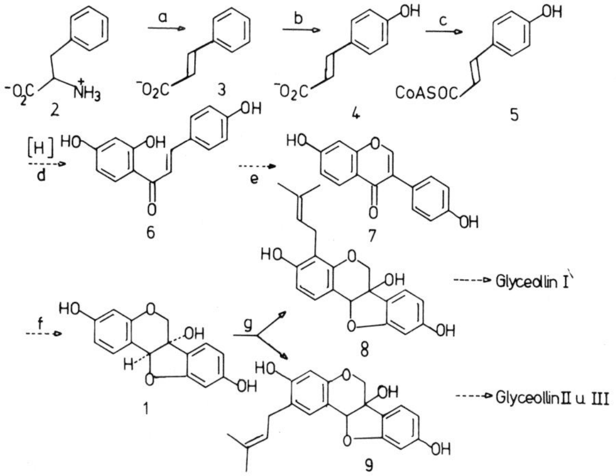

FIGURE 2. Biosynthesis of glyceollins. Known enzymes: a, phenylalanine ammonia-lyase; b, cinnamate 4-hydroxylase; c, 4-coumarate CoA: ligase; g, glycinol dimethylallyltransferase. Broken arrows indicate reactions which have not yet been found in cell-free systems.

TABLE 1
INDUCTION OF ENZYME ACTIVITIES IN SOYBEAN SEEDLINGS

Enzyme	Plant material	Induction method	lag period	Maximal activity	Remarks	References
			h after infection			
PAL	Amsoy 71 hypocotyl	mycelium of P.megasperma f.sp.glycinea	2.5	14	de novo synthesis has been proved	[3,8]
Chalcone Synthase	"	"	2.5	17	"	[3,8]
Prenyltransferase	"	"		16		[4]
Glucose-6P Dehydrogenase	"	"		16		[3]
Glutamate Dehydrogenase	"	"				[3]
HMG-CoA Reductase	Amsoy 71 cotyledons	Pmg elicitor			no induction but slight decrease in activity	[4]
Pterocarpan 6a Hydroxylase	Harosoy 63 hypocotyls	"			strong induction	[6]
Cinnamate 4 Hydroxylase	Harosoy 63 cotyledons	"			5 fold increase after 20 h	[20]
4-Coumarate: CoA Ligase	"	"			6 fold increase	[20]

have been determined. The following conclusions can be drawn from these results: a) enzymes of general phenylpropanoid metabolism (phenylalanine ammonia-lyase, cinnamate 4-hydroxylase, 4-coumarate CoA:ligase) and of the flavonoid pathway proper (chalcone synthase), as well as particular enzymes of glyceollin biosynthesis (prenyltransferase, pterocarpan 6a-hydroxylase) are induced. In contrast, 3-hydroxy-3-methyl-glutaryl coenzyme A reductase, an enzyme involved in formation of the prenyl side-chain, showed a decrease in activity after infection or elicitor treatment [4]. b) Increases in enzyme activities are not confined to enzymes involved in glyceollin biosynthesis but have been also observed for glucose 6-phosphate dehydrogenase and glutamate dehydrogenase.

In cell cultures of soybean Ebel et al. [5] have recently shown that glucan elicitor from Pmg induced a number of enzymes of the glyceollin pathway and the accumulation of glyceollin. In contrast, a xanthan from Xanthomonas campestris induced PAL and chalcone synthase but did not cause glyceollin accumulation. This result indicates that late enzymes like prenyltransferase and pterocarpan 6a-hydroxylase (Fig. 3) could have an important regulatory function in glyceollin synthesis.

The above mentioned pterocarpan 6a-hydroxylase has recently been detected for the first time in a microsomal preparation from elicitor challenged soybean cell suspension cultures [6]. It catalyzes an NADPH and dioxygen dependent 6a-hydroxylation of 3,9-dihydroxypterocarpan to 3,6a,9-trihydroxypterocarpan. The latter is a precursor for glyceollins. Optical rotatory dispersion spectra proved that the product has the natural 6aS, 11aS-configuration and that hydroxylation proceeds with retention of configuration. The 6a-hydroxylase was also found in elicitor--challenged soybean seedlings. Only very low activity of prenyltransferase(s) and 6a-hydroxylase are present in unchallenged tissue. These enzymes could therefore be rate-limiting for glyceollin synthesis.

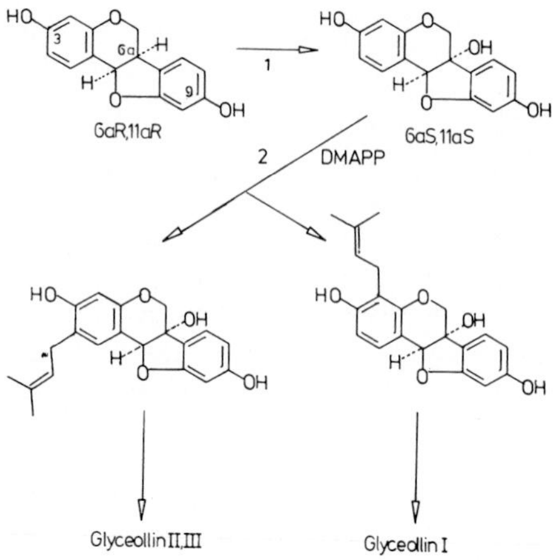

FIGURE 3:   Reaction of pterocarpan 6a-hydroxylase
and prenyltransferases in glyceollin synthesis.
Two membrane-bound prenyltransferases have been
found. DMAPP = dimethylallylpyrophosphate.

Further investigations were concerned with
the question as to whether the rapid induction of
PAL and chalcone synthase (CHS) is due to activation of inactive enzyme molecules or due to a de
novo synthesis of the enzyme protein. With both
enzymes, periods of increases in enzyme activity
coincided with high rates of enzyme synthesis, as
determined by immunoprecipitation of the in vivo
labelled proteins [3,7]. In collaboration with
E. Schmelzer and K. Hahlbrock mRNA induction in
infected soybean hypocotyls were determined by in
vitro translation in a nuclease-treated rabbit
reticulocyte lysate in the presence of L-[^{35}S]me-
thionine [8]. The translation products were immu-
noprecipitated with antisera specific for PAL,
CHS, and 4-coumarate CoA-ligase (4CL). The results
presented in Fig. 4A show rapid, transient increases in mRNA activities of the 3 enzymes.Since
the changes in mRNA translational activity can be
due to translational control of an already present
mRNA or to changes in its amount, a RNA blot hybridization experiment with a ^{32}P-labelled cDNA
probe for CHS mRNA was carried out. Fig. 4B shows

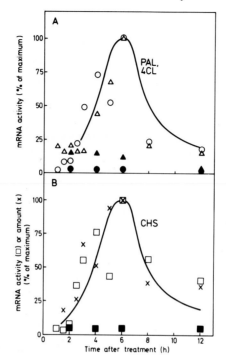

FIGURE 4. Time courses of mRNA induction in wounded (closed symbols) soybean hypocotyls and after inoculation with fungal mycelium (open symbols). A. Translational activities in vitro of PAL (o,●) and 4 CL (Δ,▲)mRNAs. B. Translational activity (□,■) and hybridizable amount (x) of CHS mRNA.

that the relative changes in the hybridizable amount of CHS mRNA after infection of soybean seedlings with the fungus coincided with the changes in CHS mRNA activity. From these results it can be concluded that glyceollin synthesis is controlled at the level of gene transcription.

CYCLIC AMP AS A POSSIBLE SECOND MESSENGER

The results discussed above raise the question as to how the plant/pathogen interaction leads to increases in the transcription of certain genes. Recently, Rosenberg et al. [9] described experiments which suggest that cAMP could play a role in mediating the production of antiviral factors in tobacco leaves infected with tobacco mosaic virus. We tested the hypothesis that cAMP could serve in an analogous role in the response of soybean to infection with P. megasperma f.sp. glycinea [10]. A radioimmunoassay (RIA) which was linear in the range between 10-400 fmole cAMP was used to quantitate cAMP in plant tissue extracts. For controls the extracts were treated with beefhart phosphodiesterase.

The amount of cAMP present in soybean seedlings infected with P. megasperma f.sp. glycinea was measured at various times after infection and compared with the amount of cAMP found in unwounded and wounded plants. Since the activities of several enzymes involved in glyceollin synthesis reach a maximum 15 h after infection one would expect to observe a rise in cAMP levels in infected plants well before this time. We were, however, unable to detect such an increase at any time after infection. In control (water-treated) or elicitor-treated soybean cell suspension cultures we were unable to detect any cAMP. According to these results cAMP is not involved in the signal cascade leading to the de novo synthesis of the enzymes mentioned above.

ACCUMULATION OF GLYCEOLLIN
IN RELATION TO HYPHAL GROWTH

To investigate the importance of phytoalexins in disease resistance it is necessary to relate its toxicity and the levels, sites and time at which it accumulates in infected tissue to the localization of fungal hyphae.

By laser microprobe mass analysis (LAMMA) it was possible to detect glyceollin at the cellular

level [11]. One hundred and fifty LAMMA spectra
taken with a 10 μm freeze microtome section of an
infected soybean cotyledon along a line perpendi-
cular to the border line of infection showed a
steep rise in glyceollin content to the border
line of infection.

Since a quantitative dertermination of glyceollin by LAMMA has not yet been developed and
since tissue preparations for LAMMA analysis pose
some difficulties, we have also developed a radioimmunoassay for glyceollin I which permits the
specific detection of glyceollin I in the range
of 1 to 100 picomoles (0.34-34 nanograms) [12].
The cross reactivity with structurally related
compounds was very low. This assay is more than
1000-fold more sensitive than other techniques
available for the quantitation of glyceollin I.

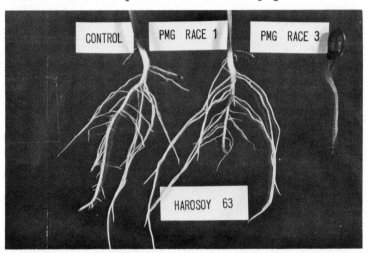

FIGURE 5.   Appearance of soybean roots after
inoculation with zoospores (about 10^4 spores) of
race 1 or race 3 of Pmg. Two day old seedlings
(cv Harosoy 63) were dip inoculated (see Fig. 6B)
for 2 h and then placed in vermiculite with sterile tap water for 4 days. Controls with water.

Furthermore we have developed an immunofluorescent technique for detection of P. megasperma f.
sp. glycinea which permits a sensitive detection
of this fungus in tissue [13]. As antigenic compound of the fungus an extracellular mannan-
-glycoprotein was used [14].

With these two methods available we set out
to analyze the accumulation of glyceollin and of
hyphal growth in the roots of soybean seedlings
(cv Harosoy 63) which had been infected with
spores of either an incompatible (race 1) or a
compatible race (race 3) of the fungus [15]. As
can be seen from Fig. 5 infection of roots of
soybean seedlings with spores shows a pronounced
difference between the incompatible (plant resistant) and the compatible (plant susceptible)
interaction.

A.   B.

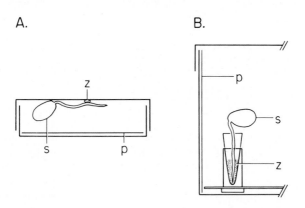

FIGURE 6A.   Droplet inoculation. Three day
old soybean seedlings were affixed to the under-
side of the lid of a petri plate and a 5 μl zoospore suspension (about 500 zoospores) was placed
between the root and the petri plate.
B.   Dip inoculation. Two day old soybean seedlings were placed in excised conical ends of
1.5 ml Eppendorf vials containing 100 μl zoospore
suspension (about 10^4 zoospores).
p = wet paper; S = cotyledon; z = zoospore suspension.

For analysis of longitutinal root sections
a droplet inoculation procedure as shown in Fig.
6A was used. Imbedded root sections were cut into
14 μm thick sections using a cryotome. Every
other section was examined by the immunohistoche-
mical technique for the presence of fungal hyphae
and each intervening section was analyzed by RIA
for glyceollin. For analysis of cross sections a
dip inoculation procedure as shown in Fig. 6B
was used. Every fourth section was analyzed for
hyphae and the three intervening sections were
combined and used for RIA analysis. The spatial
and temporal course of glyceollin accumulation
in longitudinal sections in the incompatible
(upper row) and compatible (lower row) interaction is shown in Fig. 7. Since after penetration
into the root most hyphae grow longitudinally
upwards in direction of the hypocotyl, hyphal
analysis in this case is difficult to evaluate.
The results of the glyceollin analysis allow the
following conclusions:
1.   Glyceollin which accumulates in root tissue
is over 95% isomer I (data not shown). Total

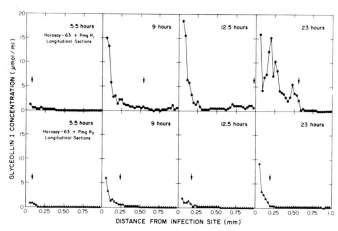

FIGURE 7.   Spacial and temporal course of
glyceollin accumulation in longitudinal root sections (14 μm) after infection with race 1 or race
3. Arrow shows position up to which the EC_{90} concentration of glyceollin was reached.

glyceollin content can therefore be quantitated
by RIA.
2.   In the early times after infection glyceollin
accumulates predominantly in the outer tissue layers (epidermis). While glyceollin accumulation
also takes place with race 3 of Pmg , the amount
is much lower and even at 12.5 and 23 h after infection no glyceollin is present in the deeper
tissue layers, although hyphae at this time have
reached the central root cylinder. This is in
sharp contrast to the result with race 1 where already at 9 h a considerable amount of glyceollin
is present at about 0.5 mm from the infection site.
     The analysis of cross sections after infec-
tion by dip inoculation with race 1 is shown in
Fig. 8. Whereas in whole root extracts glyceollin
I could be detected as early as two hours after
infection, significant phytoalexin concentration
in cross sections was detected after 5h. At this
time the glyceollin concentration reached is about
one fourth of EC_{90} (EC_{90} in vitro 0.6 µmol/ml).At
8 h the maximal penetration depth of hyphae from
race 1 has been reached. The borderline of infection corresponds to the borderline of glyceollin
accumulation which exceeds EC_{90}. At 11 and 14 h
glyceollin has further increased and was also detected in advance of the hyphae. This could mean
that a systemic signal preceeds the advancing
hyphae.
     In sharp contrast to these results in the in-
compatible reaction no glyceollin could be detected in cross sections at any time after infection
with race 3. Obviously the amount of glyceollin
which accumulates in the outer root layer (Fig.7)
is in a cross section below the detection limit
of the RIA.
     In summary, these results are strong support
for the assumption that glyceollin is an important
factor in the defense of resistant soybean cultivars against P.megasperma f.sp. glycinea. This is
corroborated by results of experiments in which
the biosynthesis of glyceollin was almost completely inhibited by L-aminooxyphenylpropionic acid
-a PAL inhibitor- resulting in loss of resistance
of Harosoy 63 against race 1 of Pmg [16].

This can bee seen from Fig. 9. At a L-AOPP concentration higher than approximately 500 μmol l^{-1} the glyceollin concentration declines drastically with a concomitant increase in infected tissue. The hypocotyls appeared dark green and water-soaked and eventually collapsed.

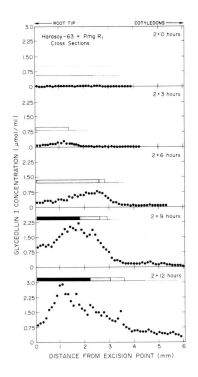

FIGURE 8. Spacial and temporal course of glyceollin accumulation in cross sections of soybean roots after infection with race 1. The extent of fungal colonization is indicated above the values for glyceollin by a broken line (single hyphae) solid line (few hyphae), open bar (weak colonization), stipled bar (stronger colonization), and solid bar (very strong colonization).

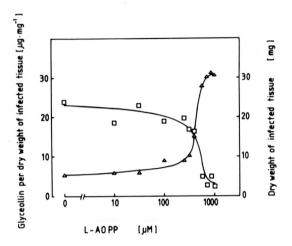

FIGURE 9.   Dependence of glyceollin content
(□) of soybean seedlings and amount of infected
tissue (Δ) on concentration of L-AOPP. Seedlings
were infected for 24 h with race 1 of Pmg by surface inoculation. From [16].

DISCUSSION

It can be estimated that about twelve enzymes
are required for the synthesis of glyceollin I,
starting from L-phenylalanine, malonyl-CoA, and
dimethylallylpyrophosphate. The plant therefore
invests considerable energy in the defense reaction. That prenylation of glycinol (Fig. 1) is of
advantage to the plant can be concluded from the
observation that it increases the toxicity against
Cladosporium cucumerinum and Ervinia carotovora
[17]. To gain deeper insight into the control
mechanisms underlying glyceollin accumulation more
knowledge on the later enzymes in this pathway are
necessary. Three of the late enzymes which are
known (6a hydroxylase and two prenyltransferases)
are membrane-bound enzymes and it could well be
that all stages from the flavanone stage on occur
on membranes with the possibility of metabolic
channeling.
The result that glyceollin synthesis is apparently controlled at the level of gene transcrip-

tion could ultimately lead to the elucidation of the molecular events that couple plant-pathogen recognition to gene activation. Is there a receptor for elicitor molecules [18], is a second messenger involved between receptor recognition and gene activation or could an alicitor-receptor complex become internalized and be directly involved in translational control, is the race specificity associated with the incompatible or the compatible interaction? These and many other questions can be asked and are now amenable to experimental investigations. The root system which we have established is probably the best system to study the natural infection process. The dramatic difference in the appearance of the soybean seedlings in the incompatible and compatible interactions (Fig. 5) must be explained,as well as, the very pronounced difference in glyceollin accumulation. One should also keep an open eye for other possible defences in soybean, such as, for example, callose formation [19].

There remain, therefore, enough problems for enthusiastic younger and older scientists to solve. The complexity of the problems involved make the combin ed effort of scientists from different fields strongly desirable.

REFERENCES

1. Baily JA, Mansfield JW (eds) (1982) "Phytoalexins", Glasgow: Blackie.
2. Moesta P, Grisebach H (1981) Arch Biochem Biophys 212:462-467.
3. Börner H, Grisebach H (1982) Arch Biochem Biophys 217:65-71.
4. Leube J, Grisebach H (1983) Z Naturforsch 38c:730-735.
5. Ebel J, Schmidt WE, Loyal R (1984) Arch BiochemBiophys 232: in press.
6. Hagmann M, Heller W, Grisebach H (1984) Eur J Biochem: in press.
7. Hille A, Purwin C, Ebel J (1982) Plant Cell Reports 1:123-127.

8. Schmelzer E, Börner H, Grisebach H, Ebel J, Hahlbrock K (1984) FEBS Lett: in press.
9. Rosenberg N, Pines M, Sela I (1982) FEBS Lett 137:105-107.
10. Hahn MG, Grisebach H (1983) Z Naturforsch 38c:578-582.
11. Moesta P, Seydel U, Lindner B, Grisebach H (1982) Z Naturforsch 37c:748-751.
12. Moesta P, Hahn MG, Grisebach H (1983) Plant Physiol 73:233-237.
13. Moesta P, Grisebach H (1983) Eur J Cell Biol 31:167-169.
14. Ziegler E, Pontzen R (1982) Physiol Plant Pathol 20:321-331.
15. Hahn MG, Grisebach H, in preparation.
16. Moesta P, Grisebach H (1982) Physiol Plant Pathol 21:65-70.
17. Zähringer U (1979) Doctoral thesis Freiburg.
18. Yoshikawa M, Keen NT, Wang M-CH (1983) Plant Physiol 73:497-506.
19. Köhle H, Young DH, Kaus H (1984) Plant Science Lett 33:221-230.
20. Ebel J (1979) Elicitor-induced phytoalexin synthesis in soybean. In Luckner M, Schreiber K (eds): Regulation of secondary product and plant hormone metabolism" Vol. 55 FEBS, Oxford: Pergamon Press.

Cellular and Molecular Biology of Plant Stress, pages 291–302
© 1985 Alan R. Liss, Inc.

THE INDUCTION OF DISEASE RESISTANCE BY HEAT SHOCK

Bruce A. Stermer and Raymond Hammerschmidt

Department of Botany and Plant Pathology,
Michigan State University,
East Lansing, Michigan 48824-1312

ABSTRACT   This paper discusses the first use of heat
shock to induce disease resistance by triggering
plant responses to stress. Immersion of cucumber
seedlings in a 50 C water bath for 40 seconds prior
to inoculation with Cladosporium cucumerinum resulted
in a 60% reduction in disease symptoms. Resistance
developed by 24 to 30 hours after heat shock and was
effective for at least 2 days. Associated with the
induction of disease resistance were plant responses
to stress that are also implicated in disease
resistance in cucurbits. The earliest response
detected in cucumber seedlings after heat shock was
a rapid increase in ethylene production. Synthesis
of 1-aminocyclopropane-1-carboxylic acid, an ethylene
precursor, was also higher in heat shocked tissues.
An increase in the soluble peroxidase isozymes
associated with cucumber cell walls were closely
correlated with the increase in resistance. In
addition, heat shock increased accumulation of
extensin (a hydroxyproline-rich glycoprotein) in
the cell wall. A possible mechanism for resistance
induced by heat shock is discussed.

INTRODUCTION

Exposure of plants to certain biological and environmental stresses may alter the outcome of a subsequent
plant-pathogen interaction. Stress of some groups of
plants caused by prior infection can render the once
susceptible plants resistant to attack from many different
pathogens. The best example of this phenomenon occurs in

cucurbits. Localized infection by viruses, bacteria, or
fungi can enhance the ability of cucumbers, muskmelons, and
watermelons to resist many different diseases (1). Tightly
correlated with the induction of disease resistance in
cucurbits is an increase in soluble, cell-wall-associated
peroxidase isozymes (2,3).

Changes in the epidermal cell wall appear to be
involved in the mechanism of induced resistance in
cucurbits against fungi. Hammerschmidt and Kuc (4) found
that the epidermal cell walls of plants with induced
resistance were lignified more rapidly and to a greater
extent than were controls in response to attack by
Cladosporium cucumerinum and Colletotrichum lagenarium; the
enhanced lignification was associated with a reduction in
successful penetrations into the tissue by the fungi. Both
cucumbers and muskmelons respond to pathogen attack with an
increase in the amount of extensin, a hydroxyproline-rich
glycoprotein that is covalently bound in the plant cell
wall (5,6). A recent study has shown that an enhanced
accumulation of bound extensin in cucumber cell walls is
associated with resistance to C. cucumerinum (6).
Ethylene, a growth regulator often produced by plants in
response to stress, was reported to increase the level of
bound extensin in the cell walls of muskmelons and also to
increase resistance to C. lagenarium (7). However,
ethylene did not induce resistance in cucumbers to the same
pathogen (8).

Heat shock has been used by many researchers to
manipulate the expression of resistance to fungi.
Generally, heat shock applied prior to inoculation has
prevented or delayed plant responses associated with
resistance. Such heat shock can inhibit cell wall
alterations such as lignification (9) and papilla
formation (10), and also reduce phytoalexin production
(11) and hypersensitive cell death (12,13). However, heat
shock can also prevent or delay susceptibility to fungi
that produce host-selective toxins; the shock reduces
plant sensitivity to the toxin (14,15,16). The common
denominator in all these effects of heat shock on disease
resistance appears to be the temporary halt of many active
processes. Thus, depending on which process requires
active metabolism, heat shock may block resistance or
susceptibility. Heat shock also has effects at the
molecular level including the de novo synthesis of "heat
shock proteins" and the reduction of normal protein
synthesis (17).

Heat shock of cucumber seedlings resistant or susceptible to C. cucumerinum immediately prior to inoculation with this pathogen allowed for the penetration and growth of the fungus in tissues (9). However, the induced susceptibility was temporary; inoculation of normally resistant seedlings 24 hours after the heat shock showed that resistance had recovered (9). Unexpectedly, the normally susceptible seedlings developed resistance to the pathogen within 24 hours after heat shock (18). In this paper, we will describe studies of heat-shock-induced resistance against C. cucumerinum in cucumber seedlings.

RESULTS

Induction of Disease Resistance by Heat Shock

Cucumber seedlings of the cultivar Marketer (normally susceptible to Cladosporium cucumerinum) were grown for five days in the dark in germination paper at 22 C. Seedlings were immersed in a 50 C water bath for 40 seconds and then returned to the germination paper for later inoculation and incubation at 22 C. Seedlings were inoculated with C. cucumerinum at various times after heat shock. Results showed a steady rise in resistance for a 12 hour period after heat treatment (18) (Fig. 1). The actual onset of resistance was 12 to 18 hours later than that indicated by the time of inoculation because the resistance continues to develop as the fungus germinates and attempts to penetrate (penetration occurs 12 to 18 hours after inoculation). Therefore, by 24 to 30 hours after heat shock the seedlings have developed their maximum disease resistance. Resistance induced by heat shock was expressed as a reduction in fungal and symptom development (18). Seedlings were rated for disease by a method modified from that of Hammerschmidt et al. (19). The disease rating for heat shocked plants was compared to that for control plants, and heat-shocked-induced resistance was expressed as percent reduction. The apparent reduction in resistance two days after heat shock is actually due to the development of maturity related resistance in controls; heat shocked seedlings remained resistant over the four day period. Heat shock also induced disease resistance in three other cultivars of cucumber and induced resistance in green cucumber plants (18).

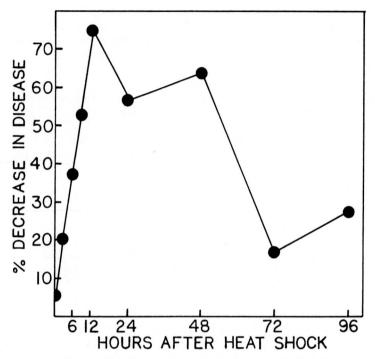

FIGURE 1. Development of resistance with time following heat shock. Plants were inoculated with Cladosporium
cucumerinum (3 X 10⁵ spores ml⁻¹) at the indicated times
after heat treatment (40 seconds at 50 C), and disease
ratings were made four days later. The disease ratings
were expressed as the percent decrease in disease rating
of heat shocked seedlings compared to unshocked control
seedlings. Data represent the mean for three experiments
(18).

Changes in Cucumber Cell Walls After Heat Shock

Etiolated seedlings were given a heat shock (40
seconds at 50 C) and the apical two cm were excised and the
cotyledons removed. The hypocotyl segments were ground in
0.5 M sucrose-phosphate buffer (.01 M, pH 6.0), centrifuged
at 10,000 xg for 20 min and the supernatant assayed for
peroxidase, using quaiacol as the hydrogen donor. Similar
to disease resistance, peroxidase activity rose quickly 24
hours after heat shock (18) (Fig. 2). Electrophoretic
analysis of soluble peroxidase showed a large increase in

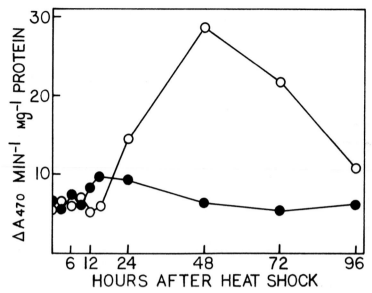

FIGURE 2. Time course for the enhancement of
peroxidase activity after heat shock. Cucumber seedlings
were heat shocked (40 seconds at 50 C) and apical
hypocotyl samples were excised at intervals. Peroxidase
activity was measured colorimetrically using guaiacol as
the hydrogen donor. Data represent the mean for two
experiments (● unshocked control; O heat shocked) (18).

the fast moving anodic isozymes (18). Other work indicates
that the fast moving anodic isozymes are located in the
cell wall (2,3). These isozymes previously were shown to
increase systemically in cucumbers with localized infection
by Colletotrichum lagenarium, i.e., these peroxidase
isozymes are associated with induced resistance in cucumber
plants (2). Further studies using heat shock at various
temperatures and durations showed a close correlation
between the level of peroxidase activity and the level of
induced resistance (18).
     Heat shock also increased the accumulation of bound
extensin, the hydroxyproline-rich glycoprotein of plant
cell walls. The apical two cm of etiolated seedlings were
excised, the cotyledons removed and the hypocotyl segments
ground with liquid N_2 and extracted with 1.0 M NaCl to
remove ionically bound substances. After acid hydrolysis

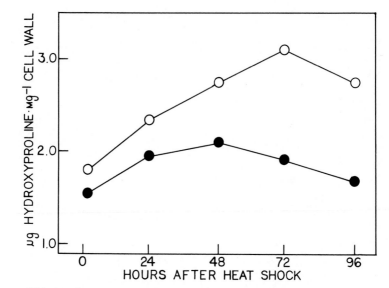

FIGURE 3.   Time course for the accumulation of cell
wall hydroxyproline after heat shock. Cucumber seedlings
were heat shocked (40 seconds at 50 C) and apical hypocotyl
samples were excised at intervals. The hydroxyproline
content of cell walls was estimated colorimetrically (6).
Data represent the mean for three experiments (● unshocked
control; ○ heat shocked).

of the cell wall powder, extensin content was estimated by
a colorimetric assay for hydroxyproline (6). By comparison
with unshocked controls, we found that there was a gradual
increase in the amount of hydroxyproline covalently bound
to the cell wall (Fig. 3). This indicates a likewise
accumulation of extensin bound in the cell wall.

Ethylene Production by Heat Shocked Cucumber Seedlings

The first response detected in cucumber seedlings
after heat shock was a rapid increase in ethylene
production. Heat shocked and control etiolated seedlings
were held in sealed flasks and ethylene accumulation was
measured. The rate of accumulation was much higher in
flasks containing the heat shocked seedlings (Fig. 4).

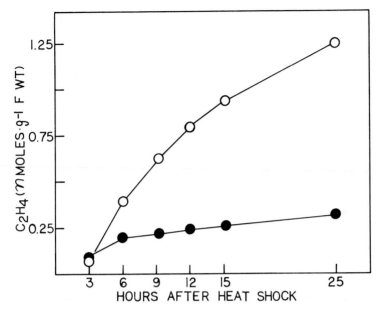

FIGURE 4. Effect of heat shock on ethylene production by cucumber seedlings. Seedlings were heat shocked (40 seconds at 50 C) and four seedlings were placed in each 50 ml sealed flask. One ml gas samples were removed at intervals and the ethylene concentration was determined by gas chromatography (20). Adjustments were made in subsequent measurements for the ethylene removed. Data represent the mean for two experiments each with four replicates per treatment (● unshocked control; O heat shocked).

Synthesis of 1-aminocyclopropane-1-carboxylic acid (ACC), an ethylene precursor, also was higher in heat shocked tissue. Amounts of ACC in heat shocked tissues were more than double those seen in control tissues by 6 hours after the shock (Table 1).

DISCUSSION

Previously, heat shock was used to modify disease resistance in plants, but only the immediate effects were determined, i.e., the inhibition of defense responses or toxin sensitivity. These temporary effects can be

TABLE 1

EFFECT OF HEAT SHOCK ON ACC LEVELS IN CUCUMBER SEEDLINGS

| Treatment | ACC (nmoles·g^{-1} fresh wt)[a] | |
| | Hours after heat shock | |
	3	6
Heat shock[b]	1.44 + .19	2.58 + .20
Unshocked control	1.26 + .11	1.20 + .22

[a]Apical hypocotyl samples were excised at three or six hours after heat shock, homogenized, and centrifuged. The supernatant was used for determination of levels of 1-aminocyclopropane-1-carboxylic acid (21) (data are means +SE).

[b]40 seconds at 50 C.

attributed to the transient reduction in protein synthesis observed after a plant is heat shocked (17). However, in this paper we discuss the first use of heat shock to induce a time dependent change in disease resistance, one that is associated with active metabolism of the plant. Along with the induction of disease resistance, heat shock also produced changes in seedlings that are commonly associated with other forms of stress.

A brief heat shock induced resistance in cucumber seedlings to disease caused by Cladosporium cucumerinum. Correlated with the rise in disease resistance was a similar increase in soluble peroxidase activity, especially in the isozymes associated with the plant cell wall. Also, heat shock led to an increase in the amount of hydroxyproline-rich glycoprotein (extensin) bound in the cell wall. However, the first observed active response to heat shock by cucumber seedlings was an increase in production of ethylene.

The in vivo production of ethylene by muskmelon plants has been implicated in the regulation of extensin accumulation in the cell wall (22). In addition, ethylene has been reported to increase peroxidase activity in various plants (23,24). Thus it is possible that ethylene induced by heat shock may be responsible for increased levels of peroxidase activity and bound extensin in shocked seedlings. The early increase in ethylene production after heat shock is consistent with such a role in peroxidase and extensin regulation.

The mechanism by which heat shock induces resistance is not known. Cell-wall-bound extensin accumulates more rapidly in heat shocked plants; this glycoprotein is also associated with resistance to C. cucumerinum in cucumber (6). Therefore, extensin accumulation may be involved in heat-shock-induced resistance. Resistance induced in cucumber by a prior infection by Colletotrichum lagenarium is associated with enhanced lignification of cell walls at points of attempted fungal penetrations (4). However, no evidence for enhanced lignification has been seen in heat shocked seedlings (Stermer and Hammerschmidt, unpublished).

Peroxidase activity is closely correlated with induced resistance whether brought on by prior localized infection or by prior heat treatment. For each method of inducing resistance the increase in peroxidase activity is at the onset of resistance, about 24 hours for heat shock and 72 hours for prior C. lagenarium inoculation. However, resistance develops at about 24 hours after stress by either method because it takes approximately 48 hours for C. lagenarium to germinate and form appressoria (i.e., to stress the plant). Furthermore, the peroxidase activity induced by C. lagenarium infection or by heat shock is in the same cell-wall-associated isozymes (18).

Crosslinking of extensin molecules in the cell wall could result from peroxidase activity, and might be a factor in resistance induced by heat shock. Recent work has shown that extensin is crosslinked in plant cell walls by the coupling of tyrosine residues, probably by peroxidase (25). The data suggests that perhaps a heat-shock-induced increase in ethylene production could regulate the accumulation of extensin and peroxidase in the cell wall. The subsequent crosslinking of extensin in the cell wall by peroxidase may confer resistance to attack by a pathogen. Of course other mechanisms of resistance should not be ruled out.

Heat shock could be a useful tool in the study of induced resistance and other plant responses linked to stress. Further studies with heat shock should increase our knowledge on the regulation of ethylene, peroxidase, and extensin, and illuminate their role in disease resistance.

ACKNOWLEDGMENTS

We wish to thank Dr. Derek Lamport for his help with
the hydroxyproline analyses and Dr. Hans Kende for his help
with ethylene measurements. Michigan Agricultural
Experiment Station Journal Article Number 11231.

REFERENCES

1. Kuc J (1982). The immunization of cucurbits against
   fungal, bacterial, and viral diseases. In Asada Y,
   Bushnell WR, Ouchi S, Vance CP (eds): "Plant
   Infection", Tokyo: Japan Scientific Societies Press,
   and New York: Springer-Verlag, p 137.
2. Hammerschmidt R, Nuckles EM, Kuc J (1982). Association
   of enhanced peroxidase activity with induced systemic
   resistance of cucumber to Colletotrichum lagenarium.
   Physiol Plant Pathol 20:73.
3. Smith JA, Hammerschmidt R (1984). Association of
   enhanced peroxidase activity with induced resistance
   of muskmelon and watermelon. Phytopathology 74:(in
   press).
4. Hammerschmidt R, Kuc J (1982). Lignification as a
   mechanism for induced systemic resistance in cucumber.
   Physiol Plant Pathol 20:61.
5. Esquerre-Tugaye MT, Mazau D (1974). Effect of a fungal
   disease on extensin, the plant cell wall glycoprotein.
   J Exp Bot 25:509.
6. Hammerschmidt R, Lamport DTA, Muldoon EP (1984). Cell
   wall hydroxyproline enhancement and lignin deposition
   as an early event in the resistance of cucumber to
   Cladosporium cucumerinum. Physiol Plant Pathol 24:43.
7. Esquerre-Tugaye MT, Lafitte C, Mazau D, Toppan A, Touze
   A (1979). Cell surfaces in plant-microorganism
   interactions II Evidence for the accumulation of
   hydroxyproline-rich glycoproteins in the cell wall of
   diseased plants as a defense mechanism. Plant Physiol
   64:320.
8. Jenns AE (1979). Induction of systemic resistance
   in cucurbits to anthracnose and tobacco necrosis
   virus. Ph.D. thesis, Univerity of Kentucky, Lexington,
   p 91.
9. Stermer BA, Hammerschmidt R (1982). Effects of
   heat-shock on varietal and nonhost resistance in
   cucumbers. Phytopathology 72:969.

10. Aist JR, Israel HW (1977). Effects of heat-shock inhibition of papilla formation on compatible host penetration by two obligate parasites. Physiol Plant Pathol 10:13.
11. Jerome SMR, Muller KO (1958). Studies on phytoalexins II Influence of temperature on resistance of Phaseolus vulgaris towards Sclerotinia fructicola with reference to phytoalexin output. Aust J Biol Sci 11:301.
12. Tomiyama K (1967). Further observations on the time requirement for hypersensitive cell death of potatoes infected by Phytophthora infestans and its relation to metabolic activity. Phytopathol Zeit 58:367.
13. Hazen BE, Bushnell WR (1983). Inhibition of the hypersensitive reaction in barley to powdery mildew by heat shock and cytochalasin B. Physiol Plant Pathol 23:421.
14. Bronson CR, Scheffer RP (1977). Heat- and aging-induced tolerance of sorghum and oat tissues to host-selective toxins. Phytopathology 67:1232.
15. Byther RS, Steiner GW (1975). Heat-induced resistance of sugarcane to Helminthosporium sacchari and helminthosporoside. Plant Physiol 56:415.
16. Otani H, Nishimura S, Kohmoto K (1974). Nature of specific susceptibility to Alternaria kikuchiana in Nijisseiki cultivar among Japanese pears III Chemical and thermal protection against effect of host-specific toxin. Ann Phytopathol Soc Japan 40:59.
17. Key JL, Lin CY, Chen YM (1981). Heat shock proteins of higher plants. Proc Natl Acad Sci USA 78:3526.
18. Stermer BA, Hammerschmidt R (1984). Heat shock induced resistance to Cladosporium cucumerinum and enhance peroxidase activity in cucumbers. Physiol Plant Pathol (accepted for publication).
19. Hammerschmidt R, Acres S, Kuc J (1976). Protection of cucumber against Colletotrichum lagenarium and Cladosporium cucumerinum. Phytopathology 66:790.
20. Kende H, Hanson AD (1976). Relationship between ethylene evolution and senescence in morning-glory flower tissue. Plant Physiol 57:523.
21. Lizada CC, Yang SF (1979). A simple and sensitive assay for 1-aminocyclopropane-1-carboxylic acid. Anal Biochem 100:140.
22. Toppan A, Roby D, Esquerre-Tugaye MT (1982). Cell surfaces in plant-microorganism interactions III in vivo effect of ethylene on hydroxyproline-rich glycoprotein accumulation in the cell wall of diseased

plants. Plant Physiol 70:82.

23. Birecka H, Briber KA, Catalfamo JL (1973). Compara-
     tive studies on tobacco pith and sweet potato root
     isoperoxidases in relation to injury, indoleacetic
     acid, and ethylene effects. Plant Physiol 52:43.

24. Ridge I, Osborne DJ (1970). Hydroxyproline and
     peroxidases in cell walls of Pisum sativum:
     regulation by ethylene. J Exp Bot 21:843.

25. Fry SC (1982). Isodityrosine, a new cross-linking
     amino acid from plant cell wall glycoprotein.
     Biochem J 204:449.

Cellular and Molecular Biology of Plant Stress, pages 303–318
© 1985 Alan R. Liss, Inc.

EXPRESSION OF LATENT GENETIC INFORMATION
FOR DISEASE RESISTANCE IN PLANTS[1]

Joseph Kuć

Department of Plant Pathology, University of Kentucky,
Lexington, Kentucky 40546

ABSTRACT Plants developed effective mechanisms for
resistance to all infectious agents in their
environment to survive the selection pressure of
evolution. Work in our laboratory provides biological
and chemical evidence to support the hypothesis that
all plants contain the genetic potential for
resistance mechanisms to fungal, bacterial and viral
diseases. The determinant of resistance would then be
the speed and magnitude with which this potential is
expressed and the effect of the environment on the
activity of gene products.

Working with green bean, tobacco, cucumber, watermelon
and muskmelon, we demonstrated that plants are
systemically immunized against diseases caused by
fungi, bacteria and viruses using restricted infection
with fungi, bacteria, or viruses. Immunization
protects cucumber, watermelon, muskmelon and tobacco
throughout the season, and a single immunization
protects cucumber against at least 12 unrelated
diseases.

Multiple mechanisms for the containment of infectious
agents are rapidly activated after infection in
immunized plants. Plants are sensitized to respond
rapidly as a result of immunization, but responses are
most apparent after challenge with the infectious
agent.

[1]The author's work reported in this paper was supported in
part by grants from the R. J. Reynolds Corp., Rockefeller
Foundation and Graduate School of the University of
Kentucky.

Immunization against blue mold also increased the growth of Burley tobacco. Immunized unchallenged plants at the start of flowering were 40% taller, had a 30% increase in fresh weight, 4-6 more leaves and 10-25% increase in marketable yield than control plants.

INTRODUCTION

Things are seldom what they seem. Preparation for this talk at Keystone reminded me of an incident which occurred several years ago during another trip to the Rocky Mountains of Colorado. High up in the mountains my family and I came across a field of lovely white flowers. The plants were no taller than 6 inches, but the flowers were at least two inches wide and held well above the foliage. Despite protests from my family, I did succeed in digging up one plant and transplanting it to my garden in Kentucky. The plant survived the winter, but during the following year it developed into an ugly, scraggly plant, over three feet high, and it had small insignificant flowers hidden by the foliage. Who could have imagined the growth potential for that Rocky Mountain wild flower while growing in its original habitat? It would clearly have been described as a dwarf alpine plant with attractive white flowers. This remarkable transformation in plant growth and development was accomplished not by the transfer of genetic information "from elephant to flea" but rather by the expression of a latent genetic potential and the environmental regulation of the activity of gene products. My story is not intended to denigrate the science of genetic engineering or the transfer of genetic information from "elephants to fleas" or for that matter, "fleas to elephants." It is clear, however, that "things are seldom what they seem." It has also been my experience in science that "things are seldom as simple as they seem." Now, I will proceed to the subject matter which I was invited to discuss at this symposium.

The cell envelopes carrying genetic information, whether they are animal or plant, are well suited for the survival and propagation of the genetic information they carry (1). Though diversity, order, and regulation are characteristic of life, change is necessary for survival. Life does not have a "status quo." In order to survive, an organism must have developed effective mechanisms to minimize damage, caused by other life forms and the environment, to its envelope and content of genetic information. It follows, that all plants have highly effective mechanisms for resistance to disease caused by infectious agents. If not, they would not have survived. It also follows that these mechanisms are subject to change and modification. The changes and modification, though perhaps slight, are sufficient for survival.

It seems a paradox, therefore, that plants growing in the wild are seldom, if ever, free of disease. The goal of producing disease-free plants originates with man and is not the rule in our ecosystem. Man has also introduced other concepts into our ecosystem which have direct bearing on plant disease and survival – the emphasis on uniformity and on maximum yield under optimal conditions for a crop's growth. Clearly, man is interested in the survival of plants only as their survival effects the survival of man, and mechanisms effective for the survival of plants may not be adequate for the intensive high-yield, high-quality and monocultural requirements placed on modern agriculture. Man has not, however, ignored the inherent mechanisms for plant disease resistance in the process of maximizing his own survival. Resistant plants are our number one defense against disease, and they are our only effective defense against diseases caused by bacteria and viruses. Until relatively recently, however, the manufacturers of pesticides seem to have established as goals the production of disease or insect-free plants. Pesticides generally have been developed with the single purpose of directly killing the disease-causing organism. Are these goals realistic? Can we make better use of disease resistance mechanisms present in all plants as a practical means for disease control?

RESISTANCE AS A FACTOR OF
DIFFERENTIATION, AGE OR TEMPERATURE

Disease resistance and susceptibility are not absolute. They are subjective evaluations within certain parameters of growth and environment. The first three or four expanding leaves of apple shoots are susceptible to apple scab, caused by <u>Venturia inaequalis</u>, and all others are resistant (2-4). From a practical viewpoint, the reaction of leaves at the growing point is determining resistance or susceptibility. Nevertheless, all apple trees have resistance to apple scab caused by all races of V. inaequalis if we consider the older leaves. The resistance is even evident in apple varieties the grower would call completely susceptible or lacking genes for resistance. Many similar situations exist. Cells of etiolated green bean hypocotyls close to the growing point are susceptible to anthracnose caused by Colletotrichum lindemuthianum. Resistance increases as the root zone is approached, and tissue near and in the root zone, as well as the roots themselves, are highly resistant (5-8). This is even true of bean cultivars which breeders or growers might consider lack genes for resistance to C. lindemuthianum.

Cucumber seedlings of cultivars susceptible to scab, caused by Cladosporium cucumerinum, are highly susceptible at the growing point and for a distance of approximately 0.5 cm below the cotyledon. Tissue beneath this zone is highly resistant (9). Tissue sections taken from the susceptible zone become macerated within one hour when placed in culture filtrates of the fungus or solutions containing a combination of proteolytic, pectinolytic and cellulolytic enzymes, whereas sections taken from the resistant zone remain intact 24 h after treatment. Temperature also has a profound effect on resistance. All cucumber cultivars are resistant to scab at temperatures of 22-24C, and some are at 20-22C. Cultivar resistance is determined at 17-20C which is the optimum temperature for scab. The pathogen grows and sporulates well in vitro at 22-24C, and spores germinate and the pathogen penetrates the foliage at these temperatures.

The fungi Physalospora obtusa, Botryosphaeria ribis, and Glomerella cingulata cause disease on ripening apple fruit late in the season (10,11). The transition from immunity to susceptibility occurs within 1 to 2 weeks in the field and the period of change varies from year to year depending on climatic conditions. The fungi are rapidly contained when they are introduced into wounds made in green apple fruit in June. However, when introduced into ripening fruit late in July, or perhaps early August, the fungi spread through the tissue and a rapidly expanding area of rot is evident within 2-3 days. Fungitoxic compounds were not detected in resistant apples and resistance was apparently due to the inability of the fungi to degrade cell walls, possibly because of the presence of pectin-protein-polyvalent cation complexes.

Leaves are generally immune to root pathogens and roots are generally immune to leaf pathogens even when the pathogens are introduced directly into tissues. The picture is made even more complex by the observation that some pathogens penetrate a host directly through the cuticle, and others penetrate through stomata or wounds.

EXPRESSION OF RESISTANCE BY
CHEMICAL AGENTS

Reports are also available of the effectiveness of chemical compounds, inactive as anti-microbial agents in vitro, as elicitors of resistance in susceptible plant hosts. The D and DL isomers of phenylalanine markedly increased the resistance of 7 apple varieties to the apple scab disease caused by Venturia inaequalis (12). A solution of the amino acid was infused into growing shoots through leaf midribs for a 24-hr period, and leaves above the points of infusion were challenged with the fungus. Solutions of D-alanine, DL-α-amino-n-butyric acid, D- and DL-leucine also elicited resistance, whereas all L isomers were inactive. In subsequent studies, α-aminoisobutyric acid (AIB) was reported to markedly increase resistance of apple foliage to scab (13). The acid was not fungitoxic at a concentration as high as 0.40 M in vitro and AIB was stable in cultures of the fungus and in apple foliage.

Injection of cell-free sonicates of virulent and avirulent isolates of Erwinia amylovora, the incitant of fire blight, or the nonpathogen, Erwinia herbicula, into "Bartlett" pear seedlings protected them against fire blight (14). Sonicates were not inhibitory to the pathogen and protection was lost when nucleic acids were precipitated from the preparations, whereas nucleic acid precipitated from sonicates protected (15). Deoxyribonuclease, but not ribonuclease, destroyed activity of nucleic acids isolated from sonicates. Activity was associated with the deoxyribonucleic acid (DNA) fraction from linear log sucrose gradients. DNA prepared by the Marmur technique protected more effectively than DNA prepared from sonicates.

Fire blight symptoms were apparent on all control seedlings, from which the radicle was excised, 4 days after they were dipped into a suspension of E. amylovora, and a few days later they were entirely destroyed. In contrast, 12 days after challenging seedlings treated with DNA, c.

50% had fire blight symptoms and only 20 to 30% of those
with symptoms were entirely destroyed. The remaining
seedlings with symptoms showed necrosis only at the
hypocotyl base and this was often dry in appearance as
compared to the normal water-soaked appearance of infected
tissues. DNA-protected seedlings, from which the radicle
had been excised, were often observed to form roots. Those
which did were transplanted into vermiculite and soil, at
high relative humidity for 1 week, and then challenged by
injecting 10^2 cells c. 0.5 cm below the cotyledons.
Control seedlings showed fire blight symptoms within 2 days
of challenge, whereas seedlings protected with DNA showed
no symptoms for at least 1 week. After 1 week, restricted
symptoms developed on seedlings treated with DNA. This
indicated the lasting nature of protection with DNA and
that protection occurs even at sites removed from the point
of initial DNA application. It is unlikely that protection
by DNA is dependent upon the specific source or nucleotide
sequence of the molecules since DNA from various bacterial
sources was active and resistance to some bacterial and
viral diseases in plants has been elicited by anionic
substances such as polyacrylic and salicylic acids.

Other chemical agents which elicit disease resistance
but are not anti-microbial in vitro include: 2,2-dichloro-
3,3-dimethyl-cyclopropane carboxylic acid, aluminum tris-
0-ethyl phosphanate and 3-allyloxy-1,2-benzisothiazole-1,1-
dioxide (16,17,18). At least in part, all of the chemicals
cited in this section appear to act by sensitizing the
plant to rapidly activate resistance mechanisms when
challenged by the pathogen. Some of the resistance-related
mechanisms activated by the chemicals after challenge,
e.g., phytoalexin accumulation, lignification, enhanced
peroxidase activity, are also common to plants systemically
protected by restricted infection. The data suggest that a
metabolic perturbation, resulting in a low level of rather
persistent stress, is responsible for the sensitization.

PLANT IMMUNIZATION

Green Bean

Protection of green bean against disease caused by C.
lindemuthianum was first demonstrated by infecting the
hypocotyls of bean cultivars, resistant to some but not all
races of the fungus, with cultivar nonpathogenic races of

the organism. Although the infected plants developed
minute lesions typical of a hypersensitive response, they
were locally and systemically protected from disease caused
by subsequent challenge with cultivar pathogenic races of
the fungus (5-8). It was apparent, therefore, that if a
cultivar had resistance to one race of a pathogen it could
be made resistant to all races of the pathogen. Further
experiments in our laboratory demonstrated that even
cultivars of bean susceptible to all races of C.
lindemuthianum were locally and systemically protected
against the disease by infection with Colletotrichum
lagenarium, a pathogen of cucurbits, prior to infection
with the pathogen of bean. C. lagenarium was equally
effective in inducing systemic protection in cultivars
susceptible to all races of C. lindemuthianum as in
cultivars resistant to one or more races of the pathogen.
The challenge fungus developed in tissue protected by C.
lagenarium to the same extent as in the tissue protected by
cultivar nonpathogenic races of C. lindemuthianum.
Protection elicited by C. lagenarium and cultivar
nonpathogenic races of C. lindemuthianum protected against
all races of C. lindemuthianum. These observations suggest
that the same mechanisms may be involved in protection
elicited by both fungi. Additional evidence for the
ability to activate a resistance mechanism in cultivars
susceptible to all races of C. lindemuthianum was obtained
by heat-attenuating cultivar pathogenic races of the
pathogen in host tissue prior to the expression of
symptoms. Such plants were protected against disease
caused by infection with the same or other cultivar
pathogenic races of the fungus. This experiment also
suggested that fungal components or metabolites, even in
cultivar pathogenic races, can elicit resistance. It was
evident, therefore, that resistance mechanisms can be
activated in bean cultivars resistant to some and
susceptible to other races of C. lindemuthianum as well as
in cultivars susceptible to all races of the pathogen. It
was also evident that even cultivar pathogenic races have
the chemical potential to elicit resistance.

   Although the accumulation of phytoalexins is
associated with the race-specific resistance of bean
cultivars to C. lindemuthianum, as well as induced local
protection by C. lindemuthianum or C. lagenarium, it is not
sufficient to explain systemic protection. Phaseollin and
other phytoalexins were not detected in systemically
protected unchallenged tissue but they rapidly accumulated

in systemically protected tissue challenged with a cultivar pathogenic race. The appearance of symptoms and the timing and magnitude of phytoalexin accumulation were as if the tissue had been challenged with a cultivar nonpathogenic race. It is possible, therefore, to separate two components of the phenomenon of induced resistance: the chemical agents, including phaseollin and other isoflavonoid phytoalexins which accumulate around the site of infection and which contribute to the inhibition of fungal development, and the signal that commits cells removed from the site of an inducing inoculation to respond rapidly when challenged.

Cucurbits

The phenomenon. Studies in my laboratory indicate that cucumber, watermelon, and muskmelon can be immunized against viral, bacterial, and fungal diseases by infection with viruses, bacteria, or fungi (5-8). Controlled infection with C. lagenarium or tobacco necrosis virus (TNV) protects cucumber, watermelon, and muskmelon against at least 12 diseases caused by a broad range of pathogens including obligate and facultative fungi, local lesion and systemic viruses, and fungi and bacteria that cause wilts and those that cause restricted and nonrestricted lesions on foliage and fruit. It is effective against foliar pathogens and a pathogen of roots. Immunization in cucurbits is systemic and requires a 3-4 day lag period between induction and challenge for expression. Mechanical injury and injuries caused by dry ice, chemicals, or fungal and plant extracts do not cause systemic immunization in cucurbits.

Removal of the first true leaf (inducer leaf) 96 hours after infection with fungi or bacteria and 72 hours after infection with TNV does not reduce the immunization of foliage above the site of induction. Similarly, removal of leaves above the inducer leaf after approximately the same lag times does not reduce the immunization of the excised leaves.

The signal for immunization in cucurbits is graft transmissible from rootstock to scion and is not cultivar, genus, or species specific. Thus, immunized cucumber rootstocks immunize cucumber, watermelon, and muskmelon scions. The experiments with grafting suggest that a chemical signal is produced at the site of induction and is translocated to other tissues where it conditions resistance. The effect of immunization is stronger for a foliar pathogen above the inducer leaf than below it; nevertheless, roots can be immunized by infecting leaves. Girdling the petiole of the inducer leaf prevents immunization above or below the inducer leaf. Girdling the petioles of leaves to be challenged, while leaving the petiole of the inducer leaf intact, prevents immunization only in challenged leaves with girdled petioles. The above data, together with those from experiments with graft transmissibility and persistence of immunization after removal of the inducer leaf, are evidence that immunization is the result of a signal transported from the inducer leaf.

Infection of the first true leaf of cucumber with C. lagenarium or TNV, followed in 2-3 weeks by a booster inoculation, immunizes cucumber against disease caused by C. lagenarium, Cladosporium cucumerinum, and Pseudomonas lachrymans through the period of fruiting. A single induction immunizes for 4-6 weeks, and without the booster inoculation, resistance is lost systemically after this period. Cucumbers can not be immunized, however, once they have started to flower and set fruit. This suggests that the programming of the plant's biological clock for reproduction turns off the ability to immunize but does not prevent the expression of the phenomenon in immunized tissues.

The concentration of inoculum used for induction and the number of lesions produced on the inducer leaf are directly related to the extent of immunization until a saturation point of inoculum or lesions is attained. A single lesion caused by C. lagenarium and as few as eight lesions caused by TNV on the inducer leaf significantly protect the tissues above the leaf.

Foliar immunization is effective against C. lagenarium or C. cucumerinum applied to the surface of foliage or infiltrated into the foliage. It reduces the multiplication of P. lachrymans and reduces the symptoms caused by TNV. It is effective against systemic cucumber mosaic virus (CMV), wilt caused by F. oxysporum f. cucumerinum, a fungal root pathogen, and by E. tracheiphila, a foliar bacterial pathogen. It is unlikely that a single mechanism for resistance would explain all the above phenomena.

Mechanisms. The penetration from appressoria of C. lagenarium into immunized cucumber is markedly reduced, whereas the germination of conidia is unaffected.

Associated with immunization is an approximately three-fold
increase in peroxidase activity. The increase is systemic
and is observed in immunized, unchallenged tissue distant
from the inducer inoculation. Since induction with C.
lagenarium, C. cucumerinum, P. lachrymans, or TNV causes
the increase, it is likely due to peroxidase activity of
host origin. As with immunization, a single lesion on the
inducer leaf results in a statistically significant
systemic increase in peroxidase activity (19). The
systemic increase in peroxidase is associated with markedly
enhanced activity of several peroxidase isozymes (19). A
booster inoculation with C. lagenarium or TNV markedly
enhances immunization and the systemic increase in
peroxidase. Peroxidase activity increases sooner in
immunized than in nonimmunized plants after challenge with
C. cucumerinum or C. lagenarium, and the enhanced activity
is also due to the increased activity of several isozymes.
Injury of a leaf with dry ice or rubbing with carborundum
enhances peroxidase activity in the injured leaf, but the
increase is not systemic. Of numerous enzymes studied,
only the increase in peroxidase activity corresponds
closely to the onset of immunization and increase in
lignification in immunized plants after challenge.

In addition to peroxidases, however, other proteins,
often referred to as "stress", "pathogenesis-related" (PR),
"b", or "E" proteins are associated with infection and
induced resistance but, unlike the peroxidases, their
increase in cucumber is limited to infected leaves whereas
the resistance induced is systemic.

Anderbrhan, et al. (20) reported that infection of
cucumber foliage with C. lagenarium or TNV caused the
accumulation of a new electrophoretically-detected protein.
Gessler and Kuc (21) reported that electrophoretic analyses
of extracts of cucumber leaves infected with C. lagenarium,
Fusarium oxysporum f. sp. cucumerinum, P. lachrymans,
Erwinia tracheiphila, TNV or CMV revealed the presence of a
protein band with an R_F value of 0.55-0.60 (based on
mobility of bromophenol blue) on 10% polyacrylamide gels.
This band was not evident in extracts of healthy or
mechanically wounded leaves. The protein was not detected
in uninfected leaves of infected plants, but it was
detected in similar amounts in infected leaves of non-
immunized plants and in challenged leaves of immunized
plants even though symptoms were not apparent on the
latter. The protein had a molecular weight of
approximately 16 kD, and was adsorbed by DEAE-cellulose,

did not react with Schiff's reagent, and did not have
ribonuclease activity. When injected into cucumber leaves,
it did not inhibit penetration of C. lagenarium or induce
resistance against disease caused by the fungus.

Investigations with cucumber and cowpea demonstrated
that the formation of local lesions by TNV led to a
localized resistance in cowpea leaves and cucumber
cotyledons and a systemic resistance induced in cucumber
leaves following homologous challenge (22). The resistance
appeared to be against symptom expression rather than
against virus multiplication. Resistance in both species
was also elicited by spraying healthy tissue with
polyacrylic acid 4 days before inoculation or osmotically-
stressing cowpea leaves 24 hr before inoculation. Virus-
elicited necrosis resulted in reproducible alterations in
the soluble protein profile of both species. Apparently
novel host proteins were induced in cowpea leaves and
cucumber tissue during necrosis. Tissue of both species
showing resistance contained either small amounts or more
often none of the apparently novel fractions and any direct
involvement in systemic resistance was suggested to be
unlikely.

The above observations indicate the pathogen-induced
"stress proteins" reported in cucumber are of host origin
and their accumulation does not depend upon the pathogen
but rather upon a general host response to infection. The
accumulation of the proteins occurs in all cucumber
cultivars tested and "stress proteins" associated with
infectious disease also accumulate in tobacco and other
plants. In tobacco, some of the "stress proteins" are
systemic. The role of the proteins in induced systemic
resistance in cucumber is unclear. If any, the role is
more likely regulatory rather than as direct inhibitors of
the development of infectious agents.

Lignification is a mechanism that has been clearly
implicated in disease resistance. In non-immunized but
genetically-resistant cucumber, localized lignification
occurs rapidly after penetration by C. cucumerinum.
Lignification also occurs in susceptible cucumbers infected
with C. cucumerinum or C. lagenarium, but the reaction is
delayed until after the pathogen has ramified through the
tissue, and the reaction is initially weak and diffuse
(23). Lignification is much more rapid, intense, and
localized in plants immunized by infection with C.
lagenarium, C. cucumerinum, or TNV and challenged with C.
cucumerinum or C. lagenarium than in unimmunized plants and

both fungi generally make little or no further growth once lignification is apparent (6-8, 23). Fungal as well as plant walls appear lignified.

Tobacco

Cruickshank and Mandryk (24) observed high foliar resistance to blue mold, caused by Peronospora tabacina, when tobacco plants were stem-infected with a spore suspension of P. tabacina and the foliage was subsequently inoculated with the same pathogen. Systemic resistance to blue mold was induced approximately 3 weeks after the stems were infiltrated with viable conidia (24,25), and lasted through flowering and seed set (25). Application of the conidia to the soil surface around stems caused restricted infection in the stem and systemically induced resistance (25). Heat-killed or sonicated conidia as well as three other pathogens and three nonpathogens of tobacco did not induce resistance (25). Induction of resistance by stem inoculations or soil treatment, however, was associated with dwarfing, reduction of leaf area, nitrogen deficiency-like symptoms and premature senescence.

Recently, a new technique for immunization by stem inoculation was developed in our laboratory. Conidia of a new isolate of the fungus were injected into tobacco stem tissue external to the cambium, and plants induced by this technique were protected 95-99%, based on the area of necrosis and amount of sporulation (26,27). The fungus injected into this tissue remained restricted in the stem. Immunized plants at the start of flowering were approximately 40% taller, had a 40% increase in dry weight, 30% increase in fresh weight and 4-6 more leaves as determined in greenhouse tests, and a 10-25% increase in marketable tobacco as determined in field tests than control plants. The first effects of induced resistance were observed 9 days after stem inoculation and 50% protection was obtained 12 and maximum protection 18-21 days after stem inoculation, respectively. Immunized plants flowered ca 2 weeks earlier than control plants. Immunized plants in the field were protected against blue mold as well or better than those treated with the best chemical control agent, Ridomil, currently available for blue mold control.

To study the movement of a protection factor produced in the stems of immunized plants, plants were girdled either 10 cm above or below the site of injection at 3 day intervals up to 21 days after injection (27,28). Plants girlded above the site of stem injection at day 6 or earlier remained susceptible to blue mold, although enhanced growth was evident with the 6 day girdling. Plants were protected when girdling occurred above the site of injection at day 9 or later. Girdling below the induction area diminished neither protection nor the increase in growth.

DISCUSSION

Immunization makes use of the plant's natural defense mechanism(s). It is systemic, can protect some annuals throughout the growing season, and it likely involves the activation of multiple resistance mechanisms which are effective against fungi, bacteria and viruses. With all plant-pathogen interactions studied in my laboratory, immunization appears to systemically sensitize plants to respond rapidly only in the presence of infectious agents, when and where needed, thereby conserving energy and metabolites. Economically-feasible chemical pesticides are not available for widespread control of bacterial diseases, and there are no effective chemicals for the control of viral diseases. In both instances, we completely rely on vector control, sanitation, and the development of disease resistance varieties.

The induction of resistance to a pathogen in plants considered completely susceptible implies that resistance is determined by the rate and magnitude of gene expression and the regulation of the activity of gene products rather than the presence or absence of genetic information for resistance. Many of the highest-yielding and quality crops become economically susceptible to disease. Breeders search throughout the world for new lines of resistance and often find them in agronomically inferior plants. Induced resistance may enable agriculture to make greater use of existing agronomically superior plants.

At this time, induced resistance is most applicable to crops that are transplanted rather than directly grown in the field, e.g., melons, tomato, tobacco. Inoculation with TNV offers promise of controlling disease on transplanted as well as field-grown plants. The virus appears inocuous, can be applied to foliage in the form of a high pressure spray, and it doesn't require the exacting conditions of moisture and temperature necessary for infection by fungal and bacterial inducers. When the natural signal for induced resistance is isolated, it may be possible to utilize it as a spray or seed treatment. Synthetic chemicals also show promise as inducers of systemic resistance.

Plant immunization would not eliminate the need for chemical pesticides. It could reduce our dependence on this form of disease control and, thereby, reduce the presence of pesticide residues on food crops and danger to the ecosystem. Of course, natural chemicals responsible for induced resistance, either as signals or agents which restrict development of pathogens, may be no safer than chemical pesticides. It seems reasonable to believe, however, that induced resistance is as safe as our first line of defense against disease – the disease resistant plants.

The ability to transmit resistance via grafts offers an added dimension to protection. Many food crops and ornamentals are propagated by grafting, and the use of immunized rootstocks may present another vehicle to get the signal into susceptible plants.

Many questions remain unanswered. Emphasis is being placed on the elucidation of the molecular mechanisms responsible for induced resistance and the testing of the concept in the field under conditions of natural infection and with different pathogens.

The implications of long range cell-to-cell communication, the possible relation of immunization to a wound response, and the role of such a response in evolutionary development, are in themselves ample reason for emphasis in this area of research. The practical control of disease on a preventive level, utilizing and perhaps improving on a mechanism that has stood the evolutionary test for survival, is a distinct possibility.

REFERENCES

1.   Dawkins R (1976). "The Selfish Gene." New York:Oxford Press, 224 pp.
2.   Biehn WL, Williams EB, Kuć J (1966). Resistance of mature leaves of Malus atrosanguinea 804 to Venturia inaequalis and Helminthosporium carbonum. Phytopathology 56:588.
3.   Grijseels AJ, Williams EB, Kuć J (1964). Hypersensitive response in selections of Malus to fungi nonpathogenic to apple. Phytopathology 54:1152.

4. Nusbaum CJ, Keitt GW (1983). A cytological study of host-parasite relations with <u>Venturia</u> <u>inaequalis</u> on apple leaves. J Agr Res 56:595.
5. Kuć J, Caruso FL (1977). Activated coordinated chemical defense against disease in plants. In Hedin PA (ed): "Host Plant Resistance to Pests." Washington DC: Amer Chem Soc, p 78.
6. Kuć J (1982). Plant Immunization. Bioscience 32:854.
7. Kuć J (1984). Phytoalexins and disease resistance from a perspective of evolution and adaptation. In "Origins and Development of Adaptation." London: Pitman, p 100.
8. Kuć J (1983). Induced systemic resistance in plants to diseases caused by fungi and bacteria. In Bailey JA, Deverall BJ (eds): "Dynamics of Host Defense." Sydney New York: Academic Press, p 191.
9. Kuć J (1962). Production of extracellular enzymes by <u>Cladosporium</u> <u>cucumerinum</u>. Phytopathology 52:961.
10. Wallace J, Kuć J, Draudt HN (1962). Biochemical changes in the water insoluble material of maturing apple fruit and their possible relationship to disease resistance. Phytopathology 52:1023.
11. Kuć J, Williams EB, Maconkin MA, Ginzel J, Ross AF, Freedman LJ (1967). Factors in the resistance of apple to <u>Botryosphaeria</u> <u>ribis</u>. Phytopathology 57:38.
12. Kuć J, Barnes A, Daftsios A, Williams EB (1959). The effect of amino acids on susceptibility of apple varieties to scab. Phytopathology 49:313.
13. MacLennan DH, Kuć J, Williams EB (1963). Chemotherapy of the apple scab disease with butyric acid derivatives. Phytopathology 53:1261.
14. McIntyre JF, Kuć J, Williams EB (1973). Protection of pear against fireblight by bacteria and bacterial sonicates. Phytopathology 63:872.
15. McIntyre JF, Kuć J, Williams EB (1975). Protection of Bartlett pear against fireblight with deoxyribonucleic acid from virulent and avirulent <u>Erwinia</u> <u>amylovora</u>. Physiol Plant Pathology 7:153.
16. Langcake P (1981). Alternative chemical agents for controlling plant disease. Phil Trans R Soc Lond B295:83.
17. Farih A, Tsao P, Menge JA (1981). Fungitoxic activity of efosite aluminum on growth, sporulation and germination of <u>Phytophthora</u> <u>parasitica</u> and <u>P.</u> <u>citrophthora</u>. Phytopathology 71:934.

18. Sekizawa Y, Mase S (1980). Recent progress in studies on the non-fungicidal controlling agent, probenazole, with reference to the induced resistance mechanism of rice plant. Rev Plant Protection Res 13:114.

19. Hammerschmidt R, Nuckles E, Kuć J (1982). Association of enhanced peroxidase activity with induced systemic resistance to Colletotrichum lagenarium. Physiol Plant Pathology 20:73.

20. Andebrhan T, Coutts RH, Wagih EE, Wood RKS (1980). Induced resistance and changes in the soluble protein fractions of cucumber leaves locally infected with Colletotrichum lagenarium or tobacco necrosis virus. Phytopathol Z 98:47.

21. Gessler G, Kuć J (1982). Appearance of a host protein in cucumber plants infected with viruses, bacteria and fungi. J Exptl Bot 33:58.

22. Coutts RH, Wagih EE (1983). Induced resistance to viral infection and soluble protein alterations in cucumber and cowpea plants. Phytopathol Z 107:57.

23. Hammerschmidt R, Kuć J (1982). Lignification as a mechanism for induced systemic resistance in cucumber. Physiol Plant Pathology 20:61.

24. Cruickshank IAM, Mandryk M (1960). The effect of stem infestation of tobacco with Peronospora tabacina Adam on foliage reaction to blue mold. J Aust Inst Agr Sci 26:369.

25. Cohen Y, Kuć J (1981). Evaluation of systemic resistance to blue mold induced in tobacco leaves by prior stem inoculation with Peronospora hyoscyami tabacina. Phytopathology 71:783.

26. Kuć J, Tuzun S (1983). Immunization for disease resistance in tobacco. Rec Adv Tobacco Sci 9:179.

27. Tuzun S, Kuć J (1983). A technique which immunizes against blue mold (Peronospora hyoscyami f. sp. tabacina) and increases growth of tobacco. Phytopathology 73:823.

28. Tuzun S, Kuć J (1983). Movement of a factor in tobacco which protects against blue mold (Peronospora hyoscyami f. sp. tabacina). Plant Physiol 72:158.

Cellular and Molecular Biology of Plant Stress, pages 319–334
© 1985 Alan R. Liss, Inc.

PECTIC FRAGMENTS REGULATE
THE EXPRESSION OF PROTEINASE
INHIBITOR GENES IN PLANTS[1]

Clarence A. Ryan, Paul D. Bishop, Mary Walker-Simmons,
Willis E. Brown and John S. Graham

Institute of Biological Chemistry and
Program in Biochemistry/Biophysics
Washington State University
Pullman, WA 99164

ABSTRACT   Proteinase inhibitor inducing activity
resides in plant cell wall-derived pectic oligomeric
fragments. The induction of proteinase inhibitors by
these fragments in excised tomato, potato and alfalfa
plants is similar to that resulting from mechanical
wounding. Activity is present in α,(1-4)galacturonic
acid oligomers and increases from DP=2 through DP=6.
Chitosan, a mixture of soluble β,(1-4)glucosamine
polymers, is also a potent inducer of proteinase
inhibitors in tomato and alfalfa leaves.
Complimentary DNAs have been constructed from
inducible tomato leaf inhibitor mRNAs and
characterized. The cDNAs have been used to
demonstrate that inhibitor mRNA levels increase within
4 hr following wounding.

INTRODUCTION
    Previous research from this laboratory has shown that
wounding of tomato plants by chewing insects or other
mechanical means releases a systemic factor, called the
proteinase inhibitor inducing factor, (PIIF), that
initiates the induction of synthesis and accumulation of
Inhibitors I and II throughout the plants (1). These two
proteins are powerful inhibitors of serine endopeptidases

1. This work was supported in part by Grants from the
National Science Foundation #PCM-8315427 and #PCM-8309344,
from the USDA Competitive Grants Program #81-CRCR-1-0697
and a Grant from the Rockefeller Foundation.

that are found in both animals and microorganisms, but not in plants (2).

Both inhibitors have been isolated from potato tubers and tomato leaves and extensively characterized (3-5). They are among the array of plant chemicals that significantly reduce the nutritional quality of plant tissues and are considered to be part of the natural defense of plants toward attacking pests (1). These proteins apparently are developmentally regulated in potato tubers where they appear beginning with tuber set and accumulate throughout development (6). On the other hand, the two inhibitors accumulate in leaves of potato and tomato plants in response to the environmental stress of wounding (1). Accumulation of inhibitor proteins has been shown to take place in the central vacuole of tomato leaf cells (7).

A major goal of our research program is the elucidation of the biochemical mechanisms that regulate systemically mediated, wound-induced proteinase inhibitor synthesis and accumulation in plant leaves. We are using two approaches to this overall goal; (1) isolating the wound signal(s) and studying its origin, release, transport and mechanism of recognition in receptor cells and (2) isolating the proteinase inhibitor genes to further understand how they are regulated in response to the wound signal(s).

In this chapter we review our recent research regarding the isolation, characterization and activity of the wound signal, PIIF, and oligosaccharide fragments prepared from PIIF (8-10). Evidence from our laboratory and others show that these pectic preparations are not only active in inducing proteinase inhibitors in tomato plants, but are active elicitors of antifungal phytoalexins in peas and beans (11-13). Conversely, evidence is presented that the phytoalexin elicitor, chitosan (14), a mixture of β,(1-4) glucosamine polymers that are derived from or are part of the cell walls of arthropods and fungi, is a powerful inducer of proteinase inhibitors in tomato leaves (11,15). We also review our recent finding that the systemically mediated wound induction of proteinase inhibitors is a property of alfalfa, a member of the Leguminosae family. A trypsin inhibitor of the Bowman-Birk family of proteinase inhibitors is induced in leaves of alfalfa plants either by wounding or by supplying excised leaves with solutions containing PIIF or chitosan (16,17). Finally, the isolation and characterization of cDNAs constructed from Inhibitor I and II mRNAs will be

presented. Their usefulness in studying the regulation of
Inhibitor I and II mRNA and in isolating and
characterizating the genes that are regulated by the plant
and fungal cell wall fragments will be discussed. The data
strongly supports a hypothesis that fragments of either the
plant's own cell walls or of cell walls of fungi, released
during pest attacks, are part of recognition systems to
activate genes involved in systemically induced natural
plant protection.

Isolation and Characterization of the Systemic Wound
Signal, PIIF. Pectic polysaccharides, cumulatively called
tomato PIIF, the proteinase inhibitor inducing factor, have
been isolated from tomato leaves. These fragments are
active inducers of both Inhibitors I and II when supplied
to excised young tomato or potato plants through the cut
petioles (1,9,10). The pectic polysaccharides are
apparently fragmented from leaf cell walls by cellular
$\alpha,1-4$ endopolygalacturonases that interact with cell walls
as a result of wounding. The isolated PIIF-active
fragments exhibit an average degree of polymerization (DP)
of about 30, with a molecular weight distribution of
5,000-10,000 daltons (8).

PIIF-active polysaccharides have been further
fragmented into oligosaccharides that retain PIIF activity
(10) by limited acid hydrolysis in 2N trifluoroacetic acid
for 1 hr at 80°C in vacuo to remove neutral side chains.
The soluble material was chromatographed on DE-52 and the
polysaccharide peak, which eluted at approximately 0.3M
ammonium bicarbonate, was lyophilized. The compositional
analysis of this polysaccharide, termed "TFA PIIF",
indicated that most of the methyl esters, neutral side
chain glycosyl residues and the rhamnose-rich backbone
segments of tomato PIIF were removed by partial acid
hydrolysis. TFA PIIF is essentially a galacturonic acid
polymer with an average DP=20, containing small amounts of
rhamnose and fucose. TFA PIIF preparations were usually
slightly more active in the biological assays than the
parent tomato PIIF (10), although in some isolations the
specific activity remained virtually unchanged from its
parent tomato PIIF. Nevertheless, it is apparent that the
polygalacturonic acid backbone present in tomato PIIF is
sufficient to induce proteinase inhibitor synthesis in
excised tomato plants.

The TFA PIIF was further fragmented into a series of
oligosaccharides in order to determine the relationship of

the DP of galacturonic acid oligomers to PIIF activity.
TFA PIIF, enzymatically hydrolyzed using a partially
purified tomato polygalacturonase preparation, was
chromatographed on DE 52 and the oligosaccharide peaks that
eluted were collected and analyzed for the average degree
of polymerization and uronic acid content as well as their
proteinase inhibitor inducing activity. The component
oligomer of each peak was confirmed by comparisons with a
series of known α,(1-4) galacturonide oligomer standards
using anion exchange HPLC (10).

The overall analyses strongly suggested that the
minimum structural moiety that exhibits PIIF activity is
4-0-α-D-galacturonisyl-D-galacturonic acid (10). To
confirm these results, a series of highly purified α(1-4)
galacturonic acid oligomers derived from citrus pectin were
tested for PIIF activity. These galacturonic acid
oligomers, a gift of C.W. Nagel, Dept. of Food Science and
Technology, Washington State University, were derived by
the action of a pure endo α,(1-4) polygalacturonase from S.
fragilis on a well characterized, acid soluble
polygalacturonic acid (>95% galacturonic acid, DP 14)
obtained from grapefruit pectin. These oligomers were
purified by anion-exchange (18) and their purity (>93%
D-galacturonic acid) was assayed by HPTLC. All of the
α,(1-4) galacturonic acid oligomers tested, from dimer
through hexamer, were found to have substantial PIIF
activities (10). These activities were quantitatively
similar to those of the 'TFA PIIF' derived galacturonic
acid oligomers. These results support our hypothesis that
4-0-α-D-galacturonisyl-D-galacturonic acid unit is required
for activity.

Evidence from several laboratories is accumulating
that supports a role for plant cell wall fragments as
chemical messengers that can activate a number of different
genes involved with natural plant protection systems
(1,8-13,19-20). Plant cell wall fragments have been shown
to activate antifungal phytoalexin synthesis in soybeans
(13), castor beans (12), and peas (11). A fungal
endopolygalacturonase that releases fragments from plant
cell walls has been demonstrated to elicit phytoalexin
synthesis in castor bean seedlings (21,22). This enzyme
has been shown to be effective in eliciting both the
phytoalexin pisatin in pea pods and proteinase inhibitor
accumulation in tomato leaves (23). Plant cell wall
fragments released by a pectic lyase are also potent
phytoalexin elicitors (24). Cell wall fragments appear to

be the 'endogenous elicitor' that had previously been reported to elicit phytoalexin synthesis in Phaseolus vulgaris (19). The ability of cell wall fragments to activate defense genes appears to be a general property of a variety of plant genera in the Solanceae and Leguminosae plant families.

Major differences are apparent however, between the systems to induce proteinase inhibitor synthesis and phytoalexin synthesis. As mentioned in this study, oligomers of DP=2 through DP=6, exhibit various capacities to induce Inhibitor I when supplied to young tomato leaves. In contrast, the phytoalexin glycinol is elicited in soybeans by fragments of about DP=12, while smaller fragments are inactive (20). In addition, wound-induced proteinase inhibitor synthesis is systemically activated, with the signal being transported throughout the plants within two hours following wounding (1). Phytoalexin synthesis, on the other hand, appears to be localized to cells near pathogenic sites of infection. It is possible that small oligosaccharides are the systemic signals that activate proteinase inhibitor genes (and possibly other defense responses) whereas the fragments of DP~10 or larger are signals of localized responses. Although the oligo- and polygalacturonic acid polymers are active inducers and elicitors of proteinase inhibitors and phytoalexins it is possible the linear pectic fragments studied here are not the only signals responsible for the biological activities.

Chitosan Derivatives as Inducers of Proteinase Inhibitors in Excised Tomato Plants. The recent finding that the phytoalexin pisatin is elicited in pea pods by plant pectic fragments (11) was of particular interest since pisatin synthesis was known to be strongly elicited by fragments derived from chitin and chitosan (14), components of cell walls of arthropods and fungi (25). We subsequently found that soluble fragments of chitosan, produced by nitrous acid hydrolysis, were potent inducers of proteinase inhibitors in tomato plants (11). Nitrous acid-modified chitosans induce maximal quantities of Inhibitor I, even when supplied at levels less than one-tenth of tomato PIIF. The accumulation of Inhibitor I induced by chitosans and PIIF is shown in Table 1. Both the time-course of accumulation and quantity of inhibitors that accumulate are identical when PIIF or chitosans are used as inducers at concentrations producing maximal responses.

TABLE 1

ACCUMULATION OF TOMATO LEAF PROTEINASE INHIBITOR I
INDUCED BY TOMATO LEAF POLYSACCHARIDES AND
SHRIMP CHITOSANS

Additions*	Accumulation of Proteinase Inhibitor I in Tomato Cotyledons (µg/g leaf tissue)
Tomato PIIF (mg/ml)	
2.0	123
1.0	143
0.5	60
Chitosans (mg/ml)	
2.0	135
0.4	143
0.08	144
0.016	115
0.003	98
Buffer Alone	23

*Tomato cotyledons were supplied with tomato
PIIF, chitosans, or buffer (0.05 M Na-phosphate, pH
6.0) for one h, followed by incubation in water for 24
h and assayed immunologically.

Loschke et al. (26) have previously shown that
chitosans induced the de novo synthesis of phenylalanine
ammonia lyase (PAL), the first enzyme required for the
synthesis of pisatin from phenylalanine. PAL synthesis, as
well as the synthesis of several other enzymes involved in
isoflavonoid biosynthesis, has been shown to be induced in
parsley and French bean suspension cultures in response to
fungal elicitors (27,28). The effect of PIIF on PAL
activity in pea pods was assayed since it elicits pisatin
synthesis (11). Although the amount of PAL induced by PIIF
was only about half that induced by chitosan, the results

clearly demonstrate that PIIF induces PAL activity in pea pods, suggesting that the entire pathway of phenylalanine to pisatin may be affected by tomato PIIF.

Phenylalanine ammonia lyase activity induction in tomato leaves by chitosans or PIIF was also studied. Neither chitosans nor PIIF elicited an increase in PAL activity in excised tomato leaves. PAL activity actually decreased when the leaves were detached and this decrease was not affected by PIIF nor chitosans (11).

Chitosan, partially hydrolyzed with HCl, provided a source of a homopolymeric series of $\beta(1-4)$glucosamine oligosaccharides that have DPs of 6 and under (15,29). Separation of these oligomers from the acid hydrolyzed chitosan was achieved on Fractogel (29). Individual peaks, containing mono- through hexasaccharides, were pooled and assayed (15) for degree of polymerization (a) by the ratio of their glucosamine residues to reducing equivalents and (b) by their linearity of R_m by TLC chromatography. The % acetylation of each oligomer was determined by NMR spectroscopy. The proteinase inhibitor inducing activity of each oligomer fraction was also assayed in excised tomato plants (15). Glucosamine was totally inactive as an inducer of proteinase Inhibitor I in leaves of excised tomato plants. The dimer through tetramer were increasingly active, but still not nearly as active as the total acid hydrolysate. The pentamer and hexamer exhibited more activity than the tetramer, but the hexamer activity dropped off slightly. The larger fragments (DP>6) were as active as the total hydrolysate (15).

Monomer and oligomers obtained from partial acid hydrolysis were all less than 2.0% acetylated, indicating that the acetyl group is not a major factor in the inducing activity. To assess the effects of acetylation on activity, the total hydrolyzed chitosan mixture was partially (13-20%) acetylated. No significant difference in the inducing activities of the acetylated and nonacetylated mixtures were observed (15).

Chitosan have now been shown to induce proteinase inhibitors (11,15), phytoalexins (14,23) and lignification (30). Thus, chitosans derived from fungal cell walls can signal the activation of a variety of plant defense responses. This role for chitosan has been strengthened by the recent finding of chitinase activities in plant tissues (31-37), even though plants do not contain any chitosan or chitin. Ethylene induces chitinase activity in bean and tomato leaves and many other plants (31). In bean tissue, chitinase is known to be sequestered in the central vacuole (32) where it is likely released upon cell damage. Tomato stem endochitinase hydrolyzes both chitin and chitosan (35) and its activity increases in tomato stems following infection with the fungus Verticillium albo-atrum (37). Hadwiger et al. (34) observed that pea tissue contained enzymes capable of releasing phytoalexin elicitors from purified Fusarium fungal cell walls and application of chitosan to pea pod tissue resulted in protection against Fusarium (14).

Wound Induction of a Trypsin Inhibitor in Alfalfa Leaves. Field-grown alfalfa plants contain a proteinaceous trypsin inhibitor that is a significant component of the soluble proteins of their leaves (38-40). On the other hand, growth-chamber grown alfalfa plants have very low levels of trypsin inhibitor activity in leaves. However, when plants are detached and supplied with solutions containing tomato PIIF (41) or chitosan (W.E. Brown and C.A. Ryan, in preparation) they exhibit a several fold increase in trypsin-inhibitory activity within two days (Table 2). We have recently observed (17) that wounding of growth-chamber grown alfalfa plants on the lower leaves also induces a striking increase in trypsin inhibitory activity in the upper leaves. The rate of induction and quantity of inhibitor accumulated in leaves of wounded alfalfa plants is similar to the wound-induced phenomenon in potato and tomato plants (17).

The trypsin inhibitor from alfalfa leaves was purified. The molecular weight of the inhibitor is about 7500 daltons and inhibits trypsin with a K_i of 1×10^{-10}M (16). Partial sequence analyses (Titani, K., Ericsson, L., Brown, W.E. and Ryan, C.A. in preparation) has shown that the inhibitor is a member of the Bowman-Birk family of proteinase inhibitors. Heretofore members of this extensively studied family have only been found in seeds of legumes where they are under developmental control. Our finding of an environmentally regulated Bowman-Birk inhibitor in alfalfa leaves closely parallels the existence of the proteinase inhibitors in potato tubers (developmentally regulated) and in tomato and potato leaves (environmentally regulated).

Alfalfa provides an additional opportunity to study the regulatory region of proteinase inhibitor genes that are under both developmental and environmental regulation. These studies should determine whether different sets of proteinase inhibitor genes are regulated, one set in

TABLE 2

INDUCTION OF ATI IN EXCISED ALFALFA LEAVES BY
TOMATO PIIF OR CHITOSAN

	mg/ml[2]	ATI INDUCED[1] (μg/mg leaf protein)
PIIF	2.0	11
	1.0	12
	0.7	6
	0.4	3
Chitosan	0.01	12
	0.005	7
	0.001	9
	0.0005	2

1. Average of 2 experiments, 3 leaves per experimental value.

2. Plants were excised and supplied with solutions of chitosan (in 5 mM sodium phosphate buffer, pH 7.0) or PIIF (in water) for 90 min, and the levels of ATI were analyzed after 48 hr in constant light.

response to development and one set in response to pest attacks, or whether the same genes are activated in different tissues in response to different stimuli.

Isolation and Characterization of cDNAs for Wound-Induced Proteinase Inhibitors I and II Genes. To examine the induction process at the gene level a cDNA library was constructed in the plasmid pUC9 with poly(A$^+$) mRNAs from leaves of tomato plants that had been enriched in Inhibitors I and II mRNA by wounding. Inhibitor I and II cDNA clones were identified among the wound-induced cDNAs by colony hybridization with ^{32}p-labeled oligonucleotide probes and by hybrid select translations. A 571 bp Inhibitor I cDNA insert was isolated and its nucleotide sequence determined (41). The open reading frame revealed the presence of a 42 amino acid sequence preceeding the N-terminal leucine found in the native protein. This sequence appears to contain a 23 amino acid segment typical of a signal sequence followed by a 19 amino acid sequence containing 9 charged amino acids. The entire 42 amino acid sequence is missing from the native Inhibitor I and it represents 38% of the translated protein. The Inhibitor I amino acid sequences contains 71% identity with potato tuber Inhibitor I sequence (41,42). The P_1-P_1-putative reactive site sequence is Leu-Asp, which appears to account for its specificity towards chymotrypsin (43).

An Inhibitor II cDNA insert of 693 bp was also isolated and its nucleotide sequence determined (44). The open reading frame codes for 148 amino acids including a signal sequence preceeding the known N-terminal lysine of the mature Inhibitor II. The amino acid sequence of Inhibitor II shows considerable homology with two small proteinase inhibitors isolated from potato tuber (45) and an inhibitor from eggplant (46). The Inhibitor II sequence apparently evolved from a smaller gene by a process of gene duplication and elongation. Two identical nucleotide sequences within the 3' noncoding regions of Inhibitor I cDNA were found to be present in a similar region of Inhibitor II cDNA. These include an atypical polyadenylation signal, AATAAG, and a 10 base palindromic sequence, CATTATAATG.

The acquisition of the inhibitor cDNAs has enabled us to study the temporal regulation of the wound induced mRNAs. Utilization of these cDNAs as hybridization probes has shown that Inhibitor I and II mRNAs exhibit rapid increases in accumulation between 4-6 hrs after an initial wound. A maximum level is attained approximately 8-12 hrs. post-wounding. The time course of mRNA accumulation appears to be identical for the two inhibitors (J.S. Graham, G. Pearce and C.A. Ryan, in preparation). The magnitude of the increases in mRNA levels is a function of the multiplicity of wounds inflicted. A series of 4 wounds spaced 1 hr. apart resulted in a dramatic increase in accumulation for both mRNA and Inhibitor I protein as compared to that of a single wound (Figure 1). Thus, repeated wounding, such as inflicted by chewing insects, may result in the delivery of more PIIF to leaves and provide a reinforcement of the mechanisms responsible for Inhibitor I and II mRNA accumulation.

Both inhibitor cDNAs have been employed in selecting genomic clones of Inhibitor I and II in tomato and potato to further probe the structure of these gene families in

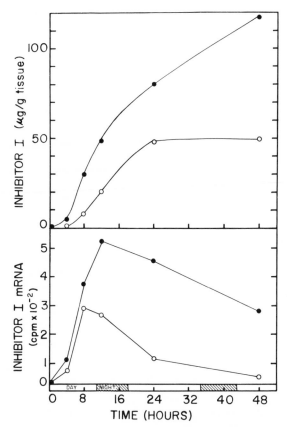

FIGURE 1. Wound induced accumulation of Inhibitor I protein (top) and mRNA (bottom) following a single wound (   ) or four wounds inflicted at 1 hr intervals (   ).

plants. This work provides the basis for defining the regulatory regions within these genes which are responsible for the modulation of Inhibitor I and II gene expression.

Concluding Remarks. The induction of proteinase inhibitors in tomato, potato and alfalfa leaves, of pisatin in pea pods and casbene in castor beans by PIIF and chitosans suggests a common mechanism may be present in a variety of plants to regulate different plant genes. This implies that different polysaccharide molecules, chitosans, and pectic fragments and β-glucans (phytoalexin elicitors in beans (20)), may participate in a common recognition system which has evolved in various plants to activate natural plant defense systems (Figure 2).

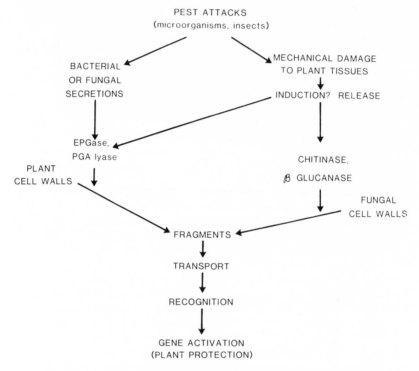

FIGURE 2. A summary of the proposed pest induced events leading to systemic defense response of plants.

The progress we have made to date, in isolating and characterizing PIIF and in obtaining wound-induced Inhibitor I and II cDNAs, will provide a basis to better understand the biochemical mechanisms by which wounding of plant leaves activates proteinase inhibitor genes. We anticipate that the complete elucidation of this system will provide entirely new concepts for plant scientists to utilize to select and fashion more pest resistant agricultural crops.

REFERENCES

1. Ryan CA (1978). Proteinase inhibitors in plant
    leaves: A biochemical model for natural plant
    protection. TIBS 5:148.
2. Ryan CA (1973). Proteolytic enzymes and their
    inhibitors in plants. Annu Rev Plant Physiol 24:173.
3. Melville JC, Ryan CA (1972). Chymotrypsin Inhibitor
    I from potatoes. Large scale preparation and
    characterization of its subunits. J Biol Chem
    247:3445.
4. Bryant, J, Green TR, Gurusaddaiah T, Ryan CA (1976).
    Proteinase Inhibitor II from potatoes: isolation and
    characterization of its protomer components.
    Biochemistry 15:3418.
5. Plunkett G, Senear DF, Zuroske G, Ryan CA (1982).
    Proteinase Inhibitors I and II from leaves of wounded
    tomato plants: purification and properties. Arch
    Biochem Biophys 213:463.
6. Ryan CA, Huisman OC, VanDenburgh RW (1968).
    Transitory aspects of a single protein in tissues of
    Solanum tuberosum and its coincidence with the
    establishment of new growth. Plant Physiol 43:589.
7. Walker-Simmons M, Ryan CA (1977). Immunological
    identification of Proteinase Inhibitors I and II in
    isolated tomato leaf vacuoles. Plant Physiol 60:61.
8. Bishop PD, Makus D, Pearce G and Ryan CA (1981).
    Proteinase inhibitor-inducing factor activity in
    tomato leaves resides in oligosaccharides enzymically
    released from cell walls. Proc Natl Acad Sci (USA)
    78:3536.
9. Ryan CA, Bishop P, Pearce G, Darvill M, McNeil M,
    Albersheim P (1981). A sycamore cell wall
    polysaccharide and a chemically related tomato leaf
    polysaccharide possess similar proteinase
    inhibitor-inducing activities. Plant Physiol 68:616.
10. Bishop PD, Pearce G, Bryant JE, Ryan CA (1984).
    Isolation and characterization of the proteinase
    inhibitor-inducing factor from tomato leaves:
    identity and activity of poly- and oligogalacturonide
    fragments. (submitted for publication).
11. Walker-Simmons M, Hadwiger L, Ryan CA (1983).
    Chitosans and pectic polysaccharides both induce
    accumulation of the antifungal phytoalexin pisatin in
    pea pods and antinutrient proteinase inhibitors in
    tomato leaves. Biochem Biophys Res Comm 110:194.

12. Bruce RJ, West CA (1981). The role of pectic fragments of the plant cell wall in elicitation by a fungal endopolygalacturonase. Plant Physiol 69:1181.

13. Nothnagel EA, McNeil M, Albersheim P, Dell A (1983). Host-pathogen interactions. XXII. A galacturonic acid oligosaccharide from plant cell walls elicits phytoalexins. Plant Physiol 71:916.

14. Hadwiger L, Beckman JM (1980). Chitosan as a component of pea-Fusarium solani interactions. Plant Physiol 66:205.

15. Walker-Simmons M, Ryan CA (1984). Protein Inhibitor I synthesis in tomato leaves: induction by chemically modified chitosan and chitosan oligomers. (submitted for publication).

16. Brown WE, Ryan CA (1984). Purification and Characterization of a Wound-Inducible Trypsin Inhibitor from Alfalfa. Biochemistry (in press).

17. Brown WE, Ryan CA (in preparation).

18. Hasagawa S, Nagel CW (1966). A. New Pectic and Transeliminase Produced Exocellularly by a Bouillus.J Food Sci 31:838.

19. Hargreaves JA, Baily JA (1978). Phytoalexin production by hypocotyls of Phaseolus vulgaris in response to constitutive metabolites released by damaged bean cells. Physiol Plant Pathol 13:89.

20. Albersheim P, Darvill AG, McNeil M, Valent B, Sharp JK, Nothnagel EA, Davis KR, Yamasaski N, Gollin D, York W, Dudman WP, Darvill J, Dell A (1983). Oligosaccharins: naturally occurring carbohydrates with biological regulatory functions. In Ciferri O, Dure L (eds): "Structure and Function of Plant Genomes," New York: Plenum, NATO Advanced Science Institute Series, vol. 63, p 293.

21. Lee S-C, West CA (1981). Polygalacturonase from Rhizopus stolonifer, an elicitor of casbene synthetase activity in castor beans (Ricinus communis L.) seedlings. Plant Physiol 67:633.

22. Lee S-C, West CA (1981). Properties of Rhizopus stolonifer polygalacturonase, an elicitor of casbene synthetase activity in castor bean (Ricinus communis L.) seedlings. Plant Physiol 67:640.

23. Walker-Simmons M, West CA, Hadwiger L, Ryan CA (1984). Comparison of proteinase inhibitor inducing activities and phytoalexin elicitor activities of pure fungal endopolygalacturonase, pectic fragments and chitosans. (submitted for publication).

24. Davis KR, Lyon GD, Darvill AG, Albersheim P (1984). Host-pathogen interactions XXV. Endopolygalacturonic acid lyase from Erwinia carotovora elicits phytoalexin accumulation by releasing plant cell wall fragments. Plant Physiol 74:52.
25. Muzzarelli RA (1977). Occurrence of chitin. In "Chitin," Oxford: Pergamon Press, p 5.
26. Loschke DC, Hadwiger LA, Wagoner W (1983). Comparison of mRNA populations coding for phenylalanine ammonia-lyase and other peptides from pea tissue treated with biotic and abiotic phytoalexin inducers. Physiol Plant Path 23:163.
27. Hahlbrock K, Lamb CJ, Purwin C, Ebel J, Fautz E, Schafer E (1981). Rapid response of suspension-cultured parsley cells to the elicitor from Phytophthora megasperma var. sojae. Plant Physiol 67:768.
28. Lawton MA, Dixon RA, Hahlbrock K, Lamb C (1983). Rapid induction of the synthesis of phenylalanine ammonia-lyase and of chalcone synthetase in elicitor-treated plant cells. Eur J Biochem 129:593.
29. Chang JJ, Hash HH (1979). The use of an amino acid analyzer for the rapid identification and qualitative determination of chitosan oligosaccharides. Anal Biochem 95:563.
30. Pearce RB, Ride JP (1982). Chitin and related compounds as elicitors of the lignification response in wounded wheat leaves. Physiol Plant Path 20:119.
31. Boller TA, Gehri A, Mauch F, Vögeli U (1983). Chitinase in bean leaves: induction by ethylene, purification, properties, and possible function. Planta 157:22.
32. Boller T, Vögeli U (1984). Vacuolar localization of ethylene-induced chitinase in bean leaves. Plant Physiol 74:442.
33. Molano J, Polacheck I, Duran A, Cabib E (1979) An endochitinase from wheat germ. J Biol Chem 254:4901.
34. Nichols EJ, Beckman JM, Hadwiger LA (1980). Glycosidic enzyme activity in pea tissue and pea-Fusarium solani interactions. Plant Physiol 66:199.
35. Pegg GF, Young DH (1982). Purification and characterization of chitinase enzymes from healthy and Verticillium albo-atrum-infected tomato plants, and from V. albo-atrum. Physiol Plant Path 21:389.

36. Young DH, Pegg GF   (1982).   The action of tomato and Verticillium albo-atrum glycosidases on the hyphal wall of V. albo-atrum.   Physiol Plant Path 21:411.
37. Pegg GF, Young DH   (1981).   Changes in glycosidase activity and their relationship to fungal colonization during infection of tomato by Verticillium albo-atrum. Physiol Plant Path 19:371.
38. Chien TF, Mitchell HL   (1970).   Purification of a trypsin inhibitor of alfalfa.   Phytochem 9:717.
39. Chang H-Y, Reed GR, Mitchell HL   (1978) Alfalfa trypsin inhibitor.   J Agric Food Chem 26:143.
40. Sukhinin, VN, Berezin, VA, Novikov, YuF, Reva, AD, Lagutenko, AV, Proidak, NI   (1981).   Purification and certain properties of the trypsin inhibitor from vegetative organs of alfalfa.   Biochemistry (USSR) 41:951.
41. Graham JS, Pearce G, Merryweather J, Titani K, Ericsson LH, Ryan CA (1984).   Wound-induced proteinase inhibitor mRNA from tomato leaves: II.   Molecular cloning and sequence analysis of Inhibitor II cDNA. J. Biol. Chem. (submitted).
42. Richardson M, Cossins L (1974).   Chymotryptic inhibitor from potatoes: the amino acid sequences of subunits B,C,D FEBS Lett 45:11.
43. Laskowski M Jr., Sealock RW   (1971).   The Enzymes. Boyer, PD (eds): "Protein proteinase inhibitors - molecular aspects." New York: p 376.
44. Graham JS, Pearce G, Merryweather J, Titani K, Ericsson LH, Ryan CA (1984).   Wound-induced proteinase inhibitor mRNA from tomato leaves: I. Molecular cloning and sequence analysis of Inhibitor I cDNA.   J. Biol. Chem. (submitted).
45. Hass MG, Hermodson MA, Ryan CA, Gentry L   (1982). Primary structures of two low molecular weight proteinase inhibitors from potatoes.   Biochemistry 21:752.

Cellular and Molecular Biology of Plant Stress, pages 335–349
© 1985 Alan R. Liss, Inc.

THE ROLE OF PECTIC FRAGMENTS OF THE PLANT
CELL WALL IN THE RESPONSE TO BIOLOGICAL STRESSES[1]

Charles A. West, Peter Moesta, Donald F. Jin,
Augusto F. Lois, and Karen A. Wickham

Department of Chemistry and Biochemistry,
University of California, Los Angeles
Los Angeles, California 90024

ABSTRACT Pectic fragments from the plant cell wall
have been shown to mediate the elicitation of the
stress metabolite casbene in castor bean (Ricinus
communis L.) seedlings during their interaction with
the fungus Rhizopus stolonifer. It has been shown
by results from several laboratories that pectic
fragments or pectin degrading enzymes can serve as
elicitors in phytoalexin-producing plants from several different Families. These findings suggest a
general role of pectic fragments as elicitors. In
castor bean seedlings pectic substances are the
most efficient oligosaccharide elicitors; chitosans
are weak elicitors and Phytophthora megasperma var.
glycinae cell wall glucans appear not to elicit.
Ophiobolin A, a phytotoxic fungal metabolite, is
also an elicitor of casbene synthetase in castor
bean seedlings. Little information is currently
available about the mechanisms involved in pectic
fragment elicitation at the molecular level.

INTRODUCTION

A variety of molecules of fungal origin, including cell
wall polysaccharides, glycoproteins and fatty acids, have
been implicated as elicitors of stress metabolite biosynthesis in responding higher plants during their interactions

[1]This work was supported by National Science Foundation
Research Grant PCM 83-02011 and United States Department of
Agriculture Research Grant 82-CRCR-1-1090.

with potentially pathogenic fungi. Recent evidence indicates that enzymes of fungal origin which catalyze the degradation of pectic substances of the plant cell wall can also stimulate stress metabolite production in plants. This response is mediated by pectic fragments released from the plant cell wall as a consequence of the action of the microbial enzymes. Therefore, plants have the capacity to recognize chemical stimuli originating both from the fungus and from their own cell walls. This paper will summarize the current status of knowledge of pectic fragment elicitors, and will consider the possible interplay of these elicitors with other types of stimuli in determining the nature of the plant's response to a stress situation.

RESULTS AND DISCUSSION

Elicitation of Casbene Synthetase Activity in Castor Bean Seedlings.

Studies of diterpene biosynthesis in castor bean (Ricinus communis L.) seedlings in our laboratory grew out of the discovery that cell-free extracts of young seedlings catalyzed the biosynthesis of a family of four polycyclic diterpene hydrocarbons plus a new diterpene hydrocarbon to which the trivial name casbene was assigned and for which the macrocylic structure illustrated in Figure 1 was proposed (1, 2). The proposed structure has since been confirmed by Crombie et al. (3) by synthesis. A mechanism for the biosynthesis of casbene is indicated in Figure 1.

P. Moesta has recently succeeded in purifying casbene synthetase to apparent homogeneity (4). A series of ion exchange and dye ligand chromatography steps permitted a 4,700-fold purification of the enzyme from the crude extract of 1,600 castor bean seedlings that had been infected with Rhizopus stolonifer spores. A single polypeptide band of apparent molecular weight 59,000 was detected by Ag-staining of an SDS-polyacrylamide gel after electrophoresis. Apparently the native enzyme also consists of a single polypeptide since a polyacrylamide gel of the partially purified native enzyme also revealed only one band of casbene synthetase activity of similar molecular weight after electrophoresis (5). Monospecific polyclonal antibodies directed against this enzyme have been raised in a rabbit. This anti-serum will be of great value for proposed studies of the transcriptional and translational regulation of the synthesis of casbene synthetase.

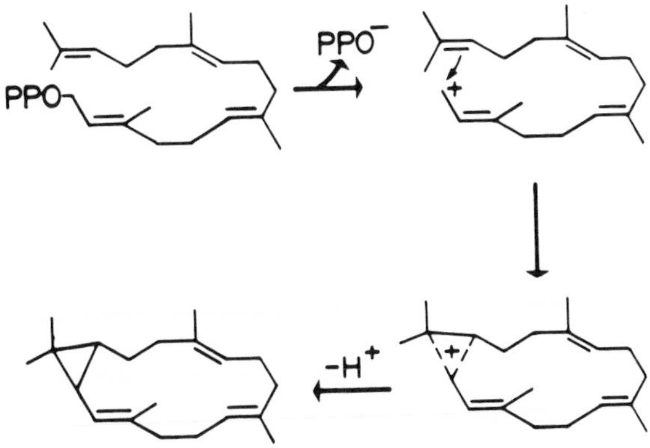

FIGURE 1. Mechanism of the reaction catalyzed by
            casbene synthetase.

Casbene synthetase is localized in the proplastids of
endosperm tissue in young seedlings. Sucrose density gradients of organelle preparations from R. stolonifer-infected
seedlings show a coincidence in the positions of the pro-
plastid marker enzyme, triose phosphate isomerase, and casbene synthetase (6).

A large variability in the levels of casbene biosynthe-
sized in cell-free extracts was observed during the early
investigations. This was traced to the degree of fungal contamination of the young seedlings (7). Seedlings intention-
ally infected with R. stolonifer or Aspergillus niger
yielded extracts with a relatively high level of casbene
synthesis, whereas seedlings maintained under aseptic conditions had little or none of this activity. Also, casbene
was shown to possess antifungal antibiotic properties. The
combination of these two characteristics led to the propo-
sal that casbene might be serving the castor bean seedling
as a phytoalexin. Even though the participation of casbene
in a hypersensitive response has not been clearly established, we elected to study the control of casbene biosynthesis as a model system for the regulation of stress metabolite production because the features of its biosynthesis
are relatively simple and well understood.
    The first objective was to study the nature of the eli-

citor molecules for casbene synthetase activity present in culture filtrates of R. stolonifer. A bioassay for elicitor was devised which depends on measuring casbene synthetase activity in cell-free extracts from a pool of half-seedlings that have been exposed under aseptic conditions to solutions of substances to be tested (8). An elicitor fraction rich in glycoproteins was partially purified from culture filtrates of R. stolonifer. The elicitor activity of this fraction was labile to heating in a manner which indicated that it depended on a native protein structure for its activity. A polygalacturonase activity was also detected in the elicitor-containing fraction. Purification of the polygalacturonase to homogeneity permitted the demonstration that both enzyme activity and elicitor activity resided in the same molecule (9). This polygalacturonase-elicitor is a mannose-containing glycoprotein of molecular weight 32,000 (10). It acts as an endohydrolase to catalyze the hydrolysis of polygalacturonic acid to a limit digest composed primarily of dimers, trimers and tetramers of α-1,4-D-galacturonide oligomers; it will also hydrolyze citrus pectin and cell wall pectic materials somewhat less efficiently (10).

The elicitor activity of the polygalacturonase depends on the presence of the catalytic activity of native enzyme; denatured enzyme appears not to serve as an elicitor (11). This finding pointed to the pectic fragments released through the catalytic action of the polygalacturonase on pectic substances of the plant cell wall as the agents responsible for the elicitor activity of the enzyme. Support for this view came from the demonstration that particulate fractions of castor bean seedlings, isolated pectic substances from castor bean cell walls, citrus pectin, and polygalacturonic acid all released heat-stable, soluble materials with elicitor activity on treatment with R. stolonifer polygalacturonase (11).

The chemical nature of pectic fragment elicitors and the structural features required for activity in the casbene synthetase elicitor assay have been investigated in our laboratory. One source of information has come from the properties of a series of α-1,4-D-galacturonide oligomers produced by partial digestion of polygalacturonic acid with R. stolonifer endopolygalacturonase and fractionation of the resulting mixture on a DEAE-Sephadex A-25 column (12). Oligomers containing eight or fewer residues were inactive, while all oligomers tested larger than this were active. Trideca-α-1,4-D-galacturonide was the most active on a weight basis of those tested. Full methyl esterification of a mixture of active oligomers with diazomethane produced a derivative which was more than twenty-times less active than an unmodified control. Similarly, reduction of the reducing termini of a mixture of active oligomers with NaBH$_4$ greatly reduced their elicitor activity. And finally, it was shown that a variety of structurally different polyanionic polymers had no detectable activity. From these results it was concluded that the size of the oligomer, its polyanionic character, the presence of a reducing terminus and other unidentified structural features of the galacturonosyl units are important structural features necessary for activity.

An incomplete study of the structural features necessary for activity has been undertaken with the complex mixture of pectic fragments released from pectic substances of the castor bean cell wall by treatment with R. stolonifer endopolygalacturonase (13). One conclusion that is clear is that the elicitor activity is associated with a diverse array of pectic fragments of different sizes, charges, and structures; however, all of them may share some necessary structural feature in common. A more tentative conclusion is that unesterified, or slightly esterified, galacturonide oligomers of sufficient length to possess elicitory activity may be present. If so, this implies that stretches of unesterified, or slightly esterified, polygalacturonic acid probably exist in the plant cell wall and can serve as a substrate for the enzyme.

These investigations of elicitation of casbene production in castor bean seedlings by the fungus R. stolonifer have demonstrated that the fungal extracellular enzyme endopolygalacturonase, which is a part of the potential pathogen's arsenal of weapons for degrading the plant cell wall, is also a major elicitor. Further, pectic fragments released from the plant cell wall through the action of the fungal enzyme are obligate intermediates in the elicitation process. Thus, rather than recognizing a fungal molecule directly, the plant is recognizing a product of the degradation of its own cell wall as a signal to initiate a defensive response.

How Generally Are Pectic Fragments Involved in the Elicitation of Stress Metabolites?

Is the utilization of pectic fragments as elicitors unique to the castor bean-R. stolonifer interaction? It soon became apparent from a parallel line of investigation in Albersheim's laboratory that this was not the case. Hahn et al. (14) reported that partial acid hydrolysis of cell walls from soybeans or other plants released substances that acted as elicitors of the phytoalexin glyceollin in soybean tissues. Nothnagel et al. (15) confirmed that the so-called 'endogenous elicitors' produced in this manner were galacturonide oligomers. The most active substance in this assay system proved to be dodeca-α-1,4-D-galacturonide. Davis et al. (16, 17) have further demonstrated that the bacterial phytopathogen, Erwinia carotovora, releases a pectate lyase-elicitor that acts on polygalacturonic acid or other types of pectic substances to release pectic fragments which serve as heat-stable glyceollin elicitors in soybeans. The close similarity of these results to those with the castor bean-R. stolonifer system summarized above is evident.

Table 1 summarizes the results we are aware of in which polygalacturonases and/or pectic fragments serve as elicitors of phytoalexins or putative phytoalexins in plants. These results are limited and, in some instances, preliminary. Nonetheless, they do suggest a widespread role for pectic fragments as elicitors. Both terpenoid and isoflavanoid phytoalexins and four different plant families are represented in these examples. The only negative result we are aware of is the failure of either polygalacturonase or pectic fragments to elicit rishitin in potato tuber tissue (Wickham, unpublished).

A different type of stress response has also been shown to involve pectic components derived from the cell wall as regulatory agents. Ryan and his associates have demonstrated the substantial accumulation of proteinase inhibitors in tomato, potato and other plants in response to the wounding of leaves mechanically or by insect feeding (20). A proteinase inhibitor-inducing factor (PIIF) released from wounded leaves transmits the response from the site of wounding to other parts of the plant. PIIF activity is associated with pectic substances of the plant cell wall (21). PIIF can be further digested by polygalacturonases to release small galacturonic acid-rich fragments with PIIF activity (22). Thus, a totally different type of response induced by another type of stress appears also to be mediated by pectic fragments originating in the plant cell wall.

TABLE 1

PHYTOALEXIN-PRODUCING PLANTS IN WHICH POLYGALACTURONASE
AND/OR PECTIC FRAGMENTS HAVE BEEN SHOWN TO SERVE AS ELICITORS

Phytoalexin	Plant	Family	Reference
Phaseollin	Phaseolus vulgaris	Leguminosae	Wickham[a]
Pisatin	Pisum sativum	Leguminosae	Hadwiger[a]
Medicarpin	Trifolium repens	Leguminosae	Gustine[a]
Glyceollins	Glycine max	Leguminosae	Hahn et al. (14)
			Davis et al. (16)
Casbene	Ricinus communis	Euphorbiaceae	Bruce and West (11)
			Lee and West (9)
Ipomeamarone	Ipomea batatas	Convolvulaceae	Wickham[a]
			Sato et al. (18)
Not identified	Beta vulgaris	Chenopodaceae	Vasil'eva et al. (19)

What is the Relationship between Pectic Fragment Elicitors and Other Types of Elicitors?

The biotic elicitors that have been described to date can be divided into two groups (23). First there is a class of fungal components that have structural features not found normally in higher plants and therefore can be recognized by the plant as a foreign molecule. Fungal glucans, chitosan and polyunsaturated fatty acids fall into this class. These substances are probably released from the microbial cell wall and surface, perhaps in part through the action of higher plant enzymes. They then may interact with specific receptors in the plant to initiate the response. A second group of elicitors are those released from the plant's own cell walls, at least in some cases through the action of enzymes from an invading microbe. It can be postulated that these, too, may bind to specific receptors in the plant to initiate the plant response. Pectic fragment elicitors are the best defined example of this class at the present time, though others may exist.

The responsiveness of castor bean seedlings to several different reasonably well-defined elicitor preparations has been evaluated with the casbene synthetase elicitor bioassay as described (12). Figure 2 presents dose response curves for four different types of elicitor preparations: an acid-hydrolyzed PIIF fraction from tomato leaves (22), a chitosan sample prepared from crab chitosan by nitrous acid hydrolysis (24), an enriched sample of Phytophthora megasperma var. glycinae glucan from a partial acid hydrolysate supplied by Sharp and Albersheim, and a mixture of α-1,4-D-galacturonide oligomers (pectic fragments) prepared by partial hydrolysis of citrus polygalacturonic acid with endo-polygalacturonase according to Jin and West (12). Table 2 lists the ED$_{50}$ values estimated from these data. The most effective elicitors are the pectic fragments and the acid-hydrolyzed PIIF sample. It is not surprising that these elicitors have comparable activities since they probably have very similar compositions. The chitosan preparation is active, but is an order of magnitude less effective. The Pmg glucan preparation is not significantly active in this assay system at the concentrations tested. These chitosan and glucan preparations would be highly active in this concentration range as elicitors of pisatin in pea tissues and glyceollin in soybean tissues, respectively. Clearly there is specificity shown toward different elicitors in the casbene synthetase elicitor assay; however, the basis for the specificity is not known.

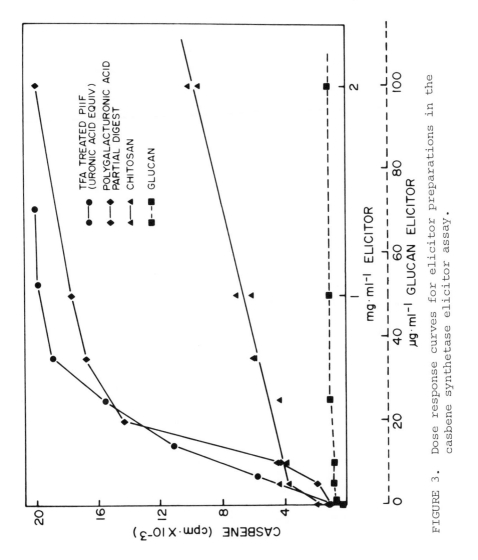

FIGURE 3.   Dose response curves for elicitor preparations in the casbene synthetase elicitor assay.

TABLE 2

ACTIVITY OF DIFFERENT ELICITOR PREPARATIONS
IN THE CASBENE SYNTHETASE ELICITOR ASSAY

Preparation	ED_{50} $mg \cdot ml^{-1}$
Pectic Fragments	0.20
PIIF, acid-hydrolyzed	0.40
Chitosan	3.2
Pmg Glucan	not active

Other types of substances can also serve as elicitors. Ophiobolin A is a sesterterpenoid metabolite produced by Helminthosporium oryzae, a fungal pathogen of rice, and related organisms. Ophiobolin A inhibits the growth of rice coleoptiles and seedling roots, and has been implicated in disease development in rice seedlings infected with H. oryzae. We were interested in examining the effect of this substance on elicitation of castor bean seedlings by pectic fragments because C. L. Tipton of Iowa State University reported to us that ophiobolin A could serve as a calmodulin inhibitor and we were interested in testing the possible role of calcium and calmodulin in elicitation by pectic fragments. To our surprise, ophiobolin A itself at relatively low concentrations elicited casbene synthetase activity. Table 3 illustrates the elicitor activity of ophiobolin A and pectic fragments for comparison. The mechanism of the effect of ophiobolin A is not clear; it could be related to the reported effects of ophiobolin A in stimulating leakage of electrolytes and organic compounds from root tissues. For example, Tipton et al. (25) have shown that ophiobolin A stimulates the net leakage of electrolytes and glucose from maize roots and inhibits uptake of 2-deoxyglucose by this tissue.

The most significant observation dealing with the interrelationship of pectic fragment elicitors with other types of elicitors comes from the work of Albersheim et al. (26). They report that a suboptimal dose of glucan elicitor, which itself gives only a very low response in the glyceollin elicitor assay in soybean cotyledons, gives a greatly in-

TABLE 3

CASBENE SYNTHETASE ELICITOR ACTIVITY
OF OPHIOBOLIN AND PECTIC FRAGMENTS

Treatment	Elicitor activity	
	casbene cpm	minus water control
Water control	1,280	–
Pectic fragments,[a] 0.5 mg/ml	8,600	7,520
Pectic fragments,[a] 1.0 mg/ml	14,060	12,780
Ophiobolin, 14µM	4,940	3,660
Ophiobolin, 30µM	13,190	11,910

[a]Pectic fragments = partial digest of PGA
with R. s. EPGase

creased response when mixed with doses of endogenous elicitor (pectic fragments) that are ineffective by themselves.
This synergism, or potentiation of the response to glucans
by pectic fragments, suggests that the two types of elicitors must be interacting at some level in vivo. Thus, the
actions of the two types of elicitors may be viewed as complementary rather than alternative.

What Are the Mechanisms Involved in Pectic Fragment Elicitation at the Molecular Level?

There is very little information available at present
to help in answering this important question. A hypothetical elicitation cascade is proposed in Figure 3 as a working model based on analogy to what is known about responses to elicitors and other stimuli in a variety of systems.
We have initiated investigations of various aspects of this

Pectic Fragment Elicitors

↓

Plasma Membrane Receptor

↓

Intracellular Messenger

↓

Activation of Transcription

↓

Specific mRNA Species

↓

Casbene Synthetase and Other
Elicitable Gene Products

FIGURE 3. Hypothetical elicitation cascade.

model in our laboratory with respect to casbene biosynthesis
in castor bean seedlings, but the results are too preliminary to be meaningful. We are particularly interested at
present in seeking evidence for specific receptors for pectic fragments and in examining the transcriptional and
translational regulation of the synthesis of casbene synthetase. The availability of antibodies to casbene synthetase
will greatly facilitate these latter investigations as
noted earlier.

What is the Physiological Importance of Elicitation by
Pectic Fragments?

Relatively high concentrations of exogenously supplied
pectic fragments are required for them to serve as elicitors
in the assay systems examined to date. It is not clear how
this relates to the concentrations required at the active
site, or what concentrations may be developed at the active
site in vivo. The observed synergism between glucan elicitors and pectic fragment elicitors in the elicitation of
glyceollins in soybean tissue observed by Albersheim and
associates is of interest with respect to the physiological
actions of these elicitors. The two types of elicitors
appear to interact to induce a large response when supplied
together at concentrations where neither alone was very effective. This suggests the possibility that the pectic
elicitors act as a general signal of wall damage which
serves to modulate both qualitatively and quantitatively
the plant's response to other types of signals reflecting
the source of wall damage e.g. the presence of a microbial

pathogen, mechanical damage, natural wall loosening associated with plant development, and so forth. Such a system would allow the plant more flexibility in responding to a variety of stimuli that are associated directly or indirectly with damage to the cell wall. Clearly, much more experimental work will be required to determine whether this suggestion has merit.

ACKNOWLEDGMENTS

We would like to thank the following individuals for making available the indicated samples for testing: Dr. Peter Albersheim of the University of Colorado for Pmg glucan elicitor; Dr. Lee Hadwiger of Washington State University for chitosan; Dr. Clarence Ryan of Washington State University for PIIF samples; and Dr. Carl Tipton of Iowa State University for ophiobolin A.

REFERENCES

1. Robinson DR, West CA (1970). Biosynthesis of cyclic diterpenes in extracts from seedlings of Ricinus communis L. I. Identification of diterpene hydrocarbons formed from mevalonate. Biochemistry 9:70.
2. Robinson DR, West CA (1970). Biosynthesis of cyclic diterpenes in extracts from seedlings of Ricinus communis L. II. Conversion of geranylgeranyl pyrophosphate to diterpene hydrocarbons and partial purification of the cyclization enzymes. Biochemistry 9:80.
3. Crombie L, Kneen G, Pattenden G, Whybrow D (1980). Total synthesis of the macrocyclic diterpene (-) casbene, the putative biogenetic precursor of lathyrane, tigliane, ingenane and related terpenoid structures. J Chem Soc Perk Trans 1:1711.
4. Moesta P, West CA (unpublished observations).
5. Dueber MT, Adolf W, West CA (1978). Biosynthesis of the diterpene phytoalexin casbene. Partial purification and characterization of casbene synthetase from Ricinus communis. Plant Physiol 62:598.
6. West CA, Dudley MW, Dueber MT (1979). Regulation of terpenoid biosynthesis in higher plants. In Swain T, Waller GR (eds): "Recent Advances in Phytochemistry" Vol 13, New York: Academic, p 163.

7. Sitton D, West CA (1975). Casbene, an antifungal diterpene produced in cell-free extracts of Ricinus communis seedlings. Phytochem 14:1921.

8. Stekoll M, West CA (1978). Purification and properties of an elicitor of castor bean phytoalexin from culture filtrates of the fungus Rhizopus stolonifer. Plant Physiol 62:571.

9. Lee S-C, West CA (1981). Polygalacturonase from Rhizopus stolonifer, an elicitor of casbene synthetase activity in castor bean (Ricinus communis L.) seedlings. Plant Physiol 67:633.

10. Lee S-C, West, CA (1981). Properties of Rhizopus stolonifer polygalacturonase, an elicitor of casbene synthetase in castor bean (Ricinus communis L.) seedlings. Plant Physiol 67:640.

11. Bruce RJ, West CA (1982). Elicitation of casbene synthetase activity in castor bean. The role of pectic fragments of the plant cell wall in elicitation by a fungal endopolygalacturonase. Plant Physiol 69:1181.

12. Jin DJ, West CA (1984). Characteristics of galacturonic acid oligomers as elicitors of casbene synthetase activity in castor bean seedlings. Plant Physiol (in press).

13. West CA, Bruce RJ, Jin DF (1984). Pectic fragments of plant cell walls as mediators of stress responses. In Dugger WM, Bartnicki-Garcia S (eds): "Structure, Function and Biosynthesis of Plant Cell Walls," Am Soc of Plant Physiologists (in press).

14. Hahn MG, Darvill AG, Albersheim P (1981). Host-pathogen interactions: XIX. The endogenous elicitor, a fragment of a plant cell wall polysaccharide that elicits phytoalexin accumulation in soybeans. Plant Physiol 68:1161.

15. Nothnagel EA, McNeil M, Albersheim P (1983). Host-pathogen interactions. XXII. A galacturonic acid oligosaccharide from plant cell walls elicits phytoalexins. Plant Physiol 71:916.

16. Davis KR, Lyon GD, Darvill AG, Albersheim P (1982). A PGA-lyase isolated from Erwinia carotovora is an elicitor of phytoalexins in soybeans. Plant Physiol 69:S-142

17. Davis KR, Nothnagel EA, McNeil M, Darvill AG, Albersheim P (1983). Endopolygalacturonic acid lyase isolated from Erwinia carotovora elicits phytoalexins in soybean by releasing an oligosaccharide elicitor from plant cell walls. Plant Physiol 72:S-161.

18. Sato K. Uritani I, Saito T (1982). Properties of terpene-inducing factor extracted from adults of the sweet potato weevil, Cylas formicarius Fabricius (Coeloptera: Brenthidae). Appl Entomol Zool 17:368.
19. Vasil'eva KV, Gladkikh TA, Davydova MA, Madosanova GA, Umralina AR (1982). Phytoalexin production in sugar beet roots by extracellular endopolygalacturonase of Rhizoctonia aderholdii (Ruhl.) Kolosh. Mikol Fitopatol 16:525.
20. Ryan CA (1980). Wound-regulated synthesis and vacuolar compartmentation of proteinase inhibitors in plant leaves. Curr Top Cell Regul 17:1.
21. Ryan CA, Bishop P, Pearce G, Darvill AG, McNeil M, Albersheim P (1981). A sycamore cell wall polysaccharide and a chemically related tomato leaf polysaccharide possess similar proteinase inhibitor-inducing activities. Plant Physiol 68:616.
22. Bishop PD, Makus DJ, Pearce G, Ryan CA (1981). Proteinase inhibitor-inducing factor activity in tomato leaves resides in oligosaccharides enzymically released from cell walls. Proc Natl Acad Sci USA 78:3536.
23. West CA (1981). Fungal Elicitors of the phytoalexin response in higher plants. Naturwissenschaften 68:447.
24. Hadwiger LA, Beckman JM (1980). Chitosan as a component of pea-Fusarium solani interactions. Plant Physiol 66:205.
25. Tipton CL, Paulsen PV, Betts RE (1977). Effects of ophiobolin A on ion leakage and hexose uptake by maize roots. Plant Physiol 59:907.
26. Albersheim P (1984). Oligosaccharins - biologically active fragments of the polysaccharides of cell walls. In Dugger WM, Bartnicki-Garcia S (eds): "Structure, Function and Biosynthesis of Plant Cell Walls," Am Soc of Plant Physiologists (in press).

Cellular and Molecular Biology of Plant Stress, pages 351–365

CHEMICAL CONJUGATION AND COMPARTMENTALIZATION:
PLANT ADAPTATIONS TO TOXIC NATURAL PRODUCTS[1]

Eric E. Conn

Department of Biochemistry and Biophysics
University of California
Davis, California 95616

ABSTRACT   Plants appear to have at least two major
mechanisms for protection against the chemically
reactive natural (secondary) compounds which they
produce. One of these is physical compartmentation
in which natural products are physically sequestered
in specific organs, tissues, cells and organelles.
The second is chemical compartmentation in which
natural products occur as derived forms or conjugates.

Higher plants produce a large variety of chemically
reactive, natural (secondary) products (1,2). In a recent
book on biochemical ecology (3), the number of those
compounds which may be involved in interactions between
plants, animals and microorganisms has been estimated as
more than 12,000. The list includes phenols, quinones,
amines, carboxylic and amino acids, cyanohydrins, alde-
hydes, ketones, diazo compounds, terpenes and alkaloids.
In some instances concentrations in the millimolar range
represent a potentially toxic environment in which the
primary metabolic processes of respiration, photosynthesis,
and the assimilation of NH_3 and H_2S must continue without
significant inhibition. For example, the concentration

[1]This work has been supported in part by continuing
grants from the U.S. Public Health Service (current grant,
GM05301-27) and the National Science Foundation (current
grant, PCM81-04497).

of the cyanogenic glycoside dhurrin is approximately 0.2 M
in the vacuoles of the epidermal layer of young, green
sorghum leaves (4). Similarly, the concentration of the
non-protein amino acid, azetidine-2-carboxylic acid, in
the tubers of some liliaceous species can be estimated as
100 mM if it is uniformly distributed (5). This value
would be increased several fold if the amino acid were
confined to a single type of tissue or cell, or to an
organelle or other intracellular location.

Plants appear to have at least two major mechanisms
for providing protection against chemically reactive
natural products. One of these, which is termed structural or physical compartmentation, refers to the physical
sequestering of a compound in a specific organ (e.g.
seeds, roots, leaves), tissue (e.g. epidermal layer or
spongy mesophyll), organelle (e.g. vacuoles, chloroplasts)
or extra cytoplasmic space (e.g. cell wall, subcuticular
space). The other protective mechanism can be termed
chemical compartmentation, for want of a better name, and
describes the fact that many natural products occur in the
plant as derived forms (e.g. amides, esters, glycosides,
ethers) in which reactive functional groups are masked.
An essential component of chemical compartmentation is the
enzymology associated with the metabolism of natural
(secondary) products. Included in this component are not
only those enzymes which catalyze either the synthesis or
degradation of the derived compounds, but also those
enzymes which are capable of directly detoxifying toxic
natural products either produced by the plant itself or
foreign compounds (i.e. xenobiotics) encountered by the
plant.

Physical Compartmentation.

The existence of specialized cells in plants which
accumulate natural products is well established. One
example is the laticifers of the Papaveraceae which contains high concentrations of alkaloids localized within
small vacuoles. Another such cell is the isodiametric
alkaloid cell of Sanguinaria canadensis. Tannin inclu-
sions have been described in the vacuoles of young tissues
of Oenothera. The production of essential oils and resins
is an important property of specialized secretory structures such as glandular trichomes and oil or resin ducts
(6,7).

Wiermann has recently reviewed the relationship between the differentiation of such specialized cells and tissues and their content of natural products (8). He points out that plants utilize two methods for storing natural products. One of these is intraplasmic involving vacuoles and plastids as storage sites; the other is extraplasmic where the cell wall, pollen wall, subcuticular space and cuticular surface are involved.

The development of methods for isolating intact vacuoles in large number from plant protoplasts (9) opened the way for examining the vacuole as a storage site and studying its role in compartmentation. Similarly, methods allowing the isolation of intact plastids, cell walls, the plasma membrane and tonoplasts have been or are being developed, making available these structures for study (10,11). Taking just the group of natural products known as glycosides, work has shown that anthocyanins (12), the cyanogenic glucoside dhurrin (4,13), 2-coumaric acid glucosides (14), coumaryl glucosides (15), and malonylated flavonoid glycosides (16) are contained in vacuoles isolated from plant tissues known to contain these natural products. Little is known, on the other hand, of where the enzymes involved in biosynthesis of these glycosides occur, although Fritz and Grisebach (17) have proposed that the glucosyl transferases presumably involved may be in close proximity but external to the vacuole. In the case of the glucoside dhurrin, the enzymes responsible for its hydrolysis are not even associated with the tissue (epidermal) from which the dhurrin-containing vacuoles were obtained (see below). Vacuoles have also been identified as the depository for the glucoside of 4-hydroxy-2,5-dichlorophenoxyacetic acid produced when the herbicide, 2,4-dichlorophenoxyacetic acid was administered to soybean callus and cell cultures (18,19).

Enzymes associated with the metabolism of natural products have been found in vacuoles (10). These include peroxidase, myrosinase, and glycosidases. However, work with the last named group of enzymes is complicated by the fact that synthetic glycosides of unphysiological aglycones (e.g. p-nitrophenol, 3,4-dinitrophenol, 4-methyl umbelliferone) have been employed as substrate in many of these studies. Such substrates have the advantage that they are commercially available and their hydrolysis readily measured. However, it is now clear that plants contain β-glycosidases which are very specific for naturally occuring glycosides found in that plant and frequently have little or no activity either towards related glycosides (e.g. a related cyanogenic glycoside) or towards the synthetic substrates. Any studies on this group of important enzymes should involve the physiological substrate (20).

Study of the cyanogenic species, Sorghum bicolor, has shown that the enzymes involved in the degradation of the cyanogenic glucoside dhurrin [4-hydroxy(S)mandelonitrile-β-D-glucoside] are effectively compartmentalized in a different tissue than that containing substrate (4). The dhurrin which occurs in 6-day old light-grown seedlings is found almost exclusively in the upper and lower epidermis of the leaf. Within the epidermal cell, the glycoside is found in the vacuole (at a concentration of 0.2 M). The β-glucosidase and hydroxynitrile lyase which accomplish the sequential decomposition of dhurrin to one mole each of 4-hydroxybenzaldehyde, glucose and HCN are predominantly (> 90%) located in the mesophyll cell. Within that cell, other work has established that most, if not all, of the β-glucosidase is associated with the chloroplast, while the hydroxynitrile lyase is located in the cytosolic fraction (21).

This intercellular compartmentation is reasonable when one considers the amount of dhurrin which has to be sequestered intact in green sorghum leaves. Such leaves contain approximately 30 μmols/g of dhurrin (5% dry wt.). The epidermal cells, which represent about 35% of the volume of the leaf, in turn consist predominantly of a large vacuole which occupies 95% of the cell. Thus the epidermal vacuoles represent approximately 30% of the leaf volume and the dhurrin concentration in these vacuoles is about 0.2 M. While the mesophyll (and vascular bundles) mainly constitute the remaining volume of the leaf, the vacuole in these cell types is significantly smaller and therefore is less suitable as a site for dhurrin storage.

The compartmentalization of the glucosides of 2-hydroxycinnamic acid and their related β-glucosidase in leaves of Melilotus alba (14) contrasts in part with that just described for dhurrin in Sorghum. First, there was no evidence for these glucosides being located exclusively in the epidermis of M. alba leaves; comparable concentrations were found in the spongy mesophyll cells of the leaf

interior. On the other hand, vacuoles isolated from protoplasts of young leaves of M. alba were shown to contain
essentially all of the glucoside present in the leaf, and
the β-glucosidase was absent from such vacuoles. While one
might have expected that the enzyme was in the cytosolic
fraction, extracts of protoplasts prepared from M. alba
leaves contained less than 1% of the β-glucosidase present
in the whole leaf extract. This in turn indicated that the
β-glucosidase is absent from the cell itself and therefore
likely to be associated with the wall itself, or alternatively with the periplasmic space surrounding the cell.
In the study on M. alba it was possible to demonstrate in
the protoplast extracts a glucosyltransferase which utilized 2-hydroxy-trans-cinnamic acid as a substrate (14).
This activity corresponded to 92% of the transferase in the
extract of an equivalent amount of leaf tissue. Thus, we
concluded that the transferase is located in the cytoplasm
of the mesophyll but external to the vacuole.

In earlier studies Marcinowski and Grisebach (22)
demonstrated that a cell-wall fraction from spruce seedlings hydrolyzed coniferin, a glucosidic precursor of
lignin. Subsequently, they were able to localize such an
enzyme at the inner layer of the secondary cell wall (23).
When isolated cells of M. alba mesophyll were prepared and
examined for β-glucosidase, only a small fraction (7%) of
the total activity was observed (E. Wurtele, unpublished).
Such results suggest either that the β-glucosidase is
loosely attached to the cell wall and easily removed during
preparation of the isolated cell, or that the enzyme is
located primarily in the intercellular spaces of the intact
leaf.

Chemical Compartmentation.

This term is used to describe the fact that numerous
natural products occur in the plant as derived forms in
which their reactive, functional groups are masked. Thus,
the carboxyl group of an acid may be masked as an ester or
amide; an amino group may be acylated with an organic acid;
a phenolic hydroxyl, which might otherwise undergo oxidation, may be methylated to form an O-methyl ether or
combined with a sugar to form a glycoside.
One has only to consult the plant herbicide literature
to encounter mechanisms by which higher plants are protected against those compounds and other xenobiotics (24,25).

Thus, herbicides with carboxyl groups (e.g. 2,4-D) are
frequently conjugated with amino acids such as aspartic or
glutamic acid. Such amide conjugates are also known for
naturally occurring acids produced by the plant. Thus,
indoleacetic acid is found as a conjugate of aspartic acid
and the epsilon amino group of lysine. Malonic acid is
frequently found in amide linkage with D-amino acids (26)
and more recently with 1-aminocyclopropane-1-carboxylic
acid (27). The biological significance of such conjugation
is not completely understood, but there is the obvious
possibility that such conjugates will have a different,
possibly lower physiological activity than the free acid
or amine.

What is being considered here is the general ability
of acids to acylate functional groups such as amines.
Other naturally occurring organic acids extensively involved in acylation are acetic acid, the several cinnamic
acids (cinnamic itself, p-coumaric, caffeic, ferulic and
sinapic) and the γ-carboxyl of L-glutamic acid. γ-Glutamyl
peptides of several non-protein amino acids are known (28)
as well as γ-glutamyl peptides of D-amino acids (29).

Acylation of hydroxyl groups, especially those of
sugars, is also a common reaction. But glucose, and other
monosaccharides, can serve as the activated conjugation
species to glycosylate hydroxyl, amino, and carboxyl
groups. The activation of the sugar is accomplished by
its conversion to a nucleoside diphosphate sugar (e.g.
uridine diphosphate glucose). The activated acylating
reagents discussed above would be the thioester (e.g.
cinnamoyl-CoA). In the case of glutamic acid, the γ-
carboxyl may be activated as the enzyme-bound acyl phosphate.

One example of a toxic natural product which is
rendered less toxic by acylation is β-nitropropionic acid.
This compound, which is a suicide inactivator of succinic
dehydrogenase (30), is found as a glucose ester in Indigofera, Astragalus and some other legumes. It can exist both
as the 1,6-diester (31) and as the 1,2,6-triester (32) of
glucose. These glucosides are responsible for various
toxic syndromes observed in livestock which feed on these
plants, especially in the western United States and in
Australia. Presumably they are hydrolyzed to form the free
acid which then can act as an inhibitor of the tricarboxy-
lic acid cycle.

An example of an extremely toxic plant product which
is rendered non-toxic by glucosylation is methylazoxyme-

thanol. This alcohol is a vigorous methylating
reagent capable of modifying the purine and pyrimidine
bases. It is found in members of the Cycadaceae as a
glycoside called cycasin. When injected intraperi-
toneally into rats, cycasin is excreted almost quantitatively in the urine. If administered orally,
however, it is highly toxic presumably due to its
hydrolysis by bacteria in the digestive tract. The
aglycone itself when administered orally is also very
toxic (33).

Acylated glycosides, involving both a naturally
occurring acid and a sugar, are known. One example
is the malonic acid hemi-ester of the β-glucoside of
the flavone apigenin which is produced in parsley
cell cultures after inactivation with UV light (34).
Another example is the 6-0-malonyl-β-D-glucoside of
betanidin (35). These complex glycosides can be
isolated from their natural sources provided care
is taken to inactivate endogenous enzymes (esterases
and glycosidases) which are present in the source
tissue. Enzymes which catalyze the synthesis of the
hemi-ester bond (36,37) and its hydrolysis (38,39)
have been described. Recently (16) a role has been
proposed for malonylation in the transport of flavonoid glycosides across the tonoplast and in storage
of the more soluble malonylated pigment in the
vacuole.

Enzymology of Chemical Compartmentation.

Plants possess an armory of enzymes useful in protecting themselves against foreign compounds to which
they are exposed. Consideration of the ways in which
plants react to and metabolize xenobiotics (e.g. herbicides) show that the basic reactions are oxidation,
reduction and hydrolysis in addition to conjunction
discussed above (24,25). The oxidation reactions are
usually catalyzed by oxygenases and peroxidases which
accomplish such reactions as aromatic hydroxylation,
N- and O-dealkylation, epoxidation and sulfur oxidation. Enzymatic reduction, as a detoxification process,
is not common, but one example is the reduction of
pentachloronitrobenzene to the corresponding aniline
in peanut (40). If the xenobiotic contains ester
or amide bonds, these can be hydrolyzed by appro-

priate esterases to the corresponding acids, alcohols,
and amines.

The several examples of chemical conjugates cited in
the previous section obviously indicate the existence of
enzymes catalyzing their synthesis (and degradation).
Those enzymes concerned with sugar conjugates (glycosyltransferases and glycosidases) have recently been reviewed
and a few generalizations have been stated (41). The
glycosyltransferases which act on simple phenols are not
able to utilize flavonoids as an alternative substrate.
On the other hand, the transferases which glucosylate the
flavonoids show a specificity for the different hydroxyl
groups on the flavonoid skeleton. In the case of a racemic
mixture of α-hydroxynitriles, an enzyme from Sorghum showed
strict specificity for the epimeric substrate possessing
the S configuration (42). Finally, UDP sugars are the best
substrates by far for these transferases and with regard to
the sugar moiety, glucose is almost exclusively preferred
(41).

The enzymatic specificity observed for most plant
glycohydrolases (glycosidases) is significantly greater
than that of the glycosyltransferases. With such specificity one wonders about the role of these enzymes in the
plant. In the great majority of cases the glycosidases are
located in plant tissues or organelles which are devoid of
the substrate. Thus, significant breakdown does not occur
until the plant tissue is disrupted and the enzyme is mixed
with its substrate. The rapid release of aglycone may well
serve to protect the plant at this point, but one can also
speculate that the glycosidase may be needed for a slower
endogenous turnover of the glucoside. Whatever their
function, it is generally true that the glycosylated aglycone usually represents a more soluble form of the natural
product. Therefore, conjugation may well be required
before the product can either be moved around the plant or
stored in significant concentrations in the vacuole.

It is also possible to cite enzymes or enzyme systems
which play a direct and obvious role in protecting the
plant from deleterious natural compounds produced by that
plant. One interesting example is the prolyl-t-RNA synthetase found in plants which produce the non-protein
amino acid, azetidine-2-carboxylic acid (A2C). This amino
acid is produced in relatively large amounts by several
legumes (e.g. Parkinsonia aculeata, Delonix regia, Conval-
laria majus) and some members of the Liliaceae, but is
never found as a component of any protein in the species

which produce A2C. On the other hand, A2C, as an analog
of proline, is incorporated into proteins when fed to seedlings of non-producer plants, and in mung beans, for example, the synthesis is lethal and the seedlings don't
survive. The incorporation into protein in non-producer
plants and the exclusion in producer plants is explained
by considering the kinetic parameters of the prolyl-t-RNA
synthetases from several species (43).

TABLE 1

KINETIC PARAMETERS OF PROLYL-tRNA SYNTHETASES FROM
A2C PRODUCER AND NONPRODUCER PLANTS[a]

Plant species	Produces large amounts of A2C	L-Proline $K_m(X\ 10^{-4}\ M)$	L-Azetidine-2-carboxylic acid $K_m(X\ 10^{-3}\ M)$	V_{max}[b]
Parkinsonia aculeata	Yes	4.35		0-5
Delonix regia	Yes	1.82		0-5
Convallaria majalis	Yes	4.5		0-5
Beta vulgaris	No	4.5	2.2	73
Hemerocallis fulva	No	6.25	5.3	75
Phaseolus aureus	No	1.37	1.43	55
Ranunculus bulbosa	No	2.9	2.0	66

[a]All data are derived from measurement of the ATP-PP$_i$
exchange reaction as described by Norris and Fowden (53).
   [b]V_{max} values expressed as a percentage of that
obtained for proline.

Table 1 shows that the enzyme from producer plants fails to
bind A2C significantly. On the other hand, the enzyme from

the non-producer plants can bind A2C and the maximal velocity of reaction is of the same order of magnitude as the
reaction with proline. The producer plant has obviously
evolved an effective mechanism for protecting itself
against a toxic amino acid which in turn may well have
provided some benefit to the plant during evolution.

　　Another interesting example of an unusual but effective protective enzyme is the enzyme β-cyanoalanine synthase which catalyzes the replacement of the thiol group
in cysteine with the nitrile group.

$$HS-CH_2-CHNH_2-CO_2H + HCN \longrightarrow NC-CH_2-CHNH_2-CO_2H + H_2S \quad (1)$$

This enzyme was discovered in experiments designed to
examine the role of HCN in the biosynthesis of cyanogenic
glycosides (44). Rather than being involved in their
biosynthesis, HCN was shown to be effectively detoxified
by its conversion to asparagine via β-cyanoalanine as an
intermediate. While such detoxification clearly was
desirable in cyanogenic species which had the capacity to
produce HCN, it was surprising at that time to find that
this enzyme was present in every plant examined, as well as
Escherichia coli and Chlorella (45). The enzyme has been
purified from lupine seedlings and extensively characterized (46).

　　Recent work on the biosynthesis of ethylene from 1-
aminocyclopropyl-1-carboxylic acid (ACC) may account for
the ubiquitous occurrence of β-cyanoalanine synthase in
plants. Peiser et al. (47) have presented data indicating
that HCN is produced in amounts equivalent to the ethylene
formed when ACC is administered to intact plants. They
have shown that $4-^{14}C$-labelled asparagine is produced when
ACC labeled in the 2 position with carbon-14 is fed to mung
bean seedlings. Thus, the synthase can serve to protect
the plant against the HCN produced in connection with
ethylene production.

　　One final example of how plants may be protected
against chemically reactive species can be cited; this is
protection from unstable or reactive intermediates produced
by multienzyme complexes or polyfunctional proteins. Such
intermediates are usually prevented from undergoing deleterious side reactions because they are produced and immediately utilized within the complex. Bifunctional
complexes of phenylalanine ammonia-lyase and cinnamic
acid-4-hydroxylase exist in several plants (48) where
they catalyze the first two steps in the formation of a

large number of phenylpropanoid compounds. Other such complexes exist in plants and their role in natural product biosynthesis has been recently reviewed (49).

In the case of cyanogenic glycosides, all but the last step in their biosynthesis is catalyzed by a membrane fraction resembling if not identical with the endoplasmic reticulum (50). This membrane fraction, obtained from several species, catalyzes a four-step biosynthetic reaction sequence which converts certain amino acids to the corresponding α-hydroxynitriles (cyanohydrins) (51,52). In Sorghum (51) the reaction sequence exhibits extensive channeling of two metabolic intermediates, namely a N-hydroxyamino acid and a nitrile. While such data might suggest the existence of two bifunctional complexes in the membrane, it has not been possible to solubilize or otherwise manipulate the membrane fraction in order to obtain information about the arrangement or organization of the individual enzymes involved. Nevertheless, the channeled nature of the reactions involved clearly prevents the two channeled metabolites (i.e. the N-hydroxyamino acid and nitrile) from undergoing deleterious side reactions. They also may facilitate the flow of carbon and nitrogen atoms in the precursor amino acids to the product cyanogenic glycosides.

CONCLUSION

Plants are protected from deleterious compounds which they either produce themselves, or encounter in their environment, by two major mechanisms. One of these is physical compartmentation in which the compound is sequestered in some structure where it minimally affects the normal metabolism of the plant. The second mechanism is chemical compartmentation whereby the deleterious compound is chemically modified, usually by plant enzymes, to a form where its potentially harmful action is reduced or eliminated.

REFERENCES

1. Bell EA, Charlwood BV (1980). Secondary plant products. In Bell EA, Charlwood, BV (eds): "Encyclopedia of Plant Physiology," Berlin: Springer-Verlag, Vol 8, p 674.

2. Conn EE (1981). Secondary plant products. In Conn EE (ed): "The Biochemistry of Plants, A Comprehensive Treatise," New York: Academic, Vol 7, p 798.

3. Harbourne JB (1982). "Introduction to Ecological Biochemistry." London: Academic, p 3.

4. Kojima M, Poulton JE, Thayer SS, Conn EE (1979). Tissue distributions of dhurrin and of enzymes involved in its metabolism on leaves of Sorghum bicolor. Plant Physiol 63:1022.

5. Fowden L (1959). Nitrogenous compounds and nitrogen metabolism in the Liliaceae. Biochem J 71:643.

6. Fahn A (1979). "Secretory Tissues in Plants." New York: Academic, p 158.

7. Schnepf E (1974). Gland cells. In Robards AW (ed): "Dynamic Aspects of Plant Ultrastructure," London: McGraw-Hill, p 331.

8. Wiermann R (1981). Secondary plant products and cell and tissue differentiation. In Conn EE (ed): "Secondary Plant Products," New York: Academic, p 85.

9. Wagner GJ, Siegelman HW (1975). Large-scale isolation of intact vacuoles and isolation of chloroplasts from protoplasts of mature plant tissues. Science 190:1298.

10. Wagner GJ (1982). Compartmentation in plant cells: the role of the vacuole. Rec. Adv. Phytochem 16:1

11. Wagner GJ (1983). Higher plant vacuoles and tonoplasts. In Hall JL, Moore AL (eds): "Isolation of membranes and organelles from plant cells," London: Academic, p 83.

12. Hrazdina G, Wagner GJ, Siegelman HW (1978). Subcellular localization of enzymes of anthocyanin biosynthesis in protoplasts. Phytochem 17:53.

13. Saunders JA, Conn EE (1978). The presence of the cyanogenic glucoside dhurrin in isolated vacuoles from Sorghum. Plant Physiol 61:154.

14. Oba K, Conn EE, Canut H, Boudet AM (1981). Subcellular localization of 2-(β-D-glucosyloxy) cinnamic acids and the related β-glucosidase in leaves of Melilotus alba Desr. Plant Physiol 68:1359.

15. Werner C, Wiemken A, Matile PH (1982). Proceedings of the Groupe Polyphenol, Toulouse, France (abstract).

16. Matern U, Heller W, Himmelspach K (1983). Conformational changes of apigenin 7-0-(6-0-malonyl glucoside), a vacuolar pigment from parsley, with solvent composition and proton concentration. Eur J Biochem 133: 439.

17. Fritsch H, Grisebach H (1975). Biosynthesis of cyanidin in cell cultures of Haplopappus gracilis. Phytochem 14:2437.
18. Schmitt R, Sandermann Jr H (1982). Specific localization of β-D-glucoside conjugates of 2,4-dichlorophenoxyacetic acid in soybean vacuoles. Zeit Natursforsch 37c:772.
19. Davidonis GH, Hamilton RH, Mumma RO (1982). Evidence for compartmentalization of conjugates of 2,4-dichlorophenoxyacetic acid in soybean callus tissue. Plant Physiol 70:939.
20. Hösel W, Conn EE (1982). The aglycone specificity of plant β-glycosidases. Trends Biochem Sci 7:219.
21. Thayer SS, Conn EE (1981). Subcellular localization of dhurrin-β-glucosidase and hydroxynitrile lyase in the mesophyll cells of Sorghum leaf blades. Plant Physiol 67:617.
22. Marcinowski S, Grisebach H (1978). Enzymology of lignification. Eur J Biochem 87:37.
23. Marcinowski S, Falk H, Hammer DK, Hoyer B, Grisebach H (1979). Appearence and localization of a β-glucosidase hydrolyzing coniferin in spruce (Picea abies) seedlings. Planta 144:161.
24. Hatzios KK, Penner D (1982). "Metabolism of Herbicides in Higher Plants." Minneaplis: Burgess
25. Fedtke C (1982). "Biochemistry and Physiology of Herbicide Action." Berlin: Springer-Verlag
26. Robinson T (1976). D-amino acids in higher plants. Life Sciences 19:1097.
27. Hoffmann NE, Yang SF, McKeon T (1982). Identification of 1-(malonylamino)cyclopropane-1-carboxylic acid as a major conjugate of 1-aminocyclopropane-1-carboxylic acid, an ethylene precursor, in plants. Biochem Biophys Res Commun 104:765.
28. Rosenthal GA (1982). "Plant nonprotein amino acids and imino acids." New York: Academic, p 273.
29. Kawasaki Y, Ogawa T, Sasaoka K (1982). Two pathways for formation of D-amino acid conjugates in pea seedlings. Agric Biol Chem 46:1.
30. Alston TA, Mela L, Bright HJ (1977). 3-Nitropropionate, the toxic substance of Indigofera, is a suicide inactivator of succinic dehydrogenase. Proc Natl Acad Sci (Wash) 74:3767.
31. Stermitz FR, Lowry WT, Ubben E, Sharifi I (1972). 1,6-Di-3-nitropropanoyl-β-D-glucopyranoside from Astragalus cibarius. Phytochem 11:3525.

32. Harlow MC, Stermitz FR, Thomas RD (1975). Isolation of nitro compounds from Astragalus species. Phytochem 14:1421.
33. Kobayashi A, Matsumoto H (1965). Studies on methyl-azoxymethanol, the aglycone of cycasin. Arch Biochem Biophys 110:373.
34. Kreuzaler F, Hahlbrock K (1973). Flavonoid glycosides for illuminated cell suspension cultures of Petroselinum hortense. Phytochem 12:1149.
35. Minale L, Piatelli M, De Stefano S, Nicolaus RA (1966). Pigments of Centrospermae. VI. Acylated betacyanins. Phytochem 5:1037.
36. Hahlbrock K (1972). Malonyl coenzyme-A: flavone glycoside malonyltransferase from illuminated cell suspension cultures of parsley. FEBS Letters 28:65.
37. Matern U, Potts JRM, Hahlbrock K (1981). Two flavonoid-specific malonyltransferases from cell suspension cultures of Petroselinum hortense: partial purification and some properties of malonyl-coenzyme A: flavone/flavonol-7-O-glycoside malonyltransferase and malonyl-coenzyme A: flavonol-3-O-glucoside malonyltransferase. Arch Biochem Biophys 208:233.
38. Davenport HE, Dupont MS (1972). The enzymic hydrolysis of malonated flavone glycosides. Biochem J 129:18P.
39. Matern U (1983). Acylhydrolases from parsley (Petroselinum hortense). Relative distribution and properties of four esterases hydrolyzing malonic acid hemiesters of flavonoid glucosides. Arch Biochem Biophys 224:261.
40. Rusners DC, Lamoureux GL (1980). Pentachloronitrobenzene metabolism in peanuts. J Agric Food Chem 28:1070.
41. Hösel W (1981). Glycosylation and glycosidases. In Conn EE (ed): "Secondary Plant Products," New York: Academic, p 725.
42. Reay PF, Conn EE (1974). The purification and properties of a UDP-glucose: aldehyde-cyanohydrin β-glucosyltransferase from Sorghum seedlings. J Biol Chem 249:5826.
43. Fowden L, Lea PJ, Bell EA (1979). The non-protein amino acids of plants. Adv Enzymol 50:117.
44. Blumenthal-Goldschmidt S, Butler GW, Conn EE (1963). Incorporation of hydrocyanic acid labelled with carbon-14 into asparagine in seedlings. Nature 197:718.

45. Dunnill PM, Fowden F (1968). Enzymatic formation of β-cyanoalanine from cyanide by Escherichia coli extracts. Nature 208:1206.

46. Akopyan TN, Braunstein AE, Goryachenkova EV (1975). Beta-cyanoalanine synthase: purification and characterization. Proc Natl Acad Sci (Wash) 72:1617.

47. Peiser GD, Wang T, Hoffman NE, Yang SF, Liu H, Walsh CT (1984). Formation of cyanide from carbon-1 of 1-aminocyclopropane-1-carboxylic acid during its conversion to ethylene. Proc Natl Acad Sci (Wash)

48. Czichi U, Kindl H (1977). Phenylalanine ammonia lyase and cinnamic acid hydroxylases as assembled consecutive enzymes on microsomal membranes of cucumber cotyledons: cooperation and subcellular distribution. Planta 134:133.

49. Stafford HA (1981). Compartmentation in natural product biosynthesis by multienzyme complexes. In Conn EE (ed): "Secondary Plant Products," New York: Academic, p 117.

50. Conn EE (1979). Biosynthesis of cyanogenic glycosides. Naturwissenschaften 66:28.

51. Møller BL, Conn EE (1980). Channeling of intermediates in dhurrin biosynthesis by a microsomal system from Sorghum bicolor (Linn.) Moench. J Biol Chem 255:3049.

52. Cutler AJ, Hösel W, Sternberg M, Conn EE (1981). The in vitro biosynthesis of taxiphyllin and the channeling of intermediates in Triglochin maritima. J Biol Chem 256:4253.

53. Norris RD, Fowden L (1972). Substrate discrimination by prolyl-tRNA synthetase from various higher plants. Phytochem 11:2921.

Cellular and Molecular Biology of Plant Stress, pages 367–380
© **1985 Alan R. Liss, Inc.**

GENETIC CONTROL OF PATHOTOXIN-INDUCED STRESS IN PLANTS[1]

David Gilchrist, Beverly McFarland[2], Deborah Siler[3],
Steven Clouse, and Ann Martensen

Department of Plant Pathology
University of California
Davis, CA   95616

ABSTRACT: Recent evidence obtained with the host-specific AAL-toxins produced by Alternaria alternata f. sp. lycopersici, causal agent of Alternaria stem canker of tomato, suggests a mechanism for genetic control of toxin-induced stress in tomato. In this system genetic control of AAL-toxin sensitivity and disease resistance is determined by a single locus which expresses complete dominance for pathogen resistance and incomplete dominance for toxin sensitivity. The presence of an aspartate-like moiety in the toxin suggested metabolite competition studies which focused attention on aspartate carbamoyl transferase (ACTase) as a possible site of action. Partial purification and detailed kinetic characterization of the enzyme from green tomato leaf tissue revealed sigmoidal substrate saturation kinetics for both substrates, aspartate and carbamyl phosphate, and inhibition by UMP and the AAL-toxins. ACTase from both resistant and sensitive (susceptible) genotypes was indistinguishable in kinetic response to the substrates and UMP but the resistant plant enzyme activity exhibited a two-fold higher toxin-enzyme dissociation constant (Kiapp). The in vitro combination

[1] Supported in part by Grant No. 59-2063-10406 CRGO, USDA, SEA, and NSF Grant No. PCM 80-1173.
[2] Present address: Chevron Biotechnology Group, Richmond, CA 94804.
[3] Present address: Department of Biochemistry and Biophysics, University of California, Davis, CA 95616.

of UMP and AAL-toxins, when assayed at variable
aspartate concentration, appeared to act synergistically to form a double dead-end inhibitor
complex which drove the ACTase reaction to zero
at concentrations of UMP that were not inhibitory
in the absence of AAL-toxins. The UMP:AAL-toxin
synergism was up to forty-fold greater with
ACTase from the susceptible compared to the
resistant genotype. Feedback loops regulating
intermediary metabolism of arginine, glutamine
and pyrimidine biosynthesis provide the basis
for a testable hypothesis of events leading from
toxin-induced stress to cell death.

INTRODUCTION

Knowledge of the molecular mechanisms of host resistance and pathogen virulence at the gene product level
is virtually nonexistant. It is true that many resistance genes have been "identified" by conventional plant
breeding techniques and introgressed into new genetic
backgrounds by sexual crossing methods (1). However,
these procedures have not provided stable resistance in
some cases while in others sexual incompatibility mechanisms prevent the movement of resistance loci outside a
species-limited sphere.

In part because genetic resistance to pathogens is
frequently associated with single gene effects, loci
controlling the host-parasite interaction are prime
candidates, economically as well as scientifically, for
genetic manipulation by recombinant DNA technology. A
major limitation to such directed genetic change at
present is that there is no known technique for isolating
resistance genes by recombinant DNA methods; only genes
whose protein product is known have been isolated so
far (2). Hence, the biochemical function of genetic
elements controlling the host-parasite interaction is
among the most eagerly sought and widely speculated
molecular mechanistic goals in contemporary plant
biology.

One of the molecular mechanisms used by fungal pathogens to induce disease is a class of metabolites referred
to as host-specific pathotoxins. A host-specific toxin

is defined as a metabolic product of a pathogen which is
toxic only to the host which is susceptible to the pathogen. Because of their host-specificity these compounds
appear to play the determinative role in host-specificity
of the pathogens and expression of disease symptoms in
the host (3, 4, 5). Lastly, if they induce nearly all
the symptoms associated with the disease such toxins are
direct tools to study the molecular development of disease
and the molecular basis of the genetic control of resistance to disease development. Even so, no single host-
parasite system has been described which permits conclu-
sive statements regarding the chemical identity of disease
determinants, their physiological and kinetic mode of
action in the host, and the molecular basis of genetic
resistance to the molecular determinants of disease.

In the example to be discussed here we reported
earlier that Alternaria alternata f. sp. lycopersici
produced a host-specific toxin fraction in culture
(6) which acted as a primary disease determinant and
that a single genetic locus (designated R) in tomato
controlled resistance to the pathogen and sensitivity
to the toxin(s) (AAL-toxins).

The AAL-toxins were characterized chemically as two
closely related esters of 1,3,3-propanetricarboxylic acid
and 1-amino-11,15-dimethylheptadeca-2,4,5,13,15-pentol
each occurring in culture as two isomers (TA and TB).
Esterification sites are the terminal carboxyl of the
acid and C_{13} (TA_1) or C_{14} (TA_2) of the amino pentol. The
TB form also consists of two components with the same carbon skeleton as TA_1 and TA_2 minus the C_5 hydroxyl (7).
All four AAL-toxin components have been separated quantitatively from culture filtrates by high performance liquid
chromatography (HPLC) (8) and recovered from infected
tomato leaves (9). Phytotoxic fractions isolated from
extracts of necrotic leaves of tomato plants infected with
A. alternata f. sp. lycopersici were indistinguishable
from AAL-toxins obtained from culture filtrates of the
pathogen based on chromatographic mobility in both thin-
layer and HPLC. Extracts from both sources possessed
equal specific activity and were equally host-specific.
Quantitative estimates indicated that toxin concentration
in diseased leaves was similar to the concentration of
AAL-toxin required to induce necrosis in detached leaves
of the same cultivar (10 ng/ml).

The presence of an L-aspartate-like moiety in the
AAL-toxins (tricarballylic acid) prompted temporal protection studies against toxin induced necrosis using aspartate and related metabolites. Preliminary results indicated that L-aspartate (at 34 μM) but not D-aspartate
reduced the sensitivity of rr leaves 100-fold in the
standard leaf bioassay. In contrast to many other host-
specific toxins ion leakage was not detected prior to the
onset of visible necrosis (10, 11). Additional protection
studies with a total of 43 structurally or metabolically
related compounds were conducted to more closely focus on
specific metabolic sites of interaction. Of all compounds
tested the most striking protectant was orotate, an intermediate unique to the orotic acid pathway of pyrimidine
biosynthesis (12). Orotate was at least 10-fold more
effective than L-aspartate, protecting at 3 μM which
suggested that a possible site of action was the step
catalyzed by aspartate carbamoyltransferase (ACTase,
E.C 2.1.2.3), an allosteric enzyme reported from several
etiolated plant sources, although it had not been studied
previously in tomato.

ACTase catalyzes the condensation reaction between
carbamyl phosphate (CAP) and aspartate (ASP) to form carbamyl aspartate (CASP) (Figure 1) which is the first committed step of pyrimidine biosynthesis in higher plants
(13). Higher plant ACTase from etiolated lettuce seedlings (14), etiolated mung bean seedlings (15) and wheat
germ (16) is regulated in vitro by the pathway end product
UMP suggesting a regulatory function for this enzyme in
higher plants (12).

Preliminary assay of ACTase activity from RR and rr
tomato and mung bean (a non-host) by several methods indicated that the enzyme was inhibited by AAL toxin with the
sensitivity order rr tomato > RR tomato > mung bean
(Figure 1). However, convincing biological arguments for
toxin induced necrosis using in vitro enzyme inhibition
data depend on experimental conditions which permit the
studies to be conducted with enzyme preparations that are
both stable and responsive to known in vivo allosteric
effectors throughout the course of the experiments. Also,
one must assume that the enzyme encounters the toxin in
the presence of the allosteric effectors in planta.
Therefore, cell-free preparations used for enzyme:toxin

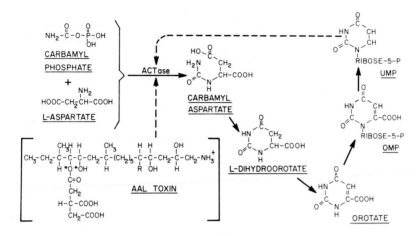

Figure 1. Orotic acid pathway of pyrimidine synthesis in plants. UMP exerts feedback inhibition at aspartate carbamoyltransferase (ACTase) in tomato. AAL-toxin inhibits ACTase and interacts with UMP to exert synergistic inhibition in the _rr_ genotype of tomato.

inhibition studies must be regulated _in vitro_ and also must be assesed in the present of the effectors and the toxins.

ACTase was partially purified from RR and _rr_ geno-types of tomato and kinetically characterized with respect to the interaction between the substrates, UMP, and the AAL-toxins. These interactions provide a bio-logical basis for genetic differences in pathotoxin induced cell stress at the level of metabolic regulation which may be applicable to other host-toxin systems.

METHODS

AAL-toxins were purified from cell-free culture fil-trates of isolate AS-273 as described elsewhere (7, 8). ACTase, purified 110 fold (29% recovery) from homozygous RR and _rr_ tomato genotypes in the presence of a stabilizing concentration of UMP (1 mM), but assayed without UMP, exhibited no metal cofactor or sulfhydral group require-

ment, had an estimated molecular weight of 79,000, and a pH
optimum of 9.9 (10). Five ACTase assay procedures were
evaluated with the most sensitive and reproducible procedure used for kinetic assays being method 2 of Prescott
and Jones (17).

RESULTS

Regulatory Properties of Tomato ACTase.

The partially purified ACTase from tomato displayed
sigmoid kinetics for both substrates, carbamyl phosphate
(CAP) and aspartate (ASP), with Hill values of 1.22 and
2.58, respectively. An ordered bi-bi mechanism for substrate addition with CAP the first ordered substrate was
suggested by these data (18). Succinate, a dicarboxylic
acid substrate analog, inhibited ACTase from both genotypes competitively with respect to ASP (Kiapp = 32 mM)
and uncompetitively with respect to CAP (I0.5 = 167 mM).

Inhibition of ACTase by UMP.

Based on analyses of Dixon plots (18), UMP appeared
to inhibit ACTase competitively with respect to CAP
(Kiapp = 0.33 mM) and uncompetitively with respect to ASP
(I0.5 = 2.4 mM), the latter may indicate UMP only binds
to the enzyme complex after the binding of ASP. No
evidence for a succinate:UMP synergism was observed when
the mixed inhibitor data was analyzed by the method of
Chou and Talaley (19). ACTase from both RR and rr tissue
was indistinguishable for all the above parameters, indicating an equivalent conservation of catalytic and regulatory properties in both genotypes in the absence of AAL-
toxins.

Inhibition of ACTase by AAL-toxins.

ACTase from both genotypes was inhibited by AAL-toxins
(TA + TB) in the absence of UMP, with the apparent dissociation constants (Kiapp) of 2.4 and 4.2 mM for rr and RR
respectively (20). Assay of ACTase at 2.0 mM AAL-toxin
decreased the Hill value for CAP to 0.72 (RR) and 0.64 (rr).

Dixon plots (1/V vs. I) (18) were linear for CAP
(2 to 16 mM) at 0, 0.5, 1.0, and 2.0 mM toxin and indicated uncompetitive binding, which presumably occurs subsequent to CAP binding. Hyperbolic velocity and linear
Dixon plots suggested either single, or multiple non-
cooperative, AAL-site(s) on the enzyme. Comparable
experiments with variable ASP at saturating CAP (10 mM)
revealed noncompetitive toxin binding with respect to ASP
for both genotypes. This type of binding differed from
succinate. The noncompetitive pattern of toxin vs ASP
is taken to indicate that the enzyme binds AAL-toxin
regardless of whether ASP is present and the substrate
saturation alone will not relieve the inhibition (18).

TABLE 1

SYNERGISTIC INHIBITION OF ASPARTATE
CARBAMOYLTRANSFERASE FROM TOMATO BY A COMBINATION
OF UMP AND AAL-TOXINS

Aspartate Concentration (mM)	ACTase Velocity Differences[a] Tomato Genotype	
	RR	rr
2.5	0.9	42.0
5.0	1.2	9.3
10.0	0.9	1.9
20.0	1.0	2.0

[a]Values are expressed as the difference between:

$$\frac{1}{V(\text{Toxin + UMP})} \quad \text{vs} \quad \left[\frac{1}{V(\text{toxin})} + \frac{1}{V(\text{UMP})} - \frac{1}{V(\text{control})} \right]$$

Inhibition of ACTase by UMP and AAL-toxins.

In the presence of UMP, which itself decreases carbamyl phosphate binding (competitive inhibition), AAL-
toxin interferes with the binding of carbamyl phosphate
in a competitive manner for ACTase from both rr and RR
tomato genotypes (Table 1). However, an interesting
synergism occurred between toxin and UMP for the rr
ACTase (19) and was twice as great as that for the
ACTase from tolerant tissue at ASP saturation. Futher-
more, when assayed at variable ASP concentration the

AAL-toxins appeared to be noncompetitive inhibitors with respect to aspartate, and, in the presence of UMP (uncompetitive with respect to aspartate), act synergistically to produce a double, dead-end inhibitor complex with the net result that ACTase activity is decreased to zero by concentrations of UMP that are not normally (i.e., in the absence of the AAL-toxins) inhibitory. Additionally, this synergism occurred in a genotype-specific manner; the synergism was forty-fold greater for the rr genotype at subsaturating aspartate (2.5 mM) (Table 1).

DISCUSSION

Regardless of the cause, the "diseased state" is the result of an unrelieved disruption in host metabolism. It seems intuitive that host metabolism will be altered first at the molecular site of action in host-toxin interactions. The ultimate objective is to describe, in a chemically defined fashion, the amplified sequence of secondary metabolic events which lead to the symptoms characteristic of disease stress. Genetic differences in host reaction to pathotoxins are key experimental variables used to study the site and molecular basis of toxin induced stress as well as to characterize the gene products controlling the qualitative or quantitative differences in disease expression.

The existing literature indicates two general ways that pathotoxins may alter metabolism; either by combining with a primary target to directly induce physiological stress or through interaction with the primary target site to initiate a subtle initial change which may be amplified through feedback loops to produce the dramatic changes characteristic of the diseased state (21). When one considers the multitude of known metabolic regulatory mechanisms the potential for sites of pathotoxin interaction appear extensive. In this context, Lardy observed that "for every enzymatic reaction the biochemist wishes to study, there is, in nature, a compound uniquely designed to serve as its inhibitor" (22). Although future studies may reveal exceptions, it is interesting to note that all of the toxin-specific sites for which solid evidence presently exists involve inhibition of a different and critical enzyme (21). Particularly important, as well as vulnerable, may be those enzymes which have roles in regulatory integrated intermediary metabolism.

The existence of multiple substrate and effector sites (inhibitors or activators) on regulatory enzymes is the basis for allosteric interactions. The concept of allostery simply emphasizes the fact that most metabolic activators and inhibitors bear little structural resemblance to the substrate of the enzyme. The existence of multiple interacting sites on regulatory enzymes also affords a subtle and potentially significant mechanism for nonsubstrate-site interactions by pathotoxins with little or no resemblence to the substrate but which have structural similarities to the effectors. Additional interactions may occur at sites on the enzyme which have no apparent relationship to either substrates or products of the enzyme. For example, Yon has reported a hydrophobic binding site on aspartate carbamoyltransferase from wheat germ where binding of compounds such as deoxycholate induces an allosteric transition state of lowered enzyme activity (23). Existence of such evolutionarily conserved, but metabolically mysterious, sites affords yet another potential site for toxin interaction.

Evidence presented in this paper suggest the existance of an allosteric site on ACTase from tomato which binds the AAL-toxins leading to a proposed conformational change with an enhanced sensitivity to the natural feedback inhibitor UMP. Based on kinetic data the metabolic consequence of the synergistic interaction between the AAL-toxins and UMP is the ability of this combination to drive the reaction velocity to zero, an effect not observed with either compound alone (Figure 1). The further observation that the pathogen susceptible and toxin sensitive genotype (rr) is up to 40-fold more sensitive to this synergistic interaction than the resistant genotype (RR) is consistent with a genotype-specific explanation of the physiological basis of AAL-toxin induced cell stress.

Precisely how the inhibition of ACTase by such a mechanism leads to cell stress and eventually cell death is unknown. However, consideration of certain alternative explanations may be relevant to other host-toxin interactions and provide a general format for stress induced in other plants by structurally unrelated pathotoxins.

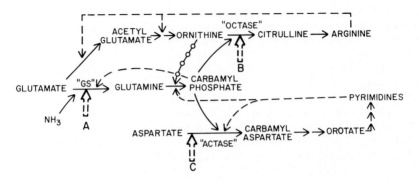

Figure 2. Feedback inhibition (− −) and activation (-o-o-o-) loops regulating glutamine, arginine, and pyrimidine synthesis. Reported sites of inhibition for tabtoxinine-β-lactam (A), phaseolotoxin (B), and AAL-toxins (C) are glutamine synthetase (GS), ornithine carbamoyl transferase (OCTase), and aspartate carbamoyltransferase (ACTase), respectively.

In the present case, inhibition of ACTase through AAL-toxin − UMP synergism could lead eventually to a limitation in essential pyrimidines causing metabolic stress. However, recycling of existing pyrimidines in the cell may forestall an immediate loss in cell integrity. Alternatively, interlocking metabolic pathways could be affected through feedback loops eventually leading to inhibition of specific processes for which no short-term compensatory mechanisms exist.

Comparison of feedback loops involving glutamine, arginine and the pyrimidines provides an example of how inhibition of different key enzymes by structurally unrelated pathotoxins could lead to physiological stress through a common fate (Figure 2). Pseudomonas syringae pv. tabaci produces a non-specific toxin, tabtoxinine-β-lactam, which specifically inhibits glutamine synthetase. Inhibition of glutamine synthetase is associated with an in vivo accumulation of ammonia, generated by photorespiration in light, to phytotoxic levels (24). Normally the high Km of the uninhibited glutamine synthetase prevents the buildup of sufficient ammonia to uncouple photophos-

phorylation (25). In the pv. tabaci case there are
literally two toxins involved in symptom expression:
tabtoxinine-β-lactam from the pathogen and ammonia from
toxin-sensitive host metabolism.

Indirect alteration in glutamine synthetase activity
through feedback mechanisms via "metabolic interlock"
(26) presumably could have a similar metabolic impact.
For example, evidence exists for direct inhibition of
ornithine carbamoyltransferase (OCTase) by phaseolotoxin
produced by P. s. pv. phaseolicola (27). In this case
OCTase inhibition is accompanied by ornithine and car-
bomoyl phosphate accumulation (28) the latter which is
reported to feedback inhibit glutamine synthetase (29).
AAL-toxin inhibition of ACTase presumably also could lead
to an accumulation of carbomyl phosphate and ultimately
ammonia by the same reasoning (Figure 2).

These interactions, if they occur in vivo, present
a scenario for common mechanisms of phytotoxicity which
would be initiated at different enzymatic sites in the
cell by structurally unrelated pathotoxins. Genetic
strategies to circumvent the effect of pathotoxins would
require modification of the pathotoxin binding sites to
effectively decrease the binding constants for the toxins
on the target enzymes. Present evidence from the AAL-
toxin/ACTase interaction in tomato suggests that such
variants may exist, at least in this system. Alternatively, introduction of genes coding for target enzymes
with conserved regulatory properties but altered sensitivity to the putative pathotoxins represent a genetic
engineering opportunity.

Based on current evidence it would appear that elucidation of the primary mechanism of pathotoxin resistance
regardless of the specific model system employed, provides
a beginning for the dissection, experimentally and conceptually, of the diseased state. Furthermore, if host-
specific toxins are factors conferring both host specificity and virulence attributes to the pathogen, as is
presently proposed, then both recognition and compati-
bility may be described. When reassembled into concepts,
information of this type can serve as a base for fundamental understanding of the developmental sequence of
symptom expression. Practical application of such information, through enlightened manipulation of physiological,

environmental, or genetic systems of the host-parasite
interaction, generates future tools in disease management
and studies of resistance gene function and metabolic
regulation.

REFERENCES

1. Day P, Barrett J, Wolf M (1983). The evaluation of
    host-parasite interaction. In Kosuge T, Meredith C,
    Hollaender A (eds): "Genetic Engineering of Plants,"
    New York: Plenum Press, p 419.
2. Chilton MD (1983). A vector for introducing new genes
    into plants. Sci Amer 248(6):51.
3. Scheffer RP (1976). Forces by which the pathogen
    attacks the host plant. Host-specific toxins in
    relation to pathogenesis and disease resistance.
    In Heitefuss R, Williams PH (eds): "Encyclopedia
    of Plant Physiology, New Series," Vol 4, Berlin:
    Springer Verlag, p 247.
4. Wheeler H (1978). Disease alterations in perme-
    ability and membranes. In Horsfall JG, Cowling EB
    (eds): "Plant Disease, An Advanced Treatise,"
    New York: Academic Press, Vol. III, p 327.
5. Durbin RD (1981). "Toxins in Plant Disease."
    New York: Academic Press, 515 p.
6. Gilchrist DG, Grogan RG (1976). Production and
    nature of a host-specific toxin from Alternaria
    alternata f. sp. lycopersici. Phytopathology
    66(2):165.
7. Bottini AT, Bowen JR, Gilchrist DG (1981). Phytotoxins
    II. Characterization of a phytotoxic fraction from
    Alternaria alternata f. sp. lycopersici. Tetrahedron
    Lett 22(29):2723.
8. Siler DJ, Gilchrist DG (1982). Determination of host-
    selective phytotoxins from Alternaria alternata f.
    sp. lycopersici as their maleyl derivatives by high-
    performance liquid chromatography. J Chromatography
    238:167.
9. Siler DJ, Gilchrist DG (1983). Properties of host
    specific toxins produced by Alternaria alternata f.
    sp. lycopersici in culture and in tomato plants.
    Physiol Plant Pathol 23:265.

10. McFarland BL, Gilchrist DG (1980). A proposed enzyme specific role for the host-specific toxins of Alternaria alternata f. sp. lycopersici. Phytopathology 71(2):240 (Abst).

11. Kohmoto K, Verma VS, Nishimura S, Tagami M, Scheffer RP (1982). New outbreak of Alternaria stem canker of tomato in Japan and production of host-selective toxins by the causal fungus. J Fac Agric, Tottori Univ 17:1.

12. Ross C (1981). In Stumpf P, Conn E (eds in-chief): "The Biochemistry of Plants: A Comprehensive Treatise," V 6 Marcus A (ed), New York: Academic Press, p 169.

13. Lovatt CJ, Albert LS, Tremblay GC (1979). Regulation of pyrimidine biosynthesis in intact cells of Cucurbita pepo. Plant Physiol 64(4):562.

14. Neuman J, Jones ME (1962). ACTase from lettuce seedlings: Case of end product inhibition. Nature 195:709.

15. Ong BL, Jackson JJ (1972). Pyrimidine nucleotide biosynthesis in Phaseolus aureus: Enzyme aspects of control of CAP synthesis and utilization. Biochem J 129:583.

16. Yon R (1972). Chromatography of lipophylic proteins on absorbants containing mixed hydrophobic and ionic groups. Biochem J 126:765.

17. Prescott LM, Jones ME (1969). Modified methods for the determination of carbamyl aspartate. Anal Biochem 32:408.

18. Analysis of Rapid Equilibrium and Steady-State Enzyme Systems." New York: Wiley and Sons.

19. Chou TC, Talalay P (1977). A generalized equation for the analysis of multiple inhibitions of Michaelis-Menten Systems. J Biol Chem 252 (18):6438.

20. McFarland BL, Gilchrist DG (1982). Synergistic inhibition of tomato Aspartate transcarbamylase by UMP and the host-select toxins produced by Altrnaria alternata f. sp. lycopersici. Plant Physiol 69(4):21 (Abst).

21. Gilchrist DG (1983). Molecular modes of action. In Daly JM, Devrall B (eds): "Toxins and Plant Pathogenesis," Australia: Academic Press, p 81.

22. Lardy H (1980). Antibiotic inhibitors of mito-chondrial energy transfer. Phamacol Ther 11:649.

23. Yon RJ (1973). Wheat-germ Aspartate transcarbamaylase: Potentiation of end-product inhibition in vitro by deoxycholate. Biochem Soc Trans 1:676.

24. Turner JG (1981). Tabtoxin, produced by Pseudomonas tabaci, decreases Nicotiana tabacum glutamine synthetase in vivo and causes accumulation of ammonia. Physiol Pl Pathol 19(1):57.

25. Miflin B, Lea P (1976). The pathway of nitrogen assimilation in plants. Phytochemistry 15:873.

26. Jensen R (1969). Metabolic interlock: regulatory interactions exerted between biochemical pathways. J Biol Chem 244:2816.

27. Mitchell RE, Johnson JS, Ferguson AR (1981). Phaseolotoxin and other phosphosulfamyl compounds: biological effects. Physiol Plant Pathol 19:227.

28. Jacques S, Sung ZR (1981). Regulation of pyrimidine and arginine biosynthesis investigated by use of phaseolotoxin and 5-flurouracil. Plant Physiol 67:287.

29. O'Neal T, Joy K (1975). Pea leaf glutamine synthetase. Regulatory properties. Plant Physiol 55:968.

Cellular and Molecular Biology of Plant Stress, pages 381–400
© 1985 Alan R. Liss, Inc.

EFFECTS OF STRESS ON
THE DEFENSIVE BARRIERS OF PLANTS[1]

P. E. Kolattukudy and C. L. Soliday

Institute of Biological Chemistry
Washington State University
Pullman, Washington

INTRODUCTION

The cuticle is the major barrier between higher plants and their environment. It is composed of a structural polymer called cutin which is embedded in a complex mixture of soluble lipids collectively called wax (1,2,3). Cutin is composed of interesterified hydroxy and hydroxyepoxy fatty acids shown in Figure 1. In the polymer most of the primary hydroxyl groups are involved in ester

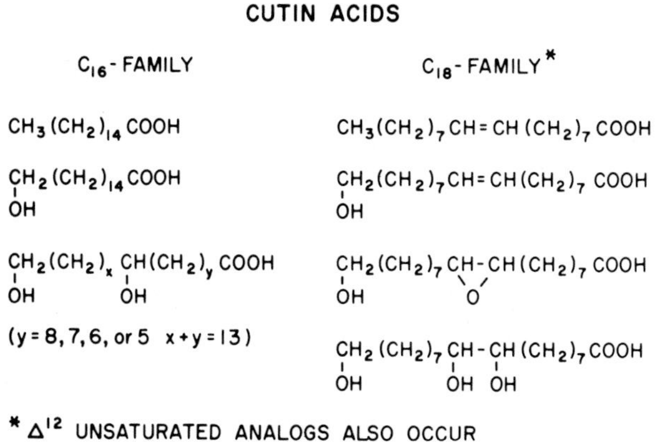

CUTIN ACIDS

C_{16}- FAMILY $\qquad$ C_{18}- FAMILY *

$CH_3(CH_2)_{14}COOH$ $\qquad$ $CH_3(CH_2)_7 CH = CH(CH_2)_7 COOH$

$\underset{OH}{CH_2}(CH_2)_{14}COOH$ $\qquad$ $\underset{OH}{CH_2}(CH_2)_7 CH = CH(CH_2)_7 COOH$

$\underset{OH}{CH_2}(CH_2)_x \underset{OH}{CH}(CH_2)_y COOH$ $\qquad$ $\underset{OH}{CH_2}(CH_2)_7 \underset{O}{CH - CH}(CH_2)_7 COOH$

(y = 8, 7, 6, or 5   x + y = 13)

$\underset{OH}{CH_2}(CH_2)_7 \underset{OH\ OH}{CH - CH}(CH_2)_7 COOH$

* Δ^{12} UNSATURATED ANALOGS ALSO OCCUR

Figure 1. Structure of the major monomers of cutin.

---

[1]Scientific paper #6860, project 2001, College of Agriculture Research Center, Washington State University, Pullman, WA 99164. This work was supported in part by grant PCM-8306835 from the National Science Foundation.

linkages and a considerable portion of the secondary hydroxyl groups are also esterified to provide crosslinking and/or branching points. In periderms and a variety of other anatomical regions of plants, another type of barrier called suberin is deposited on the cell wall and this barrier is also associated with waxes (1,2,3). The structure and composition of suberin are not well understood. According to the working hypothesis proposed on the basis of a variety of indirect evidence, suberin consists of phenolic and aliphatic domains. The phenolic domain, which might be attached to the cell wall polymers, is thought to be somewhat similar to lignin. The aliphatic components, which cannot by themselves form extensive polymers, presumably serve to crosslink the aromatic domains (Figure 2). In some cases in which suberized preparations contain a high proportion of the aliphatic components similar to those found in cutin, more extensive aliphatic domains might be present.

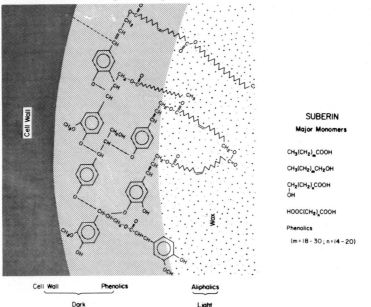

Figure 2. Tentative model proposed for the structure of suberin.

BIOSYNTHESIS OF CUTIN AND SUBERIN

The C_{16} and C_{18} family of cutin monomers are derived from the most common cellular fatty acids, namely palmitic acid, oleic acid and linoleic acid. These acids are first hydroxylated at the ω-carbon by enzymes present in endoplasmic reticulum (4). In the case of the C_{16} family of acids, an additional hydroxyl group, which is usually present at C-10, C-9 or C-8 positions, is introduced by another enzyme that is also present in the endoplasmic reticulum (5). These hydroxylases appear to have characteristics similar to other mixed function oxidases. In the case of the C_{18} family of monomers, the mid-chain functional groups are introduced at the double bonds present in the original fatty acid. The double bond is first epoxidized by a cytochrome P-450 type enzyme system with NADPH and oxygen as the required cofactors (6). The epoxide is subsequently hydrated by another enzyme to give the trihydroxy C_{18} acid (7). The polymer synthesizing enzyme, which is attached to the cuticular membranes, transfers the hydroxy and hydroxyepoxy acyl moieties from CoA to the hydroxyl groups present in the primer (8).

The aliphatic components of suberin are generated by reactions analogous to those indicated above. The unique major components of suberin, the dicarboxylic acids, are generated from the corresponding ω-hydroxy acids by an ω-hydroxy acid dehydrogenase which generates an ω-oxo acid and another dehydrogenase which converts the ω-oxo acid to the dicarboxylic acid. Since the former enzyme, but not the latter, is uniquely associated with suberization, it was purified and characterized (9,10). The aromatic components of suberin are probably generated by reactions analogous to those involved in lignification. The aromatic components are then polymerized by a process involving peroxidase. An isoperoxidase has been shown to be located specifically in suberizing tissues (11) and has been shown to be induced by abscisic acid in potato tissue cultures (12) which are known to undergo suberization under the influence of this plant hormone (13). This isoperoxidase has recently been purified as outlined in Table 1.

EFFECTS OF STRESS ON THE SYNTHESIS OF THE DEFENSIVE BARRIERS

Mechanical Injury

Very little is known about the effect of stress on the production of barrier layers. It was clearly established

TABLE 1

PURIFICATION OF PEROXIDASE FROM
WOUND-HEALED POTATO TUBER SLICES[a]

Purification Steps	Enzyme Activity $\Delta OD_{470}/min$	Specific Activity $\Delta OD_{470}/min/mg$	Recovery %
Acetone Powder Extract	14,200	43	100
NH_4SO_4 Supernatant	10,700	89	75
Sephadex G-100	6,800	523	48
Sephadex SP25	3,600	1,200	25
Sephadex Phenyl Sepharose	1,500	1,875	11

[a]K. E. Espelie and P. E. Kolattukudy, unpublished.

that wounding triggers suberization in plant organs
irrespective of the nature of the natural protective
covering of the organ (14). For example, wounding of jade
leaves, tomato fruits and bean pods which are normally
covered by a cutin polymer as well as potato tubers whose
normal protective barrier is a suberized periderm,
underwent suberization when they were wounded. Although it
has been repeatedly stated that biological stress such as
fungal attack causes lignification, it is possible that
wounding caused by fungal attacks might trigger
suberization. Since the barriers erected under such
conditions have not been fully characterized, it is
difficult to draw firm conclusions.

Cold Temperature Adaptation

     Recent experimental evidence suggested that adaptation
of rye plants to cold temperature is associated with an
increased deposition of aliphatic polymers. The thickening
of the cuticular layer and the mestome sheaths observed in
electron micrographs of cold-hardened leaves was associated
with a two- to four-fold increase in the amount of
aliphatic polymers present in the leaves (M. Griffith,
N. P. Huner, K. E. Espelie and P. E. Kolattukudy, submitted
to Protoplasma). 18-Hydroxy-9,10-epoxy C_{18} acid showed the

highest increase during cold-hardening. The increased deposition of this monomer in the polymer was confirmed by silver proteinate staining followed by electron microscopy. How the increased deposition of the polymeric material in the cuticle and the mestome sheath helps in the development of cold hardiness is not known.

Mineral Deficiency

    Mineral deficiency appears to cause dramatic changes in the deposition of the barrier layers. For example, both electron microscopic examination and chemical analysis showed that corn roots grown under magnesium deficient conditions contained several fold higher amounts of suberin than the roots grown with adequate amounts of magnesium (15). Iron deficiency in bean roots, on the other hand, caused a dramatic inhibition of suberization (Table 2). The fatty acid ω-hydroxylase level was found to be enhanced several fold under iron deficiency. However, the level of the suberization-associated isoperoxidase was found to be suppressed by iron deficiency (P. Sijmons and P. E. Kolattukudy, unpublished results). This observation is consistent with the proposed model for suberin according to which the deposition of the aromatic matrix catalyzed by this enzyme is essential for suberization.

TABLE 2
INHIBITION OF SUBERIZATION BY IRON DEFICIENCY[a]

	+ Fe	− Fe
	μg/g dry tissue	
Fatty Alcohols	63.2 (3.7)	8.0 (4.2)
Fatty Acids	83.4 (4.9)	16.0 (8.5)
ω-Hydroxy Acids	990.6 (58.6)	74.6 (48.7)
Dicarboxylic Acids	405.8 (30.0)	64.4 (33.9)
9,10,18-Trihydroxy C_{18} Acid	6.6 (0.4)	1.9 (1.0)

[a]Composition of the aliphatic components of the suberin polymer from the roots of *Phaseolus vulgaris* plants grown in iron deficient media. (P. Sijmons and P. E. Kolattukudy, unpublished results).

Induction of Suberization by Abscisic Acid

Thorough washing of potato tissue slices inhibited suberization but this inhibition was at least partially reversed by abscisic acid treatment of the washed tissue slices (13). Furthermore, abscisic acid also induced suberization in potato tissue culture. This induction was demonstrated by enhanced deposition of the aliphatic and aromatic components of suberin and associated waxes (12). Although ω-hydroxy fatty acid dehydrogenase was induced for suberization in wound-healing potato tissue slices (16), in tissue culture there was already a fairly high level of this enzyme and abscisic acid treatment resulted in only a relatively small increase in the level of this enzyme. The suberization-associated isoperoxidase, on the other hand, was clearly induced in tissue cultures as a result of abscisic acid treatment. This abscisic acid-induced isoperoxidase is quite similar to that isolated from wound-healing potato tissue slices in its properties and the two enzymes appeared to be immunologically identical (K. E. Espelie and P. E. Kolattukudy, unpublished). This system provides a means to examine abscisic acid-induced expression of the isoperoxidase gene.

DEGRADATION OF CUTIN BY PATHOGENIC FUNGI

Cutinase was purified for the first time from the extracellular fluid from *Fusarium solani pisi* grown on cutin as the sole source of carbon (17). Since then a generally applicable method has been developed for the purification of cutinase from the extracellular fluid of cutin-grown fungi (Table 3). Using this general method, cutinases have been purified from several isolates of *F. solani pisi*, *F. roseum culmorum*, *F. roseum sambucinum*, *Ulocladium consortiale*, *Helminthosporum sativum*, *Streptomyces scabies*, *Colletotrichum gloeosporioides*, *Colletotrichum capcisi*, *Colletotrichum graminicola*, *Phytothera cactorum* and *Botrytis cinerea*.
The molecular and catalytic properties of cutinases isolated from the various fungi appear to be quite similar. They all have a molecular weight near 25,000 and have very similar amino acid composition. They are all glycoproteins containing a few percent of covalently attached carbohydrates which are usually linked to the protein by *O*-glycosidic linkages. All fungal cutinases so far

TABLE 3

PURIFICATION OF CUTINASE FROM
F. SOLANI PISI[a]

Step	Total Units	% Yield
Culture Filtrate (25%) + Triton X-100 (0.1%) Wash (75%)	200,000	
Acetone (75%) Precipitate	160,000	80
QAE-Sephadex	130,000	65
Octyl-Sepharose	110,000	55
SP-Sephadex		
Isozyme C (95mg)	72,000	36
Isozyme D (45mg)	24,000	12

[a]W. Köller, C. L. Soliday and P. E. Kolattukudy, unpublished results.

examined show an optimum pH between 8.5 and 10.0. These enzymes show specificity for primary alcohol esters and generate monomers and oligomers from cutin; they also hydrolyze the oligomers to monomers. Fungal cutinases also catalyze hydrolysis of *p*-nitrophenyl esters of fatty acids often with a preference for short chain esters.

High sensitivity of cutinase to inhibition by organophosphates strongly suggested that an active serine was involved in catalysis. With radioactive diisopropylfluorophosphate it was determined that there was one active serine per mole of cutinase. A variety of organophosphates inhibit cutinase and the commonly used transition state analogs of active serine enzymes, boronic acid derivatives, also inhibit cutinase (18). Relatively specific modifying reagents were used to show that one histidine and one carboxyl group were involved in catalysis by cutinase. Thus, the involvement of the three members of the catalytic triad usually associated with an active serine catalysis was established. After inhibiting the deacylation step by carbethoxylation of the histidine, an acyl enzyme intermediate could be generated by treatment of the modified enzyme with *p*-nitrophenyl[1-^{14}C]acetate and this labeled intermediate was isolated (18).

EVIDENCE FOR CUTINASE INVOLVEMENT IN THE
PENETRATION OF PATHOGENIC FUNGI

Based on the appearance of the holes generated by the
fungi, it was suggested that the cuticular barrier was
breached enzymatically. The first direct evidence that a
fungal pathogen secretes cutinase during an actual
infection process was provided with ferretin-conjugated
antibodies prepared against cutinase (19). Cutinase was
found to be associated with the pea stem cuticle in the
region broken by invading *F. solani pisi*. Cutinase was
similarly detected in the region of attack of papaya fruits
by *C. gloeosporioides*. Cutinase was shown to be essential
for infection of pea stem by *F. solani pisi* when it was
demonstrated that specific inhibitors of cutinase prevented
infection (20,21). Subsequently, prevention of infection
by inhibition of cutinase was demonstrated for papaya
fruits against *C. gloeosporioides* (22), apple seedlings
against *Venturia inequalis* and corn seedlings against *C.
graminicola* (R. Chacko and P. E. Kolattukudy, unpublished
results). These results raised the possibility of
developing effective cutinase inhibitors as antipenetrants
to protect plants against fungal attacks.

CUTINASE BIOSYNTHESIS AND CLONING OF THE cDNA

The ability to produce cutinase can significantly
affect the pathogenicity of certain fungi on intact plant
organs (23). Therefore, knowledge of the regulation of the
biosynthesis of this enzyme could be highly relevant to an
understanding of host-pathogen interaction. Induced
cutinase production by low levels of cutin hydrolysate in
glucose-grown cultures when glucose was depleted (24)
provided a convenient method to study cutinase
biosynthesis. In such induced cultures $\sim$70% of the
extracellular protein was cutinase suggesting that
cutinase-specific mRNA might be a significant component of
the total mRNA. In fact, upon *in vitro* translation of the
poly (A)$^+$ mRNA isolated from such induced cultures, a
protein which cross-reacted with anticutinase could be
readily isolated (25). This primary translation product
was 2100 daltons larger than mature cutinase. Obviously,
cutinase synthesis involves post-translational processing
which includes proteolytic removal of a signal peptide from
the primary translation product as well as glycosylation.

The relative abundance of cutinase mRNA in the total poly(A)$^+$ RNA isolated from induced cultures made it feasible to obtain a cDNA clone containing the cutinase genetic code from a cDNA library of clones. Cloning of cDNA generated by reverse transcribing the total poly(A)$^+$ RNA isolated from induced cultures of *F. solani pisi* was accomplished in the pBR322 plasmid vector (26). The cDNA library was screened for cDNA which was unique to the induced cultures by comparing duplicate colony hybridizations with cDNA probe generated from total poly (A)$^+$ RNA, isolated from either induced cultures or uninduced cultures. The 75 colonies selected from a 600 colony cDNA library were screened in groups for clones containing the cutinase structural gene by hybrid-select translation. The groups were evaluated for efficiency in hybridizing fungal mRNA which upon cell-free translation generated a product which immunologically cross-reacted with anticutinase. Individual colonies of one such group were evaluated by hybrid-select translation. From one such colony, which showed the presence of a coding region for cutinase, the plasmids were isolated and the insert was labeled by nick-translation and this probe was used to screen the 75 clones for the presence of cutinase cDNA by colony hybridization. Thus 15 colonies were identified as containing cutinase recombinant cDNA and Southern blots of restriction endonuclease digests of isolated plasmid from each of these colonies showed cDNA inserts ranging in size from 280-950 bp. The cutinase mRNA contained 1050 nucleotides, as determined by a cutinase-cDNA hybridized "Northern blot" of electrophoretically separated poly (A)$^+$ RNA isolated from induced fungal cultures. Thus the colony (C-31) containing the largest insert (950 bp) represented a nearly full length cDNA for cutinase mRNA.

PRIMARY STRUCTURE OF CUTINASE

A strategy for nucleotide sequencing of the cutinase cDNA was devised from the restriction endonuclease map of the 950 bp insert as shown in Figure 3. After end-labeling the insert, the sequence of the first 200-250 nucleotides at the 5' and 3'-ends were determined by the method of Maxam-Gilbert (27). Treatment of the insert with either Taq I, Hpa II, or Sau 3A restriction endonuclease produced fragments which were subcloned into M13 mp10 and M13 mp11 bacteriophage vectors (28) and the extruded single-strand

DNA from these subclones was sequenced by the Sanger-dideoxynucleotide method (29). From the overlapping sequences obtained from the subclones and end-labeled inserts the complete nucleotide sequence of the cDNA insert was constructed (Figure 4). From this nucleotide sequence a 23,951 dalton protein was translated from an initiation codon to a termination codon 702 nucleotides apart in the same reading frame. The inserts from two additional cutinase-cDNA clones were also fully sequenced (C-4 and C-57 in Figure 4) in a similar manner and the sequences of both were found to be identical to that obtained with the first clone.

The primary structure translated from the nucleotide sequence was confirmed to be that of cutinase by comparing it to amino acid sequences of several tryptic peptides isolated from the protein. The enzyme was first treated with [^{3}H]diisopropylfluorophosphate, and subsequently the disulfides were reduced under denaturing conditions with dithioerythritol and the free SH groups generated were derivatized by treatment with [1-^{14}C]iodoacetamide. After succinylating this protein, tryptic fragments were generated and fractionated by HPLC. The 37 amino acid residues of one peptide containing the single tryptophan of the enzyme matched exactly with the predicted sequence near the N-terminal end of the primary structure (Figure 4). Another 31 residues of a peptide containing both the active serine (Residue 136) and a Cys residue was located in the center of the primary structure (30). A third peptide of 30 amino acid residues near the C-terminal end of the primary structure contained two additional cysteines (residues 187 and 194) and the only histidine residue (residue 204) of the enzyme which is part of the catalytic triad (W. Ettinger and P. E. Kolattukudy, unpublished results). Since all of the directly determined amino acid sequences, which comprise nearly 50% of the total protein, matched precisely with the sequence predicted from the nucleotide sequence of the cloned cDNA, there is little doubt that the cloned cDNA represents the authentic cutinase gene.

Comparison of the primary structure of the active site areas of cutinase with such regions of other enzymes containing the active serine catalyic triad showed very little homology (31). The only invariant feature which appears to be retained in cutinase is the presence of glycine residues at β positions on either side of the active serine and the fact that the active serine and

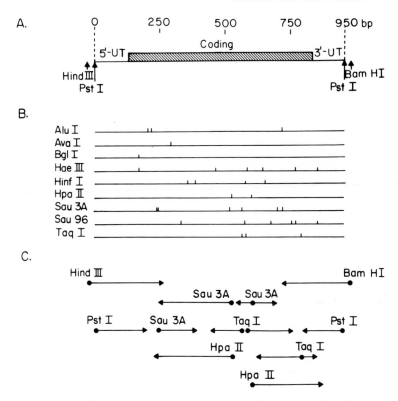

Figure 3. The restriction enzyme cleavage map of a cloned, cutinase-cDNA insert and the subcloning strategy for determining the complete cutinase-cDNA nucleotide sequence.

histidine are separated by a considerable number of amino acid residues. Since these two residues are located 67 amino acids apart it is obvious that the tertiary structure is involved in bringing the catalytic triad into the physical proximity required for catalysis. The presence of Cys residues close to the members of the catalytic triad and the observation that the reduction of the disulphide bonds in the protein results in inactivation of the enzyme (32) suggested that the disulphide bridges in the enzyme play a major role in juxtapositioning the residues involved in catalysis. Although previously it was thought that the only two Cys residues contained in cutinase from *F. solani pisi* were involved in a disulphide bridge, the nucleotide sequence of the cDNA showed that four Cys residues are present in the enzyme. Since there are no free SH groups

```
AACCACACATACCTTCACTTCATCAACATTCACTTCAACTTCTTCGCCTCTTCCTTTTCACTCTTTATCATCCTCACC
```

```
                                          10
MET LYS PHE PHE ALA LEU THR THR LEU LEU ALA ALA THR ALA SER ALA LEU PRO THR SER
ATG AAA TTC TTC GCT CTC ACC ACA CTT CTC GCC GCC ACG GCT TCG GCT CTG CCT ACT TCT
                                  ├─ C-57
                         30                   32
ASN PRO ALA GLN GLU LEU GLU ALA ARG GLN LEU GLY ARG THR THR ARG ASP ASP LEU ILE
AAC CCT GCC CAG GAG CTT GAG GCG CGC CAG CTT GGT AGA ACA ACT CGC GAC GAT CTG ATC
                         ├─ C-4
               47              50
ASN GLY ASN SER ALA SER CYS ARG ASP VAL ILE PHE ILE TYR ALA ARG GLY SER THR GLU
AAC GGC AAT AGC GCT TCC TGC CGC GAT GTC ATC TTC ATT TAT GCC CGA GGT TCA ACA GAG
                                  70
THR GLY ASN LEU GLY THR LEU GLY PRO SER ILE ALA SER ASN LEU GLU SER ALA PHE GLY
ACG GGC AAC TTG GGA ACT CTC GGT CCT AGC ATT GCC TCC AAC CTT GAG TCC GCC TTC GGC
                                  90
LYS ASP GLY VAL TRP ILE GLN GLY VAL GLY GLY ALA TYR ARG ALA THR LEU GLY ASP ASN
AAG GAC GGT GTC TGG ATT CAG GGC GTT GGC GGT GCC TAC CGA GCC ACT CTT GGA GAC AAT
                                  110
ALA LEU PRO ARG GLY THR SER SER ALA ALA ILE ARG GLU MET LEU GLY LEU PHE GLN GLN
GCT CTC CCT CGC GGA ACC TCT AGC GCC GCA ATC AGG GAG ATG CTC GGT CTC TTC CAG CAG
      125                         130                 136
ALA ASN THR LYS CYS PRO ASP ALA THR LEU ILE ALA GLY GLY TYR SER GLN GLY ALA ALA
GCC AAC ACC AAG TGC CCT GAC GCG ACT TTG ATC GCC GGT GGC TAC AGC CAG GGT GCT GCA
                                  150
LEU ALA ALA ALA SER ILE GLU ASP LEU ASP SER ALA ILE ARG ASP LYS ILE ALA GLY THR
CTT GCA GCC GCC TCC ATC GAG GAC CTC GAC TCG GCC ATT CGT GAC AAG ATC GCC GGA ACT
                                  170
VAL LEU PHE GLY TYR THR LYS ASN LEU GLN ASN ARG GLY ARG ILE PRO ASN TYR PRO ALA
GTT CTG TTC GGC TAC ACC AAG AAC CTA CAG AAC CGT GGC CGA ATC CCC AAC TAC CCT GCC
                  187             190                 194
ASP ARG THR LYS VAL PHE CYS ASN THR GLY ASP LEU VAL CYS THR GLY SER LEU ILE VAL
GAC AGG ACC AAG GTC TTC TGC AAT ACA GGG GAT CTC GTT TGT ACT GGT AGC TTG ATC GTT
            204                   210
ALA ALA PRO HIS LEU ALA TYR GLY PRO ASP ALA ARG GLY PRO ALA PRO GLU PHE LEU ILE
GCT GCA CCT CAC TTG GCT TAT GGT CCT GAT GCT CGT GGC CCT GCC CCT GAG TTC CTC ATC
                  230
GLU LYS VAL ARG ALA VAL ARG GLY SER ALA @@@ GGAGGATGAGAATTTTAGCAGGCGGGCCTGTTAAT
GAG AAG GTT CGG GCT GTC CGT GGT TCT GCT TGA
```

```
TATTGCGAGGTTTCAAGTTTTTCTTTTGGTGAATAGCCATGATAGATTGGTTCAACACTCAATGTACTACAATGCCTCCC
```

```
CCCCCCCCCCCCC
```

Figure 4. The nucleotide sequence of the cDNA showing the coding region for cutinase and the amino acid sequence deduced from the nucleotide sequence. The solid lines represent the regions for which the primary structure was confirmed by amino acid sequencing.

in the enzyme, for four SH groups must be involved in two disulphide bridges and preliminary examination of the disulphide linkages indicate that Cys-47 is linked to Cys-125 and thus the other two are probably involved in

disulphide bridge. Further work is necessary to establish
the disulphide pattern.

The N-terminal amino group of glycine was found to be
attached to glucuronic acid by an amide linkage (33). The
first glycine in the primary sequence translated from the
nucleotide sequence is at position 32 (Figure 4). The
difference in the molecular weights, as determined by SDS
electrophoresis, between the primary translation product
(24,200) and the mature extracellular enzyme (21,600) is
consistent with a 31 residue signal sequence when the 4%
carbohydrate content of the mature enzyme is taken into
account. The amino acid sequence of the signal peptide
of cutinase contains the model N-terminal hydrophilic
segment with a basic amino acid (lysine, residue 2)
preceding a central hydrophobic core of at least 12
residues (residues 3-17) as seen in most other signal
peptides (34,35). The remaining 14 residue segment is
longer than the commonly observed 4-8 residue "post-core"
region and the model signal-sequence cleavage site (37) is
not evident. Perhaps the introduction of a glucuronamide
at the N-terminus of cutinase by a transpeptidation type
reaction requires a unique recognition site in the signal
sequence. However, the post-translational processing of
cutinase is yet to be elucidated.

CUTINASE GENOMIC STRUCTURE

Since regulation of expression of the cutinase gene
could be highly relevant to pathogenesis, an understanding
of the structure of the cutinase gene and the regulatory
regions near the gene is important to the elucidation of
the early events in host-pathogen interaction. The two
isolates of *F. solani pisi*, T-8 and T-30, were chosen as
the first experimental material because the former produces
large amounts of cutinase and readily penetrates the intact
cuticle/cell wall barrier whereas the latter cannot readily
penetrate the barrier (23). The genomic DNA isolated from
the two strains was digested with Eco R1, Hind III and Pst
1 restriction endonucleases and the electrophoretically-
separated fragments were blotted and hybridized with a
[32]P-labeled cutinase cDNA probe. The results (Figure 5)
suggested that the isolate T-8 has two copies of the
cutinase gene whereas T-30 contains only one copy which is
quite similar to one of the copies found in T-8. DNA
isolated from either cutin hydrolysate-induced or uninduced

fungal cultures showed this same restriction enzyme
fragment pattern.   Perhaps the difference in expression of
cutinase gene and the difference in pathogenic virulence
between the T-8 and T-30 strains of *F. solani pisi* can be
attributed to this difference in the genomic distribution.
A comparison of the nucleotide sequence of the possible
regulatory regions flanking the cutinase genes and the
structural genes themselves may offer an explanation for
the difference in virulence.

A genomic library of randomly cleaved DNA from isolate
T-8 was constructed in a Charon 35 λ phage vector.   The
genomic library was screened for copies of the cutinase
gene by hybridizing the cloned phage DNA transferred to
nitrocellulose membranes with the cutinase cDNA probe.
From the 24,000 phage plaques screened, three hybridized
with the cDNA probe.   Considering an average cloned insert
size of 17,000 base pairs and a *Fusarium* genomic size

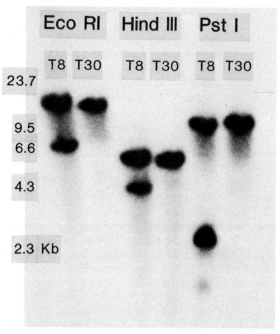

Figure 5.   A Southern blot of electrophoretically separated
restriction endonuclease digests of genomic DNA isolated
from cultures of T-8 and T-30 strains of *F. solani pisi*.
The blot was hybridized with ^{32}P-labeled cutinase cDNA.

of 2.4 x 10^7 base pairs (37), there was a 99% probability
of finding 3 copies of the cutinase gene in 20,000 phage
plaques.   Lambda DNA isolated from an amplified phage
preparation of one of the plaques containing the cutinase
gene was subjected to restriction endonuclease digestion
and electrophoretic separation.   A Southern blot of these
digests hybridized with cutinase–cDNA probe showed that an
8,200 bp piece of fungal genomic DNA containing cutinase
genetic code had been inserted into the λ vector (Figure
6).   This genomic insert is currently being subcloned for
nucleotide sequencing.   Hopefully the nucleotide sequences
of the two cutinase genes, particularly those regions
flanking the structural gene, will offer some explanation
as to the regulation of expression of cutinase.

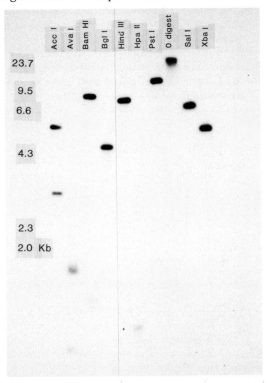

Figure 6.   A Southern blot of electrophoretically separated
restriction endoculease digests of a cutinase genomic clone
selected from the T–8 genomic library cloned into Charon 35
λ vector.

BIODEGRADATION OF SUBERIN

Since suberin is tightly attached to the cell wall, only suberin-enriched wall preparations, but not pure suberin, can be isolated. *Fusarium solani pisi, Helminthosporium solani, H. sativum, Leptosphaeria coniothyrium, Ascodichaena rugosa, Rizoctonia corcorum* and *Cyathus stercoreous* were found to grow on such a suberin-enriched preparation obtained from potato tuber periderm as the sole source of carbon. *F. solani pisi* showed the most rapid growth and therefore the extracellular fluid of suberin-grown cultures of this organism was examined for suberin-degrading enzymatic activities using labeled suberin-enriched preparations as substrates. Such preparations containing radioactive aromatic domains and aliphatic domains were biosynthetically prepared using wound-healing potato tuber slices with [U-^{3}H]cinnamic acid (38) and [9,10-^{3}H]oleic acid (39) as substrates. Incubation of these labeled insoluble polymeric materials with the extracellular fluid resulted in the release of radioactivity into soluble materials. When the

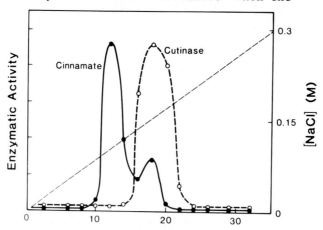

Figure 7. SP-Sephadex chromatography of an acetone precipitated fraction from the extracellular fluid of *Fusarium solani* f. sp. *pisi* T-8 grown on suberin as the sole source of carbon. Release of radioactivity from suberin biosynthetically labeled with [U-^{3}H]cinnamate (solid line) and release of aliphatic components measured with labeled cutin (broken line) are indicated (G. Fernando and P. E. Kolattukudy, unpublished results).

extracellular fluid was fractionated with this release of label as an assay for suberin-degrading activity, the enzymatic activity responsible for release of the aliphatic domains was resolved from that involved in the release of the aromatics (Figure 7). The former was purified to homogeneiety and was found to be identical to cutinase generated by the fungus when grown on cutin (40). Even after prolonged treatment of the labeled suberin with the extracellular fluid, only a small portion of the labeled aromatics could be released. Analysis of the products generated by the extracellular enzyme suggested that probably only the esterified phenolics were released by the enzyme(s), but the "core" held together by ether bonds and/or carbon-carbon bonds was not degraded (C. Woloshuk and P. E. Kolattukudy, unpublished results). Whether such enzymes are involved in the penetration of suberized tissues by pathogenic fungi is unknown.

REFERENCES

1. Kolattukudy PE (1980). Cutin, suberin, and waxes. In Stumpf PK, Conn EE (eds): "The Biochemistry of Plants Vol 4," Stumpf PK (ed): "Lipids: Structure and Function," New York: Academic Press, p 571.
2. Kolattukudy PE (1981). Structure, Biosynthesis, and Biodegradation of Cutin and Suberin. Ann Rev Plant Physiol 32:539.
3. Kolattukudy PE, Espelie KE, Soliday CL (1981). Hydrophobic layers attached to cell walls. Cutin, suberin and associated waxes. In Tanner W, Loewus FA, "Plant Carbohydrates II, Encyclopedia of Plant Physiology New Series, Vol 13B," Berlin: Springer-Verlag, p 225.
4. Soliday CL, Kolattukudy PE (1977). Biosynthesis of cutin. ω-Hydroxylation of fatty acids by a microsomal preparation from germinating *Vicia faba*. Plant Phsyiol 59:1116.
5. Soliday CL, Kolattukudy PE (1978). Midchain hydroxylation of 16-hydroxypalmitic acid by the endoplasmic reticulum fraction from germinating *Vicia faba*. Arch Biochem Biophys 188:338.
6. Croteau R, Kolattukudy PE (1975). Biosynthesis of hydroxy fatty acid polymers. Enzymatic epoxidation of 18-hydroxy oleic acid to 18-hydroxy-cis-9,10-epoxystearic acid by a particulate preparation from spinach (*Spinacia oleracea*). Arch Biochem Biophys 170:61.

7. Croteau R, Kolattukudy PE (1975). Biosynthesis of hydroxy fatty acid polymers. Enzymatic hydration of 18-hydroxy-cis-9,10-epoxystearic acid to *threo*-9,10,18-trihydroxy stearic acid by a particulate preparation from apple (*Malus pumila*). Arch Biochem Biophys 170:73.

8. Croteau R, Kolattukudy PE (1974). Biosynthesis of hydroxyfatty acid polymers. Enzymatic synthesis of cutin from monomer acids by cell-free preparations from the epidermis of *Vicia faba* leaves. Biochemistry 13:3193.

9. Agrawal VP, Kolattukudy PE (1978). Purification and characterization of a wound-induced ω-hydroxy fatty acid: NADP oxidoreductase from potato tubers (*Solanum tuberosum* L.). Arch Biochem Biophys 191:452.

10. Agrawal VP, Kolattukudy PE (1978). Mechanism of action of a wound-induced ω-hydroxy fatty acid: NADP oxidoreductase isolated from potato tubers (*Solanum tuberosum* L.). Arch Biochem Biophys 191:466.

11. Borchert R (1978). Time course and spatial distribution of phenylalanine ammonia-lyase and peroxidase activity in wounded potato tuber tissue. Plant Physiol 62:789.

12. Cottle W, Kolattukudy PE (1982). Abscisic acid stimulation of suberization. Induction of enzymes and deposition of polymeric components and associated waxes in tissue cultures of potato tuber. Plant Physiol 70:775.

13. Soliday CL, Dean BB, Kolattukudy PE (1978). Suberization: inhibition by washing and stimulation by abscisic acid in potato disks and tissue culture. Plant Physiol 61:170.

14. Dean BB, Kolattukudy PE (1976). Synthesis of suberin during wound-healing in jade leaves, tomato fruit, and bean pods. Plant Physiol 58:411.

15. Pozuelo JM, Espelie KE, Kolattukudy PE (1984). Magnesium deficiency results in increased suberization in endodermis and hypodermis of corn roots. Plant Physiol 74:256.

16. Agrawal VP, Kolattukudy PE (1977). Biochemistry of suberization. ω-Hydroxyacid oxidation in enzyme preparations from suberizing potato tuber disks. Plant Physiol 59:667.

17. Purdy RE, Kolattukudy PE (1975). Hydrolysis of plant cuticle by plant pathogens. Purification, amino acid composition, and molecular weight of two isozymes of

cutinase and a nonspecific esterase from *Fusarium solani* f. *pisi*. Biochemistry 14:2824.

18. Köller W, Kolattukudy PE (1982). Mechanism of action of cutinase: chemical modification of the catalytic triad characteristic for serine hydrolases. Biochemistry 21:3083.

19. Shaykh M, Soliday C, Kolattukudy PE (1977) Proof for the production of cutinase by *Fusarium solani* f. *pisi* during penetration into its host, *Pisum sativum*. Plant Physiol 60:170.

20. Maiti IB, Kolattukudy PE (1979). Prevention of fungal infection of plants by specific inhibition of cutinase. Science 205:507.

21. Köller W, Allan CR, Kolattukudy PE (1982) Protection of *Pisum sativum* from *Fusarium solani* f. *pisi* by inhibition of cutinase with organophosphorus pesticides. Phytopathology 72:1425.

22. Dickman MB, Patil SS, Kolattukudy PE (1982). Purification, characterization and role in infection of an extracellular cutinolytic enzyme from *Colletotrichum gloeosporioides* Penz. on *Carica papaya* L. Physiol Plant Pathol 20:333.

23. Köller W, Allan CR, Kolattukudy PE (1982). Role of cutinase and cell wall degrading enzymes in infection of *Pisum sativum* by *Fusarium solani* f. sp. *pisi*. Phys Plant Path 20:47.

24. Lin T-S, Kolattukudy PE (1978). Induction of a biopolyester hydrolase (cutinase) by low levels of cutin monomers in *Fusarium solani* f. sp. *pisi*. J Bacteriol 133:942.

25. Flurkey WH, Kolattukudy PE (1981). *In vitro* translation of cutinase in mRNA: evidence for a precursor form of an extracellular fungal enzyme. Arch Biochem Biophys 212:154.

26. Soliday CL, Flurkey WH, Okita TW, Kolattukudy PE (1984). Cloning and structure determination of cDNA for cutinase, an enzyme involved in fungal penetration of plants. Proc Natl Acad Sci USA (in press).

27. Maxam AM, Gilbert W (1980). Sequencing end-labeled DNA with base-specific chemical cleavages. In Grossman L, Moldave K (eds): "Methods in Enzymology, Vol 65," New York: Academic Press, p. 499.

28. Messing J, Vierira J (1982). A new pair of M13 vectors for selecting either DNA strand of a double-digest restriction fragments. Gene 19:269.

29. Sanger F, Nicklens S, Coulson AR (1977). DNA sequencing with chain-terminating inhibitors. Proc Natl Acad Sci USA 74:5463.
30. Soliday CL, Kolattukudy PE (1983). Primary structure of the active site region of fungal cutinase, an enzyme involved in phytopathogensis. Biochem Biophys Res Commun 114:1017.
31. Dayhoff MO, Barker WC, Hardman JK (1972). Enzyme active sites. In Dayhoff MO (ed): "Atlas of Protein Sequence and Structure, Vol 5," Washington, DC: Natl Biomed Res Foundation, p. 53.
32. Kolattukudy PE (1982). Cutinases from fungi and pollen. In Borgström B, Brockman H (eds): "Lipases." Amsterdam: Elsevier/North Holland, p. 471.
33. Lin T-S, Kolattukudy PE (1977). Glucuronyl glycine, a novel N-terminus in a glycoprotein. Biochem Biophys Res Comm 75:87.
34. Emr SD, Silhavy TJ (1982). Molecular components of the signal sequence that function in the initiation of protein export. J Cell Biol 95:689.
35. Perlman D, Halvorson HO (1983). A putative signal peptidase recognition site and sequence in eukaryotic and prokaryotic signal peptides. J Mol Biol 167:391.
36. Von Heijne G (1983). Patterns of amino acids near signal-sequence cleavage sites. Eur J Biochem 133:17.
37. Szecsi (1981). Kinetic complexity and repetitiveness of *Fusarium graminearum* nuclear DNA. Exp Mycol 5:323.
38. Cottle W, Kolattukudy PE (1982). Biosynthesis, deposition, and partial characterization of potato suberin phenolics. Plant Physiol 69:393.
39. Dean BB, Kolattukudy PE (1977). Biochemistry of suberization. Incorporation of [1-^{14}C]oleic acid and [1-^{14}C]acetate into the aliphatic components of suberin in potato tuber disks (*Solanum tuberosum*). Plant Physiol 59:48.
40. Fernando G, Zimmerman W, Kolattukudy PE (1984). Suberin-grown *Fusarium solani* f. sp. *pisi* generates a cutinase-like esterase which depolymerizes the aliphatic components of suberin. Physiol Plant Pathol 24, in press.

Cellular and Molecular Biology of Plant Stress, pages 401–412
© 1985 Alan R. Liss, Inc.

POSSIBLE MOLECULAR BASIS OF IMMUNITY OF COWPEAS TO COWPEA MOSAIC VIRUS[1]

Janet L. Sanderson, George Bruening
and Mary L. Russell

Department of Biochemistry and Biophysics
University of California, Davis, CA 95616

ABSTRACT Replication of cowpea mosaic virus strain SB
(CPMV-SB) is restricted so effectively in seedlings of
a few cowpea (Vigna unguiculata) lines that no virus
can be detected by sensitive infectivity and immunological assays performed several days after inoculation.
The cowpea line Arlington is the only immune line from
which resistant protoplasts have been isolated. CPMV-
SB yields from Arlington protoplasts are about one-
tenth the yields from protoplasts of the susceptible
line Blackeye 5. During virus replication the two
genomic RNAs of CPMV-SB are translated into polyproteins that are cleaved by virus specified protein-
ase(s) to give functional proteins. We have demonstrated in extracts of Arlington protoplasts an activity that interferes with the specific cleavage of a
precursor of the CPMV-SB capsid proteins. The properties of the observed proteinase inhibitor activity are
consistent with it having a role in the resistance of
Arlington protoplasts to CPMV-SB.

INTRODUCTION

Cowpea mosaic virus strain SB (CPMV-SB) is the type
member of the comovirus group of small RNA plant viruses.
Comoviruses have two single stranded RNA chromosomes of the

[1]Supported by the U.S. Department of Agriculture
Science and Education Administration under grant 82-CRCR-1-
1089 and the Agricultural Experiment Station of the
University of California.

messenger or (+) polarity that are separately encapsidated
in icosahedral shells composed of 60 copies each of two
coat proteins. The middle component (M) contains one copy
of RNA 2, whereas the more rapidly sedimenting bottom component (B) contains one copy of the larger RNA 1. Each of
the virion RNAs bears a 5'-linked protein and 3' polyadeny-
late. In gene order and gene expression the comoviruses
resemble the animal picornaviruses more closely than they
resemble most other plant viruses (1,2). Both comovirus
and picornavirus genes are expressed by the translation of
genomic-sized (+)RNAs into polyproteins that subsequently
are cleaved to produce functional proteins (3,4,5). For
example, RNA 1 of CPMV-SB encodes a proteinase that cleaves
a polyprotein that is translated from RNA 2 and that serves
as a precursor to the capsid proteins.

Approximately 60 lines of cowpea (Vigna unguiculata),
out of more than 1000 tested, are designated as being
operationally immune to cowpea mosaic virus strain SB
(CPMV-SB) according to specific criteria (6). The criteria
for immune seedlings require that CPMV-SB replication be
restricted so effectively that no virus can be detected by
sensitive infectivity and immunological assays performed
several days after inoculation with 100 times a level of
inoculum which will uniformly infect susceptible cowpeas.
The observed immunity has been very reliable. It was not
overcome by heat shock, hormone treatments or graft inoculation (6).

The cowpea line Arlington is the only immune line from
which resistant protoplasts have been isolated. Arlington
protoplasts accumulated only approximately one tenth the
amount of CPMV-SB accumulated by protoplasts of the susceptible line Blackeye 5 and the other immune lines when
protoplasts were inoculated with virions at 1 to 4 µg/ml
(7,8).

Results from genetic crosses of Arlington and Blackeye
5 cowpeas indicate that seedling immunity is inherited as a
simple dominant character (9). Immune progeny seedlings
are sources of resistant protoplasts, implying that both
immunity and resistance reflect a restriction of virus
replication at the cellular level. Inheritance of seedling
immunity as a dominant character indicates that immunity is
mediated by an inhibitor of virus replication rather than
by the absence of some factor required by the virus. Another comovirus, cowpea severe mosaic virus strain DG
(CPSMV-DG), infects all cowpeas and cowpea protoplasts to
which it has been inoculated, including Arlington (7;

unpublished observations). Thus the postulated inhibitor should be specific for CPMV-SB replication.

Kiefer et al. (9) compared CPMV-SB replication in Arlington and Blackeye 5 protoplasts. They found that the two systems were similar in most respects, except that Arlington protoplasts accumulated significantly less CPMV-SB (-)RNA and capsids. Similarly, intact (-)RNA could not be detected in Arlington seedlings that had been heavily inoculated with CPMV-SB (10). We postulate that Arlington restricts CPMV-SB replication by specifically reducing or preventing the production of CPMV-SB proteins, including capsid proteins and at least one polypeptide that is necessary for RNA dependent RNA polymerase activity.

The production of CPMV-SB proteins obviously requires both translation of the virus (+)RNAs and the processing of the polyproteins. Of these two steps, processing is a more likely target of an inhibitor from the Arlington cowpea. It is difficult to conceive of a mechanism that could discriminate between such similar RNAs as those of CPMV-SB and CPSMV-DG and allow only the latter to be translated in tissue that had been inoculated with both viruses (11). However, the proteinase of CPMV-SB that cleaves the precursor to the capsid proteins fails to cleave the corresponding polyprotein of CPSMV-DG (12), indicating discrimination in the processing reactions. Our working hypothesis is that the postulated inhibitor effects immunity or resistance to CPMV-SB by acting on a virus specified proteinase to reduce formation of functional virus proteins.

We demonstrate here that extracts of Arlington cowpea protoplasts interfere with the specific cleavage of polyproteins translated from CPMV-SB RNA 2. The inhibitor action was heat sensitive and was enhanced by partial fractionation of the extracts.

METHODS

Sources of cowpea lines and virus were as reported (7). Cowpea protoplasts were isolated and inoculated essentially as described (8,13,14) except that 70 mg/ml Macerase (Calbiochem) solution was chromatographed (sample approximately 0.05 of bed volume) on Sephadex G-25 (Pharmacia) cross-linked dextran resin in 2.5 mM $CaCl_2$, 0.5 mM Na_3EDTA, pH 7.2, and cellulase (Worthington, LS0002611) was dissolved at 20 mg/ml in 0.5 M mannitol, 2 mM $CaCl_2$, 4 mg/ml bovine serum albumin (BSA), 3 mM sodium mopholino-ethanesulfonate (MES), pH 6.5. Digestion of abraded cowpea primary leaves was in 0.5 M mannitol, 2 mM $CaCl_2$, 0.5 mM sodium MES, 1.5 mg/ml BSA, 3.75 mg/ml cellulase, and approximately 1.9 mg/ml Macerase for 4 to 20 hr at 30° to 37° with gentle agitation.

The source of CPMV-SB proteinase activity was extract of protoplasts that had been inoculated with 100 μg/ml of a CPMV-SB preparation that was highly enriched in component B (encapsided RNA 1). The inoculated protoplasts were incubated at a distance of 2 m from closely spaced fluorescent lamps for 2 days at 22°. Protoplasts were washed twice with 0.5 M mannitol (2 min centrifugation at 200 x g) and once with proteinase buffer (4 min at 200 x g). Proteinase buffer (adapted from the HB buffer of Franssen et al. (15)) is: 0.3 M mannitol, 50 mM Tris, 41 mM acetic acid, 10 mM dithiothreitol (DTT), 10 mM Na_3EDTA, 1 to 2 mM phenylmethylsulfonylfluoride (PMSF). DTT and PMSF were added last, the latter as a 3 M solution in dimethylformamide. Protoplasts were disrupted as a suspension in proteinase buffer in a volume of two to three times the protoplast packed volume. Thirty to 100 strokes in a glass and Teflon homogenizer produced nearly complete breakage of cells. Light microscopy revealed a suspension of chloroplasts. The extract was centrifuged for 5 min, 15 min and 15 min in an Eppendorf model 5414 centrifuge, recovering the supernatant after each centrifugation. Extracts of mock inoculated Blackeye 5 cowpea protoplasts were prepared similarly. The final supernatant was used directly or was rapidly frozen and stored at -80°.

The sources of fractions that were tested for proteinase inhibitor activity were freshly isolated, uninoculated Arlington or Blackeye 5 protoplasts that were extracted as described in the previous paragraph except that the initial centrifugation of protoplasts was at 400 x g for 4 min and the protoplasts were disrupted in either proteinase buffer or 50 mM Tris, 27 mM HCl, 0.1 mM Na_3EDTA, 12.5 mM DTT, 0.3 mM PMSF. All extracts were approximately pH 7.5 after the centrifugation steps. Extracts were either frozen and stored at -80° or subjected to fractionation. $(NH_4)_2SO_4$ was added as a pH 7.5 solution that was saturated at room temperature. The suspensions that formed were stored on ice for 18 or more hr before collecting the precipitate by centrifugation. The supernatants were brought to saturation with solid $(NH_4)_2SO_4$ and stored on ice. Precipitates

were dissolved in approximately 0.1 of their original volume of the Tris buffer discribed above and were dialyzed
against that buffer, then against 50 mM Tris, 27 mM HCl.
Alternatively, they were chromatographed on BioGel P6DG
(BioRad; sample volume 0.2 of bed volume) or Sephadex G-50
(sample volume approximately 0.1 of bed volume) in 50 mM
Tris, 27 mM HCl.

Substrates for CPMV-SB proteinase were the translation
products of CPMV-SB RNA 2. Reaction mixtures of 50 μl
contained 35 μl of micrococcal nuclease-treated rabbit
reticulocyte lysate (Promega Biotec), 50 μCi of 1200
mCi/mmole [^{35}S]-methionine, other amino acids at 20 μM, 0.5
mM MgCl$_2$, 70 mM potassium acetate, other salts and buffer
as indicated by Pelham and Jackson (16), and 1 to 5 μg of
RNA 2. The amount of RNA was adjusted to give maximum
incorporation into hot trichloroacetic acid insoluble material. RNA stimulated incorporation was 10-fold greater
than endogenous incorporation and converted 6 to 7% of the
[^{35}S]-methionine into hot acid-insoluble material. Typical
incubation mixtures for analysis of proteinase activity
contained translation reaction mixture, centrifuged extract
of protoplasts that had been inoculated with CPMV-SB component B ("proteinase") and extract of Arlington or Blackeye
5 protoplasts or buffer. Incubation was for 1 hr at 30°
(15).

Products of the reactions were analyzed by electrophoresis in sodium dodecyl sulfate (SDS)-permeated polyacrylamide gels as modified by Franssen et al. (15) from
the procedure of Laemmli (17). To avoid loss of radioactivity in the gel wells and at the top of the running gel,
samples were incubated at 100° in 10 mg/ml SDS at pH 8.8
for 2 to 4 min and titrated to pH 6.8 with HCl before
loading on 1.5 x 130 x 230 mm long 12.5% acrylamide, 0.09%
methylene-bis-acrylamide gels for electrophoresis at 200
volts. Molecular weight (12,000 to 200,000) standards were
^{14}C-labeled proteins (BRL, 6021SA). Gels were impregnated
with 2,5-diphenyloxazole from a 20% dimethylsulfoxide solution (18), washed in water and dried for autoradiography.
To determine the radioactivity of single zones a 10 mg
rectangle of excised, dry gel was dispersed in 1 ml of 1:9
water:NCS (Amersham) by incubation at 50° for 6 hr (19).
Counting was in 5 to 10 ml of toluene based scintillation
fluid. Zones from regions of the gel that showed no autoradiographic exposure were used for background subtraction.
External standard channels ratios were constant, indicating
similar counting efficiencies from vial to vial.

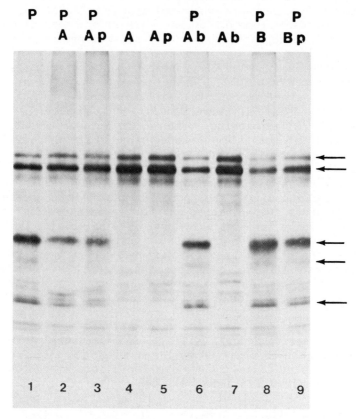

FIGURE 1.   Comparison of effects of extracts from
Arlington and Blackeye 5 cowpea protoplasts on in vitro
processing of translation products of CPMV-SB RNA 2.   (P)
indicates reaction mixtures that received proteinase rather
than an equal volume of protoplast disruption buffer.   Extract of Arlington protoplasts was used directly (A) or
after heating to approximately 90° for 5 min (Ab).   Both
Blackeye 5 (B) and Arlington protoplast extracts also were
supplied as the dissolved and dialyzed precipitate from a
solution 0.3 saturated with ammonium sulfate (Bp and Ap).
The reaction mixture for lane 1 contained 10 µl of Tris-HCl
buffer in place of extract of uninoculated protoplasts.
The arrows indicate, from top to bottom of the gel, zones
designated as containing polypeptides 105K, 95K, 60K, 58K
and 48K on the basis of comparisons with mobilities of
standards and published assignments (15).

RESULTS AND DISCUSSION

The reaction mixtures for the assays of CPMV-SB proteinase that are presented in Fig. 1 were 2 μl of reticulocyte extract in which CPMV-SB RNA 2 had been translated
("substrate") plus 5 μl of extract from Blackeye 5 cowpea
protoplasts that had been inoculated with CPMV-SB component
B ("proteinase") plus 10 μl of extract of uninoculated
protoplasts. Fig. 1, lane 1 (and Fig. 2, lane 3) demonstrates the expected proteinase activity by the appearance
of the zones for polypeptide cleavage products designated
60K, 58K and 48K and a decrease in the intensity of zones
for translation products 105K and 95K (compare lanes 1 and
4 of Fig. 1). The patterns of zones resemble published
patterns for processing catalyzed by CPMV-SB proteinase
from protoplasts (15) and from an in vitro translation
reaction (3). When extract of uninoculated Blackeye 5
protoplasts was substituted for the extract of inoculated
protoplasts, the 105K and 95K zones remained undiminished,
and we observed no 60K, 58K or 48K zones (data not shown).
Extract of freshly isolated Arlington protoplasts
reduced the extent of processing (Fig. 1, lane 2), whereas
heated extract was less effective (lane 6). Heating the
extract to 100° for longer periods resulted in greater loss
of the proteinase inhibiting activity (data not shown).
Quantitative analyses of the radioactivity associated with
the precursor and product zones from electrophoresis gels
(Table 1) showed that extract of freshly isolated, uninfected Blackeye 5 protoplasts did not reduce the extent of
processing of CPMV-SB RNA 2 translation products. Our
yields of both proteinase activity and proteinase inhibitor
activity have varied from preparation to preparation. However, each strongly active inhibitor preparation from Arlington protoplasts effectively reduced the extent of processing catalyzed by every active proteinase preparation
against which it was tested. When we added extract of
Arlington cowpea protoplasts after, rather than before, a
one hour period of incubation of "proteinase" with "substrate," the products of the processing reaction were present in the expected amounts (data not shown).
A portion of the proteinase inhibitor activity from
Arlington cowpea protoplasts precipitated at 0° in the
presence of a concentration of $(NH_4)_2SO_4$ that corresponded
to 0.3 of saturation at room temperature (Fig. 1, lane 3;
Table 1). At 0.5 of saturation the activity completely
precipitated (Fig. 2, lanes 1 and 2). Most of the

TABLE 1

EFFECT OF PROTOPLAST EXTRACTS ON THE PROCESSING
OF CPMV-SB TRANSLATION PRODUCTS

Additions to RNA 2 translation products	Extent of processing[a]
Experiment 1[b]	
"Proteinase"	0.48
"Proteinase" plus Blackeye 5 protoplast extract	0.59
"Proteinase" plus Arlington protoplast extract	0.32
"Proteinase" plus Arlington protoplast extract, heated	0.41
"Proteinase" plus $(NH_4)_2SO_4$ insoluble fraction Blackeye 5 protoplast extract	0.44
"Proteinase" plus $(NH_4)_2SO_4$ insoluble fraction Arlington protoplast extract	0.20
Arlington protoplast extract, heated	-0.02
$(NH_4)_2SO_4$ insoluble fraction Arlington protoplast extract	-0.02
Experiment 2[c]	
"Proteinase"	0.61
"Proteinase" plus Void volume fraction, Arlington protoplast extract	0.22

[a]Extent of processing is defined as the ratio of cpm from the 60K and 48K zones divided by the sum of cpm from 105K, 95K, 60K and 48K zones.
[b]Zones indicated by arrows in Fig. 1 were excised and radioactivity determined.
[c]Conditions like those of Fig. 2, lanes 3 and 4.

inhibitor activity emerged at the void volume during gel
exclusion chromatography on cross-linked polyacrylamide gel
(Fig. 2, lanes 3,4; Table 1). In other experiments the
inhibitor activity emerged after the void volume but before
the total volume from Sephadex G-50 columns eluted with
Tris-HCl buffer (data not shown).

    CPSMV-DG provides an opportunity to test the specifi-
city of the Arlington proteinase inhibitor _in vitro._
CPSMV-DG proteinase activity has proved to be more difficult to demonstrate than that of CPMV-SB. In recent preliminary experiments we have found sufficient activity in
extracts of CPSMV-DG-infected Blackeye 5 seedlings, using
the _in vitro_ translation products of CPSMV-DG RNA 2 as
substrates. As expected, the proteinase inhibitor activity
of Arlington cowpea protoplasts was ineffective against the
CPSMV-DG proteinase activity (data not shown).

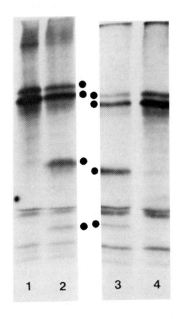

FIGURE 2. Precipitation and
chromatography of proteinase
inhibitor from Arlington cowpea
protoplasts. The precipitate
from 0.6 of saturation (lane 1)
and the supernatant from 0.5 of
saturation (lane 2) ammonium
sulfate precipitation steps
were analyzed for proteinase
inhibitor activity. Reaction
mixtures had the composition
indicated in the legend for
Fig. 1. The precipitate from
an Arlington protoplast extract
that had been brought to 0.5 of
saturation with ammonium sulfate was chromatographed on
BioGel P6DG. Reactions mixtures contained 3 µl of "proteinase," 2 µl of "substrate,"
and 5 µl of either column buffer (lane 3) or void volume
fraction from the column (lane 4). Chromatography fractions that preceded or succeeded the void volume fractions
gave results similar to those seen in lane 3. Dots between
lanes 2 and 3 indicate, from top to bottom, zones for 105K,
95K, 60K and 48K. The 58K zone was not apparent.

Extracts of Arlington cowpea protoplasts apparently contain a substance or substances that reduced the extent of specific cleavage of CPMV-SB RNA 2 translation products by CPMV-SB proteinase. The activity appears to be mediated at least in part by a high molecular weight material(s), presumably a protein(s), because the activity was heat sensitive, was precipitated by ammonium sulfate, and emerged before the total volume fractions during gel exclusion chromatography on two types of resins. The results are consistent with a simple model for the mechanism of resistance of Arlington cowpea protoplasts, and the immunity of Arlington cowpea seedlings, to CPMV-SB. In this model Arlington cowpeas form a protein that binds to and inhibits a proteinase that is specified by RNA 1 of CPMV-SB. Thus the translation of either RNA 1 or RNA 2 in inoculated Arlington cells will result in the formation of reduced amounts, or perhaps none, of the functional CPMV-SB proteins that are required for its replication.

ACKNOWLEDGMENTS

J. L. Sanderson is the recipient of a fellowship from the McKnight Foundation training program at the University of California, Davis.

REFERENCES

1. Goldbach R, Rezelman G (1983). Orientation of the cleavage map of the 200-kilodalton polypeptide encoded by the bottom-component RNA of cowpea mosaic virus. J Virol 46:614.
2. Franssen H, Leunissen J, Goldbach R, Lomonossoff G, Zimmern D (1984). Homologous sequences in non-structural proteins from cowpea mosaic virus and picornaviruses. EMBO J 3:855.
3. Pelham HRB (1979). Synthesis and processing of cowpea mosaic virus proteins in reticulocyte lysates. Virology 96:463.
4. Franssen H, Moerman M, Rezelman G, Goldbach R (1984). Evidence that the 32,000-Dalton protein encoded by bottom-component RNA of cowpea mosaic virus is a proteolytic processing enzyme. J. Virol. 50: 183.

5. Franssen H, Goldbach R, van Kammen A (1984).
    Translation of bottom component RNA of cowpea mosaic
    virus in reticulocyte lysate: faithful proteolytic
    processing of the primary translation product. Virus
    Res 1:39.
6. Beier H, Siler DJ, Russell ML, Bruening G (1977).
    Survey of susceptibility to cowpea mosaic virus among
    protoplasts and intact plants from Vigna sinensis
    lines. Phytopathology 67:917.
7. Beier H, Bruening G, Russell ML, Tucker CL (1979).
    Replication of cowpea mosaic virus in protoplasts
    isolated from immune lines of cowpeas. Virology
    95:165
8. Bruening G, Kiefer M (1981).   Virus RNA synthesis in
    resistant protoplasts inoculated with cowpea mosaic
    virus. Proc Symp IX Internat Congr Plant Protection
    1:225.
9. Kiefer MC, Bruening G, Russell ML (1984). RNA and
    capsid accumulation in cowpea protoplasts that are
    resistant to cowpea mosaic virus strain SB. Virology
    (in press).
10. Eastwell KC, Kiefer MC, Bruening G (1983). Immunity
    of cowpeas to cowpea mosaic virus. In Goldberg RB
    (ed): "Plant Molecular Biology," UCLA Symposia on
    Molecular and Cellular Biology, New Series, Vol. 12.
    New York: Alan R. Liss, p 201.
11. Bruening G, Lee S-L, and Beier H (1979). Immunity to
    plant virus infection. In Sharp WR, Larsen PO,
    Paddock EF, Raghavan V (eds): "Plant Cell and Tissue
    Culture -- Principles and Applications," Ohio State
    University, Biological Sciences Colloquia, Chapter 22,
    p 421.
12. Goldbach R, Krijt J (1982). Cowpea mosaic virus-
    encoded protease does not recognize primary
    translation products of M RNAs from other comoviruses.
    J Virol 43:1151.
13. Beier H, Bruening G (1975). The use of an abrasive in
    the isolation of cowpea leaf protoplasts which support
    the multiplication of cowpea mosaic virus. Virology
    64:272.
14. Cassells AC, Barlass M (1978). The initiation of TMV
    infection in isolated protoplasts by polyethylene
    glycol. Virology 87:459.

15. Franssen H, Goldbach R, Broekhuijsen M, Moerman M, van Kammen A (1982). Expression of middle-component RNA of cowpea mosaic virus: In vitro generation of a precursor to both capsid proteins by a bottom-component RNA-encoded protease from infected cells. J Virol 41:8.
16. Pelham HRB, Jackson RJ (1976). An efficient mRNA-dependent translation system from reticulocyte lysates. Eur J Biochem 67:247.
17. Laemmli UK (1970). Cleavage of structural proteins during the assembly of the head of bacteriophage T4. Nature (London) 227:680.
18. Bonner WM, Laskey RA (1974). A film detection method for tritium-labelled proteins and nucleic acids in polyacrylamide gels. Eur J Biochem 46:83.
19. Zaitlin M, Hariharasubramanian V (1970). Proteins tobacco mosaic virus-infected tobacco plants. Bioch Biophys Res Commun 39:1031.

Cellular and Molecular Biology of Plant Stress, pages 413–433
© 1985 Alan R. Liss, Inc.

PLANT DEFENSE RESPONSES TO VIRAL INFECTIONS

G. Loebenstein and Adina Stein

Virus Laboratory, Agricultural Research Organization,
Bet Dagan, Israel

INTRODUCTION

The interaction of a virus with a plant may vary
between immunity and susceptibility, with most plants being
resistant to infection thanks to the process of natural
selection during evolution. The resistance of many plants
to viruses is probably of a passive preformed type,
prepared in advance of the infection, as mechanical
barriers, lack of compatible infection sites or metabolites
required for virus multiplication (38). In animals, and
especially mammals, two potent mechanisms of resistance
developed during the evolutionary route: the immune
reaction-based on antibodies, and the interferon system.
Both are due to active host responses induced by the
pathogen. In plants, even though apparently the preformed
passive mechanisms- resulting in an escape from infection-
are dominant, active host resistance responses are known
and are often used in breeding programs.

An active response may be defined as a mechanism
whereby, after the introduction of the virus into the
plant, virus multiplication or spread is inhibited, and
which requires transcription of cellular DNA. Such an
activation of host genes may give rise to structural
changes forming physical barriers to virus spread, or to
functional changes that interfere in the viral replication
process. Although quite a range of active resistance
response phenomena is known, understanding the latters
mechanism is still often speculative. This review will be
centered around the local lesion response, including
subliminal infections, effect of the localized infection on
non-invaded tissues (induced resistance), and several
resistance responses in systemic infections. It is expected
that knowledge of the molecular basis of these phenomena

may be of major value in developing novel ways of crop
protection.

THE LOCAL LESION RESPONSE

This is probably the most notable active resistance
phenomenon, where, after inoculation, the virus invades and
multiplies in up to several hundred cells, but does not
spread to other tissues.

The types of localized infections known are: (a)
necrotic lesions of the nonlimited type; (b) self-limiting
necrotic local lesions; (c) ring-like patterns or
ringspots; (d) starch lesions; (e) microlesions; and (f)
subliminal infections (42). In the local lesion a
discontinuous and uneven distribution of virus between
cells is observed and the estimated number of virus
particles per cell, averaged over all cells in the lesion,
is about two to four orders of magnitude lower than in a
comparable systemic infection. In starch lesions a gradient
of particle density from the center to the periphery of the
lesion is observed.

In subliminal infections, such as TMV in cotton
cotyledons, only minuscle amounts of virus, $<$ 0.1 ug/g of
fresh tissue, are recovered, compared with $>$ 1000 ug/g from
highly susceptible hosts (11). No symptoms are observed and
virus can be recovered only from directly inoculated
tissue. The virus-replicating capacity in cotton can be
increased by dark treatment and by application of several
plant hormones, such as kinetin, indole-3 -acetic acid,
2,4-dichloro phenoxyacetic acid, and Etephon (10). The
subliminal infection seems to be one of the most efficient
resistance responses.

Activation of the Localizing Mechanism by Non-Viral Agents.

The localizing mechanism can be activated also by
non-viral-agents. Various polyanions such as yeast-RNA,
poly I: poly C (34), ethylene maleic anhydride copolymer
(EMA 31) (60), and polyacrylic acid (24), have been
reported to activate resistance. In general the resistance
developed gradually and was sensitive to actinomycin D. In
the antiviral state induced by EMA-31, good correlations
were observed between resistance and a specific ribosomal

fraction (61). The resistance induced by polyacrylic acid
was associated with the appearance of three new proteins.
These three proteins were similar to those produced in
TMV-infected leaves of cv. 'Xanthi' plants that also are
resistant to infection (69).

Hypotheses Explaining Localization.

     Death of cells. This explanation is not satisfactory
even in necrotic local lesion hosts and still less so in a
chlorotic or starch-lesion host, as virus particles are
found outside the necrotic area in viable cells (44).
     Structural changes in advance of the infection
leading
to cell death. Around the core of necrotic cells of the
lesion in tobacco cv. 'Samsun NN', there is a zone of cells
that stain darkly. Seven days after inoculation with TMV,
the mesophyll cells surrounding the lesion showed
structural features suggestive of metabolic activity.
Compared with cells of normal mesophyll, those within this
zone had smaller vacuoles, more cytoplasm and ribosomes,
and a marked system of endoplasmic reticulum. It was
suggested that structural changes induced in advance of
infection leading to cell collapse - may be instrumental in
virus localization (53). These changes may progress to a
level that causes collapse of a narrow ring of cells not
yet invaded by virus, resulting in a barrier to further
movement of virus. However, in N. glutinosa, the peripheral
tissue was found to consist of necrotic cells, intermingled
with healthy cells and others showing various levels of
disorganization (27).
     Barrier substances. Several barrier substances in
cells surrounding the local lesion have been implicated as
possible factors preventing virus movement.
     Lignification and suberin deposition. Lignin
deposition has been observed at the borders of necrotic
lesions induced by TMV U-5 in N. glutinosa and in the lumen
of tracheids situated close to them (16). Lignin formation
seems to be associated with wound-healing mechanisms, when
the wound undergoes necrosis. Lignin may be involved with
cell-wall thickening, postulated to be a barrier to the
spread of virus. However, in tomato bushy stunt-induced
necrotic lesions in Gomphrena globosa, conspicuous
thickenings of lignin were deposited outside the cells of
the halo surrounding the necrotic area, but these did not

follow the cell perimeter (4). Deposition of suberin was
also observed around these lesions, as well as around TMV-
and TNV-induced local lesions on bean (15). Apparently, the
modification of cell walls by lignin and suberin occurs as
a general response to injury, amplified by necrotic lesion
formation.

Paramural bodies, and callose deposition. Blocking of
plasmodesmata or rupture of plasmodesmatal connections by
paramural bodies, and callose deposition along cell walls,
have been studied in relation to virus localization.
Membrane-bound vesicular (paramural) bodies were observed
between the plasma membrane and the cell wall at the
periphery of TMV-induced lesions on Pinto bean leaves (59)
and in cells of PVX-induced lesions on Gomphrena globosa
(1). However, paramural bodies associated with cell wall
protrusions have also been observed in systemic infections,
as in barley infected with barley stripe mosaic (43). Their
specific association with virus localization is therefore
questionable.

The deposition of callose has received much attention
in recent years. Wu and colleagues observed callose
deposition around TMV-induced lesions on Pinto beans (72).
Large depositions of callose were also observed in
association with cell walls surrounding the necrotic zone
of PVX-incited local lesions on G. globosa (1). These
caused a protrusion of the arms of the plasmodesmata until
finally the increased callose deposition ruptured
plasmodesmatal connections. Membrane-bound bundles of virus
particles embedded in callose were observed outside the
protoplast. Allison and Shalla (1) suggested that the
callose deposition and exclusion of virus particles from
the protoplast may limit the spread of virus in a
hypersensitive host.

However, heavy callose depositions were also observed
in tobacco infected with TSW and with tobacco rattle virus,
which produce systemic necrosis where the virus does not
remain localized (57). On the other hand, no callose
deposition was observed around TMV-induced starch lesions
on cucumber cotyledons, where the infection remains
localized without necrosis (12).

It seems, therefore, that the various barrier
substances, as well as ultrastructural changes in advance
of the infection, could be responses to necrotization and
not necessarily responsible for localization.

To obtain a better insight into localization, it is
therefore necessary to select a system where the infection

remains localized but without necrotization, so that any
response can be related to the localizing process, without
having to consider the possible association with the
necrotic reaction. The starch lesion system, such as TMV in
cucumber cotyledons, seems to be the most suitable, as here
the infection remains localized without necrotization. No
callose deposition was observed around the starch lesion
(35), and no ultrastructural changes were observed at the
periphery of the lesion (12). Thus, except for the fact
that the cells at the periphery of the lesion contained
virus, they were otherwise normal. These virus-containing
cells were connected through plasmodesmata with normal
non-infected cells. Apparently, in this starch lesion host,
localization of TMV is not due to ultrastructural changes
in advance of the infection, or to blocking of
plasmodesmata. It was suggested that these cells at the
periphery of the lesion do not support virus synthesis
(34).

     In conclusion, ultrastructural changes and various
barrier substances associated with necrotic lesions could
be a response of the host to necrotization, but not
necessarily responsible for localizing the infection,
although they may help in slowing down virus movement.

     Inactivation of a "transport protein". Recently it was
suggested that restriction of virus in a subliminal
infection is associated with the inactivation of a
"translocation protein" (62). Thus, by evaluating
protoplasts from TMV-inoculated cowpea leaves, it was
deduced that only those cells that received virus during
mechanical inoculation supported virus replication, and
that virus was unable to move from these original infection
centers. That in subliminal infections a putative
"transport protein" is destroyed or inactivated is in line
with suggestions based on observations with certain
temperature sensitive (ts) mutants of TMV (64). These
mutants, although replicating normally in protoplasts, do
not spread systemically at restrictive temperatures. It was
suggested that a specific transport function is coded by
the TMV genome, which is essential for the systemic
spreading of the infection. At high temperatures it is not
produced or inactivated. It was further suggested that a
30,000 MW protein, coded by TMV, is a possible candidate
for the role of this "transport protein" (33). Such a
hypothetical "transport protein" might increase the number
of plasmodesmata, prevent plasmodesmatal loss, or

participate in the formation of a special transport complex
with the virus or viral - RNA.

However, whether resistance in subliminal infections
is associated with inactivation of a "transport protein",
thereby limiting virus spread, or due to the effective
production of an inhibitor of virus replication (IVR) (37),
will have to be seen. That initially infected cells are
able to support virus replication with no virus in adjacent
cells could also be due to a time lapse required for
activation of IVR production and not necessarily due to
lack of a transport function. It seems relevant to mention
in this connection that, in cucumbers inoculated with TMV,
causing starch lesions, no differences were observed in the
number or shape of plasmodesmata connecting infected to
neighboring non-infected cells to those between two healthy
control cells (12).

Localization due to reduced virus multiplication - a
substance(s) that inhibits virus multiplication.
Metabolic inhibitors and ultraviolet irradiation.
Indications of a host-mediated process that suppresses
virus multiplication in a starch lesion host were obtained
in studies using antimetabolites (41) and UV-irradiation
(35). Virus concentration in cucumber cotyledons increased
markedly when cotyledons were treated with actinomycin D,
chloramphenicol or UV, one day after inoculation with TMV.

These results suggest that the localizing mechanism
operates mainly via inhibition of virus multiplication.
This is an active process which requires new m-RNA
synthesis, and therefore treatments which inhibit
transcription suppress localization and increase virus
multiplication in such hosts. The virus itself activates
this mechanism.

SUBSTANCES THAT INHIBIT VIRUS MULTIPLICATION IN
RESISTANT TISSUE

Antiviral or interfering agents have been extracted
from both infected and uninfected-resistant tissue.
However, their connection with localization or induced
resistance is still an open question, especially as they
have been tested as inhibitors of infection given during
inoculation (by mixing with the test virus) and not as
inhibitors of virus multiplication, which should be applied
to the infected cell 5-12 hours (or later) after
inoculation. An antiviral factor (AVF) developing in local

and sytemically infected plants has been reported by Sela
and co-workers (56). The active principle extracted from
leaves of N. glutinosa 48 hours after inoculation with TMV
was first considered to be RNA but later to be a
phosphoglycoprotein with a molecular weight of about 22,000
(45). However, AVF was tested mainly as an inhibitor of
infection, and its association with localizing the
infection through inhibition of virus multiplication has
yet to be shown.

Inhibitory substances have also been extracted from
resistant uninfected tissues (40). However, as activity was
tested by mixing the substance with the virus, the effect
on virus multiplication and its role in localization and
induced resistance have yet to be determined.

For reliable evaluation of substances as inhibitors of
virus multiplication, it is necessary to have a test system
where cells that have been synchronously infected can be
treated uniformly - after different times of inoculation -
with the test substance. The protoplast system seems to be
most suitable, as it is possible to inoculate
simultaneously all cells and to obtain "one-step growth
curves," in contrast to the intact plants where only a
small percentage of the cells becomes infected during the
primary inoculation. It is also possible to apply the test
material uniformly to every protoplast, in contrast to a
parallel application in the intact plant. The use of the
protoplast system is therefore obvious.

Furthermore, the protoplast system can be used to
evaluate antiviral compounds associated with resistance,
which are released from the protoplast into the medium,
thereby obviating the necessity of obtaining and testing
such substances from crude plant homogenates, which contain
many other compounds.

Inhibitor of Virus Replication (IVR) from TMV-infected
Protoplasts of a Local Lesion-responding Tobacco Cultivar.

Recently, we reported that a substance(s) inhibiting
virus replication (IVR) is released into the medium from
tobacco mosaic virus-infected protoplasts of a cultivar in
which infection in the intact plant is localized. IVR
inhibited virus replication in protoplasts from both
local-lesion-responding resistant Samsun NN and
systemic-responding susceptible Samsun plants, when applied
up to 18 h after inoculation. IVR was not produced in

protoplasts from susceptible plants or from non-inoculated
protoplasts of the resistant cultivar. IVR was partially
purified using ZnAc$_2$ precipitation, and yielded two
biologically active principles with molecular weights of
about 26,000 and 57,000, respectively, as determined by gel
filtration (37).

IVR was also recovered from the protoplasts
themselves, and not only from the incubation medium. It did
not affect TMV directly. IVR was found to be sensitive to
trypsin and chymotrypsin but not to RNAse, and its activity
was abolished by incubation at 60 C for 10 min, suggesting
that it is of a proteinaceous nature (21).

Actinomycin D and chloramphenicol, added up to 24 h
after inoculation of tobacco NN protoplasts with TMV,
suppressed production of IVR almost completely. This was
concurrent with a marked increase in TMV replication (22).

IVR was found also to inhibit virus replication in
leaf tissue disks (21), with inhibition rates ranging
between 60% and 90%, as determined by both local lesion
assay and ELISA. Similar results were obtained when IVR was
applied to TMV-infected tomato leaf disks. IVR also
inhibited replication of CMV and PVX in different host
tissues, indicating that it is neither virus- nor
host-specific.

IVR also inhibited TMV replication when applied
through cut stems and by spray. Application of IVR through
cut stems of Samsun plants 2 h before or 5 h after
inoculation reduced extractable TMV by about 80% and 70%,
respectively. Spraying intact plants with IVR 2 h before
or 5 h after inoculation, gave inhibition rates of about
78% and 56%, respectively, when inoculated leaves were
assayed 7 days after inoculation. In non-inoculated leaves
one spray gave inhibition rates of 81% and 43%,
respectively, which are still far from being of practical
value, although it should be noted that only one spray,
with relatively low concentrations of IVR (2 or 3 units),
was given.

Recently an antiserum specific to IVR was prepared.
The serum gave positive reactions with IVR in both ELISA
and gel-diffusion tests and also neutralized IVR's
biological activity (unpublished results). The serum did
not react with purified TMV, TMV coat protein or against α-
and γ-interferon. It gave a weak reaction with the
antiviral factor (AVF) obtained by Sela et al. (56). Such
an antiserum is a prerequisite for the future isolation of
the IVR-specific m-RNA.

These studies with IVR, associated with localization -
which is a true inhibitor of virus replication, may open up
new approaches for the control of virus diseases, in
addition to providing a better understanding of the
mechanisms of resistance to viruses in plant.

SYSTEMIC INDUCED RESISTANCE

Induced resistance is apparently also due to
activation of the localizing mechanism. Thus, resistance in
uninoculated parts of a hypersensitive host can be induced
by inoculating other parts of the plant with necrotic local
lesion producing viruses, fungi (34) and bacteria (29). For
example, inoculation of the lower leaves of tobacco NN
plants with TMV or TNV induced resistance in the upper
uninoculated leaves; and inoculation of primary leaves of
Pinto beans with AMV, TMV or TNV induced resistance in the
opposite leaf. Lesions developing after challenge
inoculation of the resistant tissue are consistently
smaller and usually fewer in number than those formed on
previously uninoculated control plants. In tobacco NN
plants induced resistance, measured by reduction in number
and size of lesions, was found to be closely correlated to
virus concentration determined by ELISA (unpublished
results). This indicates that in the resistant tissue virus
replication and not only the development of necrotic local
lesions is suppressed (17).
Localized necrotic infection with fungi and bacteria
also induces resistance to local lesion producing viruses.
Thus, Thielaviopsis basicola induced resistance in tobacco
(28), and Colletotrichum lagenarium and Pseudomonas
lachrymans (29) in cucumber plants. Although in general
resistance is expressed against necrotic localized
challenge infection, inoculation of cucumbers with C.
lagenarium, P. lachrymans and TMV also induced systemic
resistance against systemic infection by CMV (8). Induced
resistance was expressed as a decrease in the number of
chlorotic, primary lesions and as a delay in the time of
appearance of systemic mosaic symptoms, virus titer not
being measured.
The development of induced systemic resistance has
been observed mainly with pathogens that produce necrotic
local lesions. Recently, however, it was reported that
inoculation of cucumber cotyledons and first foliage leaves
with TMV, which remains localized (starch lesions), induces

systemic resistance in other leaves, when challenged 2
weeks later with TNV (51). Conversely, systemic resistance
can be induced by necrotic spots that develop after
pricking tobacco leaves with an ethylene-releasing compound
(67), or by spraying the lower leaves with mercuric
chloride (68) although in general mechanically or
chemically caused necroses do not induce resistance (52).

Induced resistance seems to be due to an active
cellular process, depending on the transcription mechanism
from DNA to RNA, as its development is markedly inhibited
in the presence of actinomycin D (39). It was suggested
that after the initial inoculation a substance(s) is
produced that induces resistance in uninoculated tissues
(34).

In induced resistant tobacco NN tissues at least four
(and possibly ten) new host-coded proteins become
detectable (68) Similar proteins have also been observed in
other Nicotiana species and in virus-infected cowpea
leaves (13). The tobacco "pathogenesis-related proteins"
(PR's) are host specific soluble proteins, preferentially
extracted at low pH. They differ from the low
pH-extractable proteins of non-infected leaves by their
resistance to proteases. It has been suggested that these
PR's are involved in localization and induced resistance
(30). However, several lines of evidence negate this
suggestion. PR's accumulate to high levels after necrosis
is caused by the systemically spreading viruses TRV and
PVY^n, although the spread of these viruses is not impeded.
On the other hand, spraying with mercuric chloride that
induces systemic resistance was not accompanied by new PR's
(65). Furthermore, no dose quantitative relationship
between the amount of PR and the intensity of systemic
induced resistance was observed and resistance could be
demonstrated before detection of PR's (19). Also, no novel
soluble protein fraction was found in the upper leaves of
cucumber plants when lower leaves were locally infected
with TNV or Colletotrichum lagenarium, with the appearance
of a new PR in the acidic leaf extract of the inoculated
leaves (14). Also, the presence of PR's in tobacco NN
plants with TMV necrotic lesions did not prevent systemic
spread of the virus when these plants were transferred from
$20^{o}C$ to $30^{o}C$ (66); and watering 'Xanthi-nc' tobacco with
low doses of methyl benzimidazole-2yl-carbamate caused
accumulation of PR's but not resistance (19). Large amounts
of PR's were also detected in healthy flowering 'Xanthi-nc'
tobaccos (18). However, treatments such as inflorescence

removal drastically reduced PR's without affecting
susceptibility to TMV. Large amounts of similar proteins
were also found in senescing non-infected tissue of
Gomphrena globosa (49) and in osmotically stressed tissues
(70). Furthermore, partially purified PR's did not induce
resistance and did not reduce virus replication when added
to tobacco protoplasts (31).

The N gene controlling localization and probably
induced systemic resistance is not directly responsible for
the synthesis of PR's. Thus, the tobacco variety 'Judy's
Pride Burley', which lacks the N gene, nevertheless
produces PR's when local necrotic lesions are induced by
the fungus Thielaviopsis basicola (23). It seems,
therefore, that PR's are associated mainly with pathogen-
induced necroses, osmotic stresses or other signals but not
with resistance.

An alternative explanation for induced resistance
based on changes in host growth regulator metabolism, has
recently been reviewed (20). Upper leaves of 'Xanthi-nc'
tobaccos showing induced resistance had a higher (though
not very great) cytokinin content 8 days after the lower
leaves were inoculated with TMV (7), treated with a leaf
spot inducing bacterium, or treatment with mercuric
chloride causing chemical necrosis (63).

As virus content in resistant leaves was not
decreased, it was suggested that cytokinins inhibit
necrotization and not virus replication. However, it should
be noted that virus content in the resistant leaves was
assayed only 48 h after the challenge inoculation, at a
time when virus replication was probably still continuing.
Application of exogenous cytokinin (Kinetin) by spraying
also suppressed the number of visible necroses in
TMV-infected 'Xanthi-nc' tobacco, without, however,
reducing virus replication (6) As induced resistance is
generally associated with reduced virus- replicating
capacity and not only with suppression of necrotization,
cytokinin does not seem to be the major factor.

A higher content of abscisic acid (ABA) has been
observed in the upper resistant leaves of N. tabacum 'White
Burley', using the TMV-flavum strain that forms local
necrotic lesion on this tobacco (17, 72). Using labeled ABA
it was shown that increased ABA in the upper resistant
leaves was partly due to transport of ABA from the lower
leaves. Resistance in this host, which lacks the N gene, is
manifested by a decrease in lesion size and number, but
accumulation of viral RNA (replication) was not affected.

It would therefore be of interest to see if in tobaccos
having the N gene, induced resistance will also be
correlated with an increase in ABA. The marked reduction in
lesion number, with only a minor decrease in lesion size,
which is observed when healthy 'Xanthi-nc' tobaccos are
sprayed with ABA before inoculation (19), suggests that ABA
mainly alters the susceptibility of the plant to infection,
while its effect on virus replication (if at all) was
minor. This change in susceptibility to infection may be
associated with a change in the leaf water status (9).
However, as induced resistance in general is manifested by
a decrease in lesion size, it does not seem at present that
ABA, which affects mainly infection establishment, is a
major factor in induced resistance.

A greater capacity to convert the ethylene precursor
1-aminocyclopropane-1-carboxylic acid (ACC) to ethylene was
observed recently in resistant 'Samsun NN' leaves (32). An
earlier increase of ethylene in challenge-inoculated
resistant leaves was not observed, in contrast to the
findings of Pritchard and Ross (50). It was suggested (32)
that increased ethylene production contributes, at least
initially, to virus localization. However, it remains to
be seen if other factors that cause changes in membrane
permeability and necrotization, especially systemic
necrotic virus infections, do not result in a similar
increase in ACC conversion capacity. This ACC conversion
enzyme may be peroxidase (3) the activity of which
increases with severity of symptoms (34).

So far systemic resistance can be induced mainly by
pathogen-induced necroses, while in directly treated
tissues the localizing mechanisms can be activated by a
variety of chemicals. It may therefore be of interest to
learn whether the signal for systemic induction is similar
to the fungal and plant cell wall fragment elicitors,
suggested as activating phytoalexin production. Thus, a
number of elicitors, which are or contain complex
polysaccharides, have been obtained from the breakdown of
the cell wall of several fungi (2, 5). Further studies
showed that plant cells contain elicitors that are capable
of inducing phytoalexin accumulation (26, 25). Galacturonic
acid was found to be an essential constituent of the
elicitor-active wall fragment. Elicitors were also released
from citrus pectin by partial acid hydrolysis and the
active component was suggested to be dodeca α -1,4 D
galacturonide (47).

It is therefore proposed that degradation products
released from the plant cell wall by the action of pectic
enzymes such as endopolygalacturonase act, in turn, to
elicit phytoalexin production in the plant (71). A further
suggestion proposed by West implicated Ca^{++} ions
sequestered in pectin, as the ultimate messenger. The
pectin fragments may serve as carriers to bring the ions to
neighboring cells, where they activate the genes that
promote phytolexin production. In this connection it is of
interest to mention the results of Ryan et al. (55), with
the proteinase inhibitor inducing factor (PIIF). This
factor induces the synthesis of the proteinase inhibitor
that interferes with digestive enzymes of insects, which
break down and digest plant tissues. The proteinase
inhibitors make their appearance in response to stress such
as wounding, and probably serve as protectants of the
plants against the insect. Thus, proteinase inhibitors
accumulate in potato and tomato leaves when they are
damaged by beetles and when infected by Phytophthora
infestans (48) in both damaged and unwounded leaves of the
plant. Such increases also occur in tomato leaves after
mechanical wounding (58). Ryan et al. (55) suggested that
PIIF, a hormone-like factor which signals a wound-induced
response throughout the plant (54), is probably a wall
fragment related to pectic substances. It may therefore be
of interest to see whether pectic (oligosaccharide)
fragments, released after necrosis-forming virus infection,
might not act as an elicitor which is systemically
transported to upper leaves, and activates there the
localizing mechanism.

ACTIVE DEFENSE RESPONSES IN SYSTEMIC INFECTIONS

Resistant Cultivars.

        In some "resistant" cultivars resistance has been
shown to be due to an active mechanism, developing in the
plant after inoculation. Thus, resistance in cucumber
cultivars to cucumber mosaic virus, depends on three genes
and gives a stable type of resistance, whereby virus
multiplication is markedly reduced. Injection of leaves
with actinomycin D within 48 h after inoculation, or
ultraviolet irradiation of leaves, led to a substantial
increase in virus multiplication, indicating that the

resistance mechanism is activated by the virus and depends
on transcription from cellular DNA (46).

"Green Islands".

In several systemic infections virus is unevenly
distributed, forming in several virus-host combinations
"green islands" that contain extremely low concentrations
of virus and are resistant to further infection. Especially
in tobaccos infected with cucumber mosaic virus (CMV)
strain 6, marked "green islands" are observed (36). When
protoplasts were prepared from these "green islands," an
IVR-like compound was found to be released from them during
incubation (unpublished results).

The compound was found to inhibit replication of TMV
and CMV in isolated tobacco protoplasts as well as in leaf
tissue disks. The inhibition rates varied between 55% and
80%. The inhibitor could be obtained from the "green
islands" leaf tissue directly, using the $ZnAc_2$
precipitation.

A preliminary estimation of the molecular weight of the
inhibitor was obtained by gel-filtration on Sephadex G-75.
Peaks of activity were eluted at two positions, equivalent
to molecular weights of 24-26,000 and 55-58,000 daltons.

The antiserum prepared against IVR gave a partial
reaction (about 30%) when evaluated against IGI - the
inhibitory substance from the "green islands" (unpublished
results).

It seems, therefore, that in "green islands" an active
resistance mechanism is operative, which may be associated
with an inhibitor of virus replication.

PERSPECTIVES

Evidence is beginning to emerge that natural active
resistance mechanisms are associated with specific
substances that inhibit virus replication. A better
understanding of these mechanisms as well as the mode of
action of the inhibitory antiviral substances, may lead to
novel approaches for the control of plant virus diseases.
Thus, it should be possible in the future to locate the
m-RNA associated with the production of the antiviral
substance and resistance, and to use this RNA for
identifying the DNA sequences coding for resistance. Future
"genetic engineering" techniques might then be used to
introduce these DNA sequences into susceptible plants.

ACKNOWLEDGMENTS

This paper constitutes contribution No. 1036-E (1984 series) from the Agricultural Research Organization, The Volcani Center, Bet Dagan, Israel.

This work was supported in part by the Gesellschaft fur Strahlen - und Umweltsforschung (GSF), Munich, West Germany; the National Council for Research and Development (NCRD), Jerusalem, Israel; and the United States - Israel (Binational) Agricultural Research and Development Fund (BARD).

REFERENCES

1. Allison AV, Shalla TA (1974). The ultrastructure of local lesions induced by potato virus X: a sequence of cytological events in the course of infection. Phytopathology 64: 784-793.
2. Anderson AJ (1978). Isolation from three species of Colletotrichum of glucan containing polysaccharides that elicit browning and phytoalexin production in bean. Phytopathology 68: 189-194.
3. Apelbaum A, Burgoon AC, Anderson JD, Solomos T, Lieberman M. (1981). Some characteristics of the system converting I-aminocyclopropane-1-carboxylic acid to ethylene. Plant Physiol. 67: 80-84.
4. Appiano A, Pennazio S, D'Agostino G, Redolfi P, (1977). Fine structure of necrotic local lesions induced by tomato bushy stunt virus in Gomphrena globosa leaves. Physiol. Pl. Path. 11: 327-332.
5. Ayers AR, Ebel J, Valent B, Albersheim P (1976). Host-pathogen interactions. X. Fractionation and biological activity of an elicitor isolated from the mycelial walls of Phytophthora megasperma var. Sojae. Pl. Physiol. 57: 760-765.
6. Balazs E, Berina B, Kiraly Z (1976). Effect of kinetin on lesion development and infection sites in Xanthi-nc tobacco infected by TMV: single-cell local lesions. Acta Phytopath. Acad. Scient. Hung. 11: 1-9.
7. Balazs E, Sziraki I, Kiraly Z (1977). The role of cytokinins in the systemic acquired resistance of tobacco hypersensitive to tobacco mosaic virus. Physiol. Pl. Path. 11: 22-37.

8. Bergstom GC. Johnson MC, Kuc J (1982). Effects of local infection of cucumber by <u>Colletotrichum lagenarium</u>, <u>Pseudomonas</u> <u>lachrymans</u>, or tobacco necrosis virus on systemic resistance to cucumber mosaic virus. Phytopathology 72: 922-926.

9. Cassells AC, Barnett A, Barlass M (1978). The effect of polyacrylic acid treatment on the susceptibiliy of <u>Nicotiana tabacum</u> cv. Xanthi-nc to tobacco mosaic virus. Physiol. Pl. Path. 13: 13-22.

10. Cheo PC (1971). Effect of plant hormones on virus-replicating capacity of cotton infected with tobacco mosaic virus. Phytopathology 61: 869-872.

11. Cheo PC, Gerard JS (1971). Differences in virus-replicating capacity among plant species inoculated with tobacco mosaic virus. Phytopathology 61: 1010-1012.

12. Cohen J, Loebenstein G (1975). An electron microscope study of starch lesions in cucumber cotyledons infected with tobacco mosaic virus. Phytopathology 65: 32-39.

13. Coutts RHA (1978). Suppression of virus induced local lesions in plasmolysed leaf tissue. Pl. Sci. Lett. 12: 77-85.

14. Coutts RHA, Wagih EE (1983). Induced resistance to viral infection and soluble protein alterations in cucumber and cowpea plants. Phytopath. Z. 107:57-69.

15. Faulkner G, Kimmins WC (1975). Staining reactions of the tissue bordering lesions induced by wounding, tobacco mosaic virus and tobacco necrosis virus in bean. Phytopathology 65: 1396-1400.

16. Favali MA, Bassi M, Conti GG (1974). Morphological cytochemical and autoradiographic studies of local lesions induced by the U5 strain of tobacco mosaic virus in <u>Nicotiana glutinosa</u> L. Riv. Patol. Veg. Ser. IV 10: 207-218.

17. Fraser RSS (1979). Systemic consequences of the local lesion reaction to tobacco mosaic virus in a tobacco variety lacking the N gene for hypersensitivity. Physiol. Pl. Path. 14: 383-394.

18. Fraser RSS (1981). Evidence for the occurrence of the "Pathogensis-related" proteins in leaves of healthy tobacco plants during flowering. Physiol. Pl. Path. 19: 69-76.

19. Fraser RSS (1982). Are "Pathogenesis-related" proteins involved in aquired systemic resistance of tobacco plants to tobacco mosaic virus? J. gen. Virol. 58: 305-313.

20. Fraser RSS, Whenham RJ (1982). Plant growth regulators and virus infection: A critical review. Pl. Growth Regulation 1: 37-59.
21. Gera A, Loebenstein G (1983). Further studies of an inhibitor of virus replication from tobacco mosaic virus-infected protoplasts of a local lesion responding tobacco cultivar. Phytopathology 73: 111-115.
22. Gera A, Loebenstein G, Shabtai S (1983). Enhanced tobacco mosaic virus production and suppressed synthesis of a virus inhibitor in protoplasts   exposed to antibiotics. Virology 127: 475-478.
23. Gianinazzi S (1982). Antiviral agents and inducers of virus resistance: Analogies with interferon. in: Wood, R.K.S. (Ed.) Active Defence Mechanisms in Plants p. 285 Plenum Publishing Co.
24. Gianinazzi S, Kassanis B (1974). Virus resistance induced in plants by polyacrylic acid. J. gen. Virol. 23: 1-9.
25. Hahn MG, Darvill AG, Albersheim P (1981). Host-pathogen interactions XIX. The endogenous elicitor, a fragment of a plant cell wall poly-saccharide that elicits phytoalexin accumulation in soybeans. Pl. Physiol. 68: 1161-1169.
26. Hargreaves JA, Baily IA (1978). Phytoalexin production by hypocotyls of Phaseolus vulgaris in response to constitutive metabolites released by damaged bean cells. Physiol. Pl. Path. 13: 89-100.
27. Hayashi T, Matsui C (1965). Fine structure of lesion periphery produced by tobacco mosaic virus. Phytopathology 55: 387-392.
28. Hecht EI, Bateman DF (1964). Non-specific acquired resistance to pathogens resulting from localised infection by Thielaviopsis basicola or viruses in tobacco leaves. Phytopathology 54: 523-530.
29. Jenns AE, Caruso FL, Kuc' J (1979). Non specific resistance to pathogens induced systemically by local infection of cucumber with tobacco necrosis virus, Colletotrichum lagenarium or Pseudomonas lachrymans. Phytopath. medit. 18: 129-134.
30. Kassanis B, Gianinazzi S, White RF (1974). A possible explanation of the resistance of virus-infected tobacco plants to second infection. J. gen. Virol. 23: 11-16.
31. Kassanis B, White RF (1978). Effect of polyacrylic acid and b-proteins on TMV multiplication in tobacco protoplasts. Phytopath. Z. 91: 269-272.

32. de Laat AMM, van Loon LC (1983). The relationship between stimulated ethylene production and symptom expression in virus-infected tobacco leaves. Physiol. Pl. Path. 22: 261-273.

33. Leonard DA, Zaitlin M (1982). A temperature sensitive strain of tobacco mosaic virus defective in cell to cell movement generates an altered viral coded protein. Virology 117: 416-424.

34. Loebenstein G (1972). Localization and induced resistance in virus-infected plants. A. Rev. Phytopath. 10: 177-208.

35. Loebenstein G, Chazan R, Eisenberg M (1970). Partial suppression of the localizing mechanism to tobacco mosaic virus by UV irradiation. Virology 41: 373-376.

36. Loebenstein G, Cohen J, Shabtai S, Coutts RHA, Wood KR (1977). Distribution of cucumber mosaic virus in systemically infected tobacco leaves. Virology 81: 117-125.

37. Loebenstein G, Gera A (1981). Inhibitor of virus replication released from tobacco mosaic virus infected protoplasts of a local lesion- responding tobacco cultivar. Virology 114: 132-139.

38. Loebenstein G, Gera A, Stein A (1983). Plant defense mechanisms to viral infections. in: Kurstak, E. (Ed.) Control of Virus Diseases. Marcel Dekker Inc., New York, N.Y.

39. Loebenstein G, Rabina S, van Praagh T (1968). Sensitivity of induced localized acquired resistance to actinomycin D. Virology 34: 264-268

40. Loebenstein G, Ross AF (1963). An extractable agent induced in uninfected tissues by localized virus infections that interferes with infection by tobacco mosaic virus. Virology 20: 507-517.

41. Loebenstein G, Sela B, van Praagh T (1969). Increase of tobacco mosaic local lesion size and virus multiplication in hypersensitive hosts in the presence of actinomycin D. Virology 37: 42-48.

42. Loebenstein G, Spiegel S, Gera A (1982). Localized resistance and barrier substances. in: Wood, R.K.S. (Ed.) Active Defense Mechanisms in Plants. Plenum Publishing Co. p. 211-230.

43. McMullen CR, Gardner WS, Myers GA (1977). Ultrastructure of cell- wall thickenings and paramural bodies induced by barley stripe mosaic virus. Phytopathology 67: 462-467.

44. Milne RG (1966). Electron microscopy of tobacco mosaic virus in leaves of Nicotiana glutinosa. Virology 28: 527-532.

45. Mozes R, Antignus Y, Sela I, Harpaz I (1978). The chemical nature of an antiviral factor (AVF) from virus-infected plants. J. gen. Virol. 38: 241-249.

46. Nachman I, Loebenstein G, van Praagh T, Zelcer A (1971). Increased multiplication of cucumber mosaic virus in a resistant cucumber cultivar caused by actinomycin D. Physiol. Pl. Path. 1: 67-71.

47. Nothnagel EA, McNeil M, Albersheim P, Dell A (1983). Host-pathogen interactions. XXII. A galacturonic acid oligosaccharide from plant cell walls elicits phytoalexins. Pl. Physiol. 71: 916-926.

48. Peng JH, Black LL (1976). Increased proteinase inhibitor activity in response to infection of resistant tomato plants by Phytophthora infestans. Phytopathology 66: 958-963.

49. Pennazio S (1981). Changes in the soluble protein constitution of Gomphrena globosa leaves showing spontaneous local lesions. Riv. Patol. Veg. 17: 127-130.

50. Pritchard DW, Ross AF (1975). The relationship of ethylene to formation of tobacco mosaic virus lesions in hypersensitive responding tobacco leaves with and without induced resistance. Virology 64: 295-307.

51. Roberts DA (1982). Systemic acquired resistance induced in hypersensitve plants by nonnecrotic localized viral infection. Virology 122: 207-209.

52. Ross AF (1961). Systemic acquired resistance induced by localized virus infection in plants. Virology 14: 340-358.

53. Ross AF, Israel HW (1970). Use of heat treatments in the study of acquired resistance to tobacco mosaic virus in hypersensitive tobacco. Phytopathology 60: 755-770.

54. Ryan CA (1978). Proteinase inhibitors in plant leaves: A biochemical model for pest-induced natural plant protection. Trends Biochem. Sci. 3: 148-150.

55. Ryan CA, Bishop P, Pearce G (1981). A sycamore cell wall polysaccharide and a chemically related tomato leaf polysaccharide possess similar proteinase inhibitor inducing activity. Pl. Physiol. 68: 616-618.

56. Sela I, Harpaz I, Birk Y (1966). Identification of the active component of an antiviral factor isolated from virus-infected plants. Virology 28: 71-78.

57. Shimomura T, Dijkstra J (1975). The occurrence of callose during the process of local lesion formation. Neth. J. Pl. Path. 81: 107-121.
58. Shumway LK, Yang VV, Ryan CA (1976). Evidence for presence of proteinase inhibitor I in vacuolar protein bodies of plant cells. Planta 129: 161-165.
59. Spencer DF, Kimmins WC (1971). Ultrastructure of tobacco mosaic virus lesions and surrounding tissue in Phaseolus vulgaris var. Pinto. Can. J. Bot. 49: 417-421.
60. Stein A, Loebenstein G (1972). Induced interference by synthetic polyanions with the infection of tobacco mosaic virus. Phytopathology 62: 1461-1466.
61. Stein A, Loebenstein G, Spiegel S (1979). Further studies of induced interference by a synthetic polyanion of infection by tobacco mosaic virus. Physiol. Pl. Path. 15: 241-255.
62. Sulzinski MA, Zaitlin M (1982). Tobacco mosaic virus replication in resistant and susceptible plants: in some resistant species virus is confined to a small number of initially infected cells. Virology 121: 12-19.
63. Sziraki L, Balazs E, Kiraly Z (1980). Role of different stresses in inducing systemic acquired resistance to TMV and increasing cytokinin level in tobacco. Physiol. Pl. Path. 16: 277-284.
64. Talianski ME, Malyshenko SI, Pshennikova ES, Kaplan IB, Ulanova EF, Atabekov JG (1982). Plant virus specific transport function: 1. Virus genetic control required for systemic spread. Virology 122: 318-326.
65. Van Loon LC (1975a). Polyacrylamide disc electrophoresis of the soluble leaf proteins from Nicotiana tabacum var. "Samsun" and "Samsun NN", similarity of quantitative changes of specific proteins after infection with different viruses and their relationship to acquired resistance. Virology 67: 566-575.
66. Van Loon LC (1975b). Polyacrylamide disk electrophoresis of the soluble leaf proteins from Nicotiana tabacum var "Samsun" and "Samsun NN". III. Influence of temperature and virus strain on changes induced by tobacco mosaic virus. Physiol. Pl. Path. 6: 289-300.
67. Van Loon LC (1977). Induction by 2-chloroethylphosphonic acid of viral-like lesion, associated proteins, and systemic resistance in tobacco. Virology 80: 417-420.

68. Van Loon LC (1982). Regulation of changes in proteins and enzymes associated with active defence against virus infection. in: Wood, R.K.S. (Ed.) Active Defence Mechanism in Plants. Plenum Publishing Co.) p. 247-274.

69. Van Loon LC, Van Kammen A (1970). Polyacrylamide discelectrophoresis of soluble leaf proteins from <u>Nicotiana tabacum</u> var. "Samsun" and "Samsun NN". II. Changes in protein constitution after infection with tobacco mosaic virus. Virology 40: 199-211.

70. Wagih EE, Coutts RHA (1981). Similarities in the soluble protein profiles of leaf tissue following either a hypersensitive reaction to virus infection or plasmolysis. Pl. Sci. Lett. 21: 61-69.

71. West CA (1981). Fungal elicitors of the phytoalexin response in higher plants. Naturwissenschaften 68: 447-457.

72. Whenham RJ, Fraser RSS (1981). Effect of systemic and local-lesion forming strains of tobacco mosaic virus on abscisic acid concentration in tobacco leaves: consequences for the control of leaf growth in tobacco. Physiol. Pl. Path. 18: 267-278.

73. Wu JH, Dimitman JE (1970). Leaf structure and callose formation as determinants of TMV movement in bean leaves as revealed by UV irradiation studies. Virology 40: 820-827.

Cellular and Molecular Biology of Plant Stress, pages 435–446
© **1985 Alan R. Liss, Inc.**

TMV COAT PROTEIN ENCAPSIDATES SPECIFIC SPECIES OF HOST RNA BOTH in vivo AND in vitro[1]

D'Ann Rochon and Albert Siegel

Department of Biological Sciences, Wayne State University
Detroit, Michigan 48202

ABSTRACT Despite the specificity of the tobacco mosaic virus assembly mechanism, virus preparations contain a small proportion of pseudovirions, particles that contain host rather than virus RNA. The encapsidated host RNA is composed of non-ribosomal RNA chloroplast transcripts. We suggest that virus-induced stress may result, at least in part, from the sequestering of chloroplast messenger RNAs during critical stages of leaf development. The results suggest that chloroplast transcripts contain a sequence that resembles the virus RNA encapsidation initiation site. This suggestion is reinforced by the observation that chloroplast transcripts are also encapsidated when TMV capsid protein is reacted with leaf RNA in vitro. Other leaf RNA species are also encapsidated in vitro, most notably a portion of 18S ribosomal RNA but the reaction is specific because other RNA species, such as the 25S cytoplasmic ribosomal RNA are not. A number of host species may contain a sequence resembling the encapsidation initiation site but only those in organelles to which the capsid protein can gain entry are encapsidated in vivo.

[1]This work was supported in part by the National Science Foundation (PCM-8104496), USDA-SEA (59-2264-1-
-1-607-0) and Agriculture Canada.

INTRODUCTION

Viruses constitute one class of agent that cause
plant stress. Just what it is about the virus-host interaction that causes stress is poorly understood. Stress
is usually expressed as a symptom typical of a particular
host-virus combination but symptoms do not necessarily result from the mere accumulation of a substantial number of
virus particles in cells of an infected plant. There are
numerous reports in the literature of "latent" or "masked"
infections. As a specific example, the strains U1 (common) and U4 (masked) of tobacco mosaic virus (TMV) are
considered to be closely related because they are indistinguishable on the basis of serology and a number of
other physical and chemical criteria and, yet, they induce
substantially different symptoms in a systemic host (Van
De Walle and Siegel, 1982). Both strains replicate to a
fairly high level in the plant but the masked strain, as
the name implies, induces only a barely detectable,
transient symptom, whereas leaves of plants infected with
the common strain exhibit the typical light green, dark
green mosaic pattern. Even in the case of the common
strain, however, it is only the leaves that were young or
not yet formed at the time of inoculation that exhibit
symptoms of infection; inoculated fully expanded leaves
show no apparent sign of infection despite supporting the
replication of 30 or more mg of virus per leaf. There are
several distinctly different, genetically inherited types
of virus induced stress response, the physiological bases
of which are poorly understood. Again using TMV as an
example for two such stress responses, some host species
become systemically invaded by many of the strains. Virus
infection spreads to many parts of the plant in a predictable fashion and a mosaic of strain dependent severity
appears on leaves that were young or not yet formed at the
time of infection. In contrast, other host species
exhibit a hypersensitive stress response in which a restricted group of cells around an infection site collapse
to form a necrotic spot (local lesion) and thereby limit
virus spread. A third group of host species and cultivars
distinguish between strains, becoming systemically invaded
by some and producing local lesions when inoculated with
the others.

We report here experiments to define the nature of an
in vivo reaction between virus capsid protein and host RNA
and to determine whether this reaction may at least in

part be responsible for the nature and severity of virus
induced plant stress. The reaction results in the
presence of a small proportion of pseudovirions in TMV preparations (Siegel, 1971; Rochon and Siegel, 1984). These
are virus-like particles that contain host rather than
virus RNA and in which, as in virions, the RNA is protected from nucleases. Viruses are assembled from components that are synthesized independently in host cells
and mechanisms exist to ensure that only virus-specific
components are assembled into virions. This mechanism, in
the case of TMV, lies in the initiation event of nucleoprotein rod formation. This is a specific reaction between a capsid protein 34 sub-unit, double disc oligomer
and an internal viral RNA nucleotide sequence called the
encapsidation initiation site (Zimmern and Butler, 1977).
Nevertheless, the existence of pseudovirions has been
demonstrated, indicating that the selectivity of the assembly process is imperfect. The encapsidated host RNA
present in pseudovirions is not representative of the
diversity of RNA species present in infected tissue.
Rather, it consists primarily of most, or all, of the
chloroplast DNA transcripts except, perhaps, the 23 and
16S chloroplast ribosomal RNAs. These data, taken together with the observation that chloroplasts of infected
tissue contain less than full length, non-infectious rods
(Shalla et. al., 1975), suggests that cytoplasmically synthesized capsid protein gains access to chloroplasts
where, upon reaction with resident RNA molecules, pseudovirion assembly occurs. Perhaps the mosaic symptoms
exhibited by infected plants results at least in part from
the sequestering in pseudovirions of chloroplast messenger
RNAs during a critical stage of leaf development. A study
was undertaken of the in vitro reaction of capsid protein
with host RNA in order to explore whether conditions in
the chloroplast might be such as to encourage non-specific
assembly of capsid protein with any available RNA or
whether pseudovirion formation might be quite specific and
depend on the presence of a nucleotide sequence or structure common to chloroplast transcripts which resembles the
viral RNA encapsidation initiation site. Fraenkel-Conrat
(1959) determined conditions for in vitro assembly of TMV
nucleoprotein rods and this reaction has been studied and
defined by several groups of investigators (for reviews
see Butler, 1984). Because of the way the reaction is
initiated, it is quite specific for TMV RNA with only a
few observed exceptions (Fritsch et al., 1973).

MATERIALS AND METHODS

TMV capsid protein was prepared by the acetic acid method of Fraenkel-Conrat (1957). The reconstitution reaction was carried out as follows. Capsid protein in H_2O was adjusted to 0.1M Na pyrophosphate, pH 7.25 and centrifuged to remove denatured, aggregated protein. Further incubation of the supernatant solution at a concentration of 1 mg/ml for 20 hr did not lead to appearance of additional insoluble protein and was considered to contain primarily the disc aggregate. Reconstitution reactions contained 1 mg/ml of disc aggregate and 1 mg/ml of leaf RNA or 20 μg/ml of virus RNA. After overnight incubation at room temperature, the solution was made to a density of 1.5 g/ml with CsCl and centrifuged for 20 hr in the SW 50.1 rotor. The upper third of the solution was collected, diluted with 2 volumes of H_2O and the nucleoprotein and protein aggregates collected by sedimentation. The pellet was resuspended in H_2O and treated with 0.1 μg/ml pancreatic ribonuclease A for 30 min at room temperature followed by treatment with proteinase K for 30 min and the RNA extracted by phenol-SDS treatment.

Other techniques used in this paper are as described (Rochon and Siegel, 1984).

RESULTS

Ribonuclease Protection of Leaf RNA by Reaction with Capsid Protein.

TMV capsid protein was reacted with leaf RNA under conditions used for TMV reconstitution in order to determine whether any of the leaf RNA species might react with the capsid protein in a manner similar to virus RNA. It was found that after reaction as described in the methods section that ca. 1% of the leaf RNA became resistant to ribonuclease treatment and was recovered by phenol treatment. In similar experiments performed with capsid protein and virus RNA from 10 - 30% of the RNA became protected. Other workers have been unsuccessful in attempts to reconstitute leaf RNA with capsid protein (Fritsch et al., 1973). We suspect that their lack of success can be attributed to either not preincubating their capsid protein to form discs or to not having a high enough ratio of RNA to protein in the reaction mix.

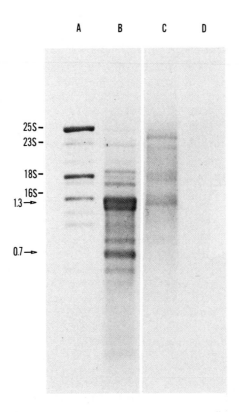

Fig. 1 Ethidium bromide stained CH₃HgOH agarose gel of in vitro reconstituted RNA species. TMV capsid protein disks were incubated with different RNAs under reconstituting conditions. Lanes B, C, and D contain material extracted from ribonuclease treated, purified reconstituted tobacco leaf, turnip root and E. coli RNAs, respectively and lane A contains unreacted tobacco leaf RNA used as a molecular weight standard. The arrows point to the major reconstituted RNA components, the sizes of which are indicated in kb.

Specificity of the Reaction of Capsid Protein with Heterologous RNAs.

  The ability of capsid protein to react with some of
the leaf RNA so as to protect it from nuclease degradation
suggested that the reaction might be non-specific in that
capsid protein might have the capability of reacting with
all RNA species under appropriate conditions. That this
is not so can be seen in fig. 1; the capsid protein reacts
with a number of leaf RNAs and with turnip root RNA but
not with a bacterial RNA. The fact that capsid protein
was able to react with an appreciable amount of root RNA
so as to render it resistant to nuclease was surprising
because of our previous observation that pseudovirions
contain much less material homologous to root than to leaf
RNA. Thus, the present data suggests that capsid protein
might react with different RNA species in vitro than in
vivo.

In vitro Encapsidated Leaf RNA Contains Chloroplast
Transcripts.

  Our previous work has demonstrated that pseudovirions
formed in vivo contain primarily non-ribosomal RNA
transcripts of chloroplast DNA. The following experiment
was performed to test whether the same species would be
encapsidated under in vitro reconstituting conditions.
Chloroplast DNA fragments were released from the
individual components of a petunia chloroplast plasmid
clone bank and these were probed with a cDNA prepared to
the in vitro encapsidated leaf RNA. As seen in fig. 2A
the in vitro encapsidated material resembles pseudovirion
RNA in containing material that hybridizes to all of the
chloroplast fragments indicating that transcripts from
many parts of the chloroplast DNA were encapsidated. A
test was made of the condition of several of the
encapsidated chloroplast DNA transcripts by incubating
nick translated cloned fragments of chloroplast DNA to
immobilized encapsidated leaf RNA. As seen in figure 2B
the messenger RNA for the large fragment of RuBC and
transcripts which map to a 9.2 kb petunia chloroplast DNA
restriction fragment are preserved intact.

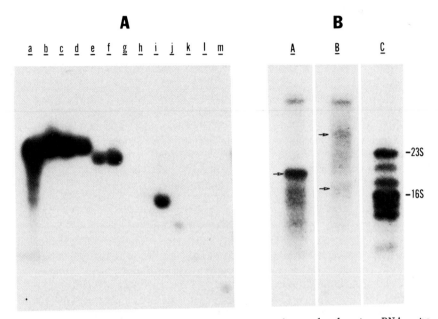

Fig. 2 Hybridization of reconstituted host RNA to chloroplast DNA restriction fragments. (A) Four hundred nanograms each of 13 different chimeric plasmids containing petunia chloroplast DNA Pst 1 restriction fragments (Palmer et al., 1983) were treated with Pst 1, size fractionated in a 1% agarose gel, transferred to GeneScreen (New England Nuclear) and incubated with cDNA made to reconstituted tobacco leaf RNA. Lanes a–m contain the 13 individual petunia chloroplast DNA fragments which together represent approximately 80% of the genome. (B) Electrophoretically separated reconstituted leaf RNA was transferred to diazotized paper and incubated with nick-translated cloned chloroplast DNA fragments. Lane A was hybridized to the 4.1 kb RuBC containing chloroplast fragment and lane B to the 9.2 kb fragment. Lane C was used as a molecular weight standard. Marker 23S and 16S rRNAs are indicated at the right of the figure and transcripts detected by the chloroplast DNA probes pointed to by arrows.

Further Characterization of in vitro Encapsidated RNA.

Leaf and root in vitro encapsidated RNAs were subjected to electrophoresis in a denaturing agarose gel with the result shown in fig. 1. A number of discrete sized bands are present ranging from ca. 0.1 to 6 kb with the most prominent one at 1.3 kb. This component was identified as a portion of 18S ribosomal RNA by the following series of experiments.

A cDNA prepared to the in vitro encapsidated leaf RNA was incubated with immobilized leaf RNA and was found to react with a number of discrete RNA species but most strongly with one of the same size as 18S (ca. 2.0kb) ribosomal RNA (fig. 3A). The fact that the 1.3 kb component in the in vitro encapsidated RNA preparation was a ribosomal RNA was confirmed by observing that it reacted with a cloned plant (Cucurbita pepo) ribosomal DNA (rDNA) repeat unit (fig. 3B). The encapsidation of 18S RNA proved quite specific in that no 25S RNA could be detected in the encapsidated RNA preparation. The presence of 25S RNA was ruled out by incubating a cDNA to the encapsidated leaf RNA with a restriction endonuclease fragmented cloned rDNA repeat unit. The results (fig. 4) show clearly that hybridization is only to fragments containing 18S coding sequences. Although it is clearly demonstrated that 18S ribosomal RNA is encapsidated in vitro, attempts to demonstrate its presence in in vivo formed pseudovirions were unsuccessful.

It is peculiar that the 18S homologous in vitro encapsidated material is shorter than 18S RNA. Its estimated size is only 3/5 of 18S RNA. We note that the cloned rDNA repeat unit hybridizes not only to the 1.3 kb component present in immobilized leaf RNA but also to a second, shorter RNA species of 0.7 kb. The two of these together add up to 2.0 kb, the size of 18S RNA. We speculate that either the 18S RNA is subject to a specific cleavage during the incubation of the encapsidation reaction or that rods containing full length 18S RNA are formed with a site specific imperfection where the RNA is exposed and subject to cleavage by the ribonuclease treatment given to the in vitro formed rods before RNA extraction. To distinguish between these possibilities, RNA was extracted from in vitro formed, non-nuclease treated rods that had been separated from unencapsidated RNA by CsCl flotation and two cycles of sedimentation. This RNA preparation contained material of 18S size but

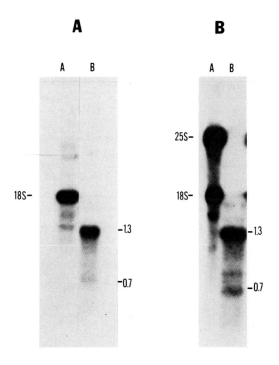

Fig. 3   Hybridization of reconstituted leaf RNA to 18S
ribosomal RNA.   (A) Size fractionated and paper immobilized total tobacco leaf (lane A) and reconstituted leaf
RNA (lane B) hybridized to ^{32}P-labeled cDNA prepared to
reconstituted leaf RNA.   (B)   The lanes are the same as in
fig. 3A but in this case the probe is a cloned nick-
translated pumpkin rDNA repeat unit.   18S   and   25S
ribosomal RNAs are indicated at the left of the figures
and the prominent 1.3 kb and less prominent 0.7 kb RNA
species are pointed out on the right (see text).

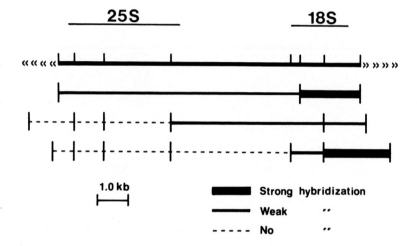

Fig. 4   Diagram to indicate regions of a cloned rDNA repeat unit which hybridize to _in vitro_ reconstituted tobacco leaf RNA.   18s and 25S refer to regions of the cloned pumpkin rDNA repeat unit (Siegel and Kolacz, 1983) which map to the respective ribosomal RNAs.   The thickest bars show regions of the repeat unit which hybridized most strongly with a cDNA to reconstituted leaf RNA; the thinner bars, regions which hybridized only weakly; and the dashed lines indicate no hybridization.

very little, if any 25S RNA or the 1.3 kb component present in the material extracted from nuclease treated rods. It would thus appear that assembly of capsid proteins with 18S RNA is imperfect in that the rods formed have a specific exposed region.

DISCUSSION

This study was undertaken to test the specificity of pseudovirion formation during infection. We presume that they are formed as a consequence of entry of capsid protein molecules into chloroplasts where they encapsidate chloroplast DNA transcripts. Evidence is supplied here that this reaction is specific and probably depends on the presence of a nucleotide sequence or structure that resembles the virus RNA encapsidation initiation site. The evidence is based on the specificity of the in vitro reaction of capsid protein with host RNA. As in the in vivo reaction, chloroplast DNA transcripts are also encapsidated in vitro. Surprisingly, one of the cytoplasmic ribosomal RNAs is encapsidated in vitro but not in vivo. This occurs only to 18S ribosomal RNA and not to the 25S RNA, indicating that of the major component RNAs in a leaf or root extract, only the 18S RNA can initiate an assembly reaction with capsid protein and, thus, probably has a structure resembling the encapsidation initiation site. What the nature of this site may be in either the chloroplast transcripts or the 18S RNA remains to be determined.

It is interesting to note that chloroplast DNA transcripts are encapsidated both in vivo and in vitro, whereas 18S RNA is encapsidated only in vitro. We speculate that the reason for this is that capsid protein enters the chloroplast during infection where reaction can take place with the chloroplast transcripts, but the capsid protein does not enter the nucleolus where it might have access to 18S RNA or its precursor before assembly into ribosomes. Evidence for the entry of capsid protein into chloroplasts is provided by the work of Schalla et al. (1975) who noted the presence of short, non-infectious rods in the chloroplasts of infected tissue.

It is possible that the symptoms exhibited by infected plants may in part result from pseudovirion formation. The light green areas of leaves exhibiting a mosaic pattern obviously suffer from a chloroplast defect or deficiency and this may result from the sequestering of chloroplast transcripts into pseudovirions.

ACKNOWLEDGEMENTS

We thank David Gilley for technical assistance and Michael Weis for his help in preparing photographs.

REFERENCES

1. Butler JG (1984). The current picture of the structure and assembly of tobacco mosaic virus. J of Gen Virol 65:253.
2. Fraenkel-Conrat H (1957). Degradation of tobacco mosaic virus with acetic acid. Virology 4:1.
3. Fraenkel-Conrat H, Singer B (1959). Reconstitution of tobacco mosaic virus. III. Improved methods and use of mixed nucleic acids. Biochimica et Biophysica Acta 33:359.
4. Fritsch C, Stussi C, Witz J, Hirth L (1973). Specificity of TMV RNA encapsidation: in vitro coating of heterologous RNA by TMV protein. Virology 56:33.
5. Palmer JD, Shields CR, Cohen DB, Orton TJ (1983). Theor Appl Genet 65:181.
6. Rochon D, Siegel A (1984). Chloroplast DNA transcripts are encapsidated by tobacco mosaic virus coat protein. Proc Nat'l Acad of Sci (in press).
7. Shalla TA, Petersen CJ, Giunchedi, L (1975). Partial characterization of virus-like particles in chloroplasts of plants infected with the U5 strain of TMV. Virology 66:94.
8. Siegel A (1971). Pseudovirions of tobacco mosaic virus. Virology 46:50.
9. Siegel A, Kolacz K (1983). Heterogeneity of pumpkin ribosomal DNA. Plant Physiol. 72:166.
10. Van De Walle MJ, Siegel A (1982). Relationships between strains of tobacco mosaic virus and other selected plant viruses. Phytopath 72:390.
11. Zimmern D, Butler PJG (1977). The isolation of tobacco mosaic virus RNA fragments containing the origin for viral assembly. Cell 11:455.

Cellular and Molecular Biology of Plant Stress, pages 447–457
© 1985 Alan R. Liss, Inc.

IMMUNOCHEMICAL IDENTIFICATION OF ANTIGENS INVOLVED IN PLANT/PATHOGEN INTERACTIONS[1]

Jody J. Goodell, Paul L. DeAngelis and Arthur R. Ayers

Department of Cellular and Developmental Biology,
Harvard University, Cambridge, MA 02138

ABSTRACT Genetic studies of plants and their pathogens
have provided many examples indicating that genes for
disease resistance in the host are complemented by
corresponding genes for avirulence in the pathogen.
Both host and pathogen must have at least one of these
pairs of corresponding genes in order for the pathogen
to be detected. In the absence of detection, which
normally leads to defensive responses, the pathogen can
spread through host tissue and produce characteristic
disease symptoms.
    Polyspecific and monoclonal antibodies are being
used to search for products of genes for resistance and
avirulence. Antibodies specific for glycomoities of
extracellular glycoproteins secreted by different races
of a fungal pathogen have been identified to examine
the possibility that avirulence genes code for glycosyl
transferases. A search for host membrane receptors for
pathogen glycoproteins is in progress, aided by
antibodies raised to plant membrane components.
    Receptors for pathogen components are also being
investigated using synthetic compounds that mimic the
structure and activity of their natural counterparts.
Elicitors, pathogen molecules that elicit the
accumulation of toxic secondary metabolites
(phytoalexins) in host tissues, are being examined by
this approach.

[1]This work was supported by grants from the National
Science (PCM 83-02789) and Maria Moors Cabot Foundations.

INTRODUCTION

Plants protect themselves from microorganisms with a variety of physical and chemical defenses. Both preformed, passive barriers and dynamic responses to the presence of potential invaders are used. In our laboratory we have focused on the active defenses, e.g. phytoalexin accumulation, invoked when a pathogen is detected. Phytoalexins, secondary plant metabolites toxic to microorganisms, accumulate in response to non-pathogens and particular ('incompatible') races of pathogens. Race-specific defenses are of special interest, because they are controlled by resistance genes in the plant and a corresponding set of complementary avirulence genes in the pathogen (a 'gene-for-gene' relationship). Each plant species devotes dozens of genes to the detection of its pathogens, and dozens more to the synthesis of phytoalexins (Fig. 1).

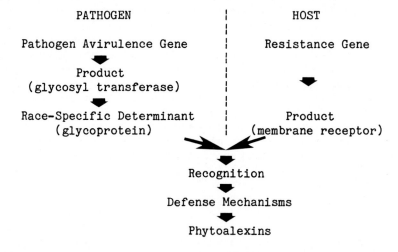

FIGURE 1. Model for interaction of host and pathogen gene products.

Clearly, disease resistance is a fundamental plant phenomenon that is both of practical and theoretical importance, and yet surprisingly little is known about the genes or gene products involved. The project outlined here is designed to examine some of the current theories of plant disease resistance and to shed some light on the host and pathogen molecules involved.

Several hypotheses have implicated polysaccharides and the carbohydrate portion of glycoproteins as the pathogen components that mediate disease resistance in race-specific host-pathogen interactions. Three of these hypotheses are: 1) race-specific elicitors of phytoalexin accumulation (1), 2) race-specific suppressors of phytoalexin accumulation (2), and 3) race-specific interactions that precede and may lead to phytoalexin accumulation. Glucans or mannans are implicated in each case. Preliminary experiments using immunochemical approaches to characterize pathogen components (especially extracellular glycoproteins) involved in disease resistance are described here. In addition the elicitor-phytoalexin system (3,4,5) is examined as a model of plant recognition of pathogen components using a facile scheme to synthesize molecules that mimic the pathogen's components in both structure and biological activity. Elicitor derivatives have also been synthesized for receptor binding studies.

METHODS

Pathogens and Plants.

The host-pathogen system in this study was soybean (<u>Glycine max</u>) and its pathogen, the causal agent of root and stem rot, <u>Phytophthora megasperma</u> f.sp. <u>glycinea</u> (Pmg). The soybean varieties used were from the Williams series (Williams, Williams '79, and Williams '82). These varieties provide a closely related genetic background with three different resistance genes to Pmg: -, Rps1^c, Rps1^k, respectively. Pmg races 1, 4 and 12 were used (Table 1).

TABLE 1
RACE-SPECIFIC INTERACTIONS OF SOYBEAN CULTIVARS
WITH RACES OF Pmg

Pmg Race	Soybean Cultivar		
	W	W-79	W-82
1	C	I	I
4	C	C	I
12	C	C	C

Polyclonal Antibodies.

Polyclonal antibodies were raised in rabbits by multiple injections of antigens (extracellular culture filtrates or purified walls of three races of Pmg, or purified membranes of soybean roots). Antibodies were partially purified by ammonium sulfate precipitation and DEAE chromatography.

Monoclonal Antibodies.

Monoclonal antibodies were produced from hybridoma cell lines derived from the fusion of myeloma line P3AG865.3 with spleen cells from BalbC mice challenged with antigen mixtures (extracellular culture filtrates or glycoproteins or purified walls from the three races of Pmg, or soybean root membranes). Cell lines were maintained in DMEM-FCS medium.

Screening Approaches for Monoclonal Antibodies.

Standard techniques of enzyme-linked immunosorbant assays (ELISA) using a second antibody linked to alkaline phosphatase were used to quantitate immunoglobulins (Ig) in general and Ig specific for antigens of interest. Competition experiments were performed by comparison of Ig binding in the presence and absence of an antigen mixture related to that immobilized on the ELISA well plates.
Inactivation of particular enzymes, e.g. B-glucosidase, present in preparations of fungal extracellular glycoproteins was used as an indication of the presence of Ig specific for the protein portion of an enzyme. Antibodies specific for glycomoieties common to several different glycoproteins were sought by simultaneous precipitation of glycoproteins responsible for at least two different activities, e.g. B-glucosidase and alkaline phosphatase. Race-specific antibodies are being identified by ability to precipitate enzymes from the same race, but not from other races.
Antibodies binding immobilized fungal extracellular glycoproteins were further analyzed for their ability to bind to bands on Western blots of glycoproteins separated by SDS-PAGE. Binding was visualized by the peroxidase-antiperoxidase method or with fluorescent second antibodies.

RESULTS

Extracellular Culture Filtrates of Pmg.

The extracellular filtrates derived from cultures of
different races of Pmg grown on a defined medium with
sucrose as the carbon source (1), or with other carbon
sources (maltose, arabinogalactan, lactose, soybean root
cell walls) show a wide variation in the quantities of
individual proteins appearing on SDS-PAGE. Each race of Pmg
could be identified by its characteristic pattern of
proteins. In addition, enzymes such as B-glucosidase and
alkaline phosphatase that had been assigned to particular
bands could be enhanced in prominance by growth of the fungi
on plant cell walls. Predictable increases in enzymes
related to carbon source metabolism, e.g. a-glucosidase
stimulation in cultures grown on starch, were also observed.
Protein patterns also changed dramatically over the period
in which the cultures were grown. Thus, extracellular
proteins reflect the dynamic state of the fungal metabolism
in adjusting to the changing concentration of nutrients in
their medium.

Polyclonal Antibodies to Extracellular Culture Filtrates.

Polyclonal antibodies were raised in rabbits to
antigens (concentrated, extracellular culture filtrates, and
purified walls) from each of the Pmg races 1, 4 and 12. In
general antibodies raised to glycoproteins (predominant
components of the culture filtrates) from one race bind
glycoprotein of that race (homologous antigens) more avidly
than glycoproteins from the other races (heterologous
antigens) in ELISA competition assays. As expected based on
the characterization of the protein mixtures by PAGE, there
are antigens from each race that are in common with other
races, and there are also antigens that are unique to each
race.

Polyclonal antibodies bound to Western blots of
glycoproteins separated by SDS-PAGE can be visualized by
binding of a second fluorescent antibody (goat-antirabbit
Ig) or by binding of the second antibody with the addition
of a peroxidase-antiperoxidase complex and development with
the peroxidase substrate. In both cases, the pattern
observed differs greatly from the pattern obtained by

directly staining the glycoproteins. The immunodominant proteins are not the major proteins present in the preparations. These observations confirm the impression that antibodies are quite often specific for minor components of an immunizing mixture -- predominance does not imply immunodominance.

Mononclonal Antibodies for Extracellular Glycoproteins.

Monoclonal antibodies have been produced by hybridoma clones derived from fusions with spleen cells challenged with a variety of antigen mixtures. The major focus of this study has been the characterization of the extracellular glyoproteins produced by Pmg races 1, 4 and 12 (Table 2).

TABLE 2
GENERAL CHARACTERISTICS OF TYPICAL FUSIONS

Antigen Race	Predominant Ig Class	% of wells producing Ig	% producing Ig to fungal glycoproteins
1	IgG	97	47
12	IgG	77	30
4	IgG	93	27
12	IgM	92	6
1	IgG	72	48
4	IgM	91	19

Clones producing antibodies specific for fungal glycoproteins have been identified using immobilized glycoproteins in ELISA. Target glycoproteins for each monoclonal are being identified by binding of the antibodies to Western blots of the glycoproteins. Monoclonal antibodies have been identified that either inactivate or precipitate fungal B-glucosidase. Screening for monoclonal antibodies capable of precipitating multiple enzymes of the same race only (race-specific) or from different races (glyco-specific) is in progress.

The characterization of soybean membranes has been initiated. Hybridoma cultures derived from mice challenged with soybean root membrane preparations (containing plasma membrane, ER, Golgi and some organelles) have been obtained and are now being characterized.

Monoclonal antibodies specific for elicitor are being sought by challenging mice with purified fungal walls, which contain abundant insoluble forms of elicitor, and screening first for wall-specific antibodies and subsequently for elicitor inactivation. Elicitor derivatives (see below) will facilitate these studies.

Synthesis of Oligosaccharide Conjugates.

Several approaches for the synthesis of oligosaccharides with terminal functional groups appropriate for labeling and synthesis of more complex structures have been examined. Aminoglycosides have been obtained from oligosaccharides by incubation (16 hr, RT) in aminophenylethylamine (6). The aminoglycosides were subsequently reduced to stable secondary amines with sodium borohydride to yield oligosaccharides with effectively one less glucosyl residue and a free amino group (aryl amine). This approach was used with laminarin oligosaccharides and resulted in quantitative derivatization with small oligosaccharides (dp 2-5) and much less efficient derivatization (1-10%) with large oligosaccharides (dp>10, laminarin). One major disadvantage to this approach is the difficulty of coupling the aryl amine to other molecules using convenient N-hydroxysuccinimide derivatives.

Another approach to the introduction of amino functional groups into oligosaccharides and direct coupling of oligosaccharides to useful backbone structures is reductive amination using cyanoborohydride (7,8,9). Oligosaccharides are incubated with cyanoborohydride and reactants containing one or more free amino groups (primary or secondary). The resulting reductive amination leads to a stable linkage of the oligosaccharide through the amine. Using this approach, lysine derivatives containing one or two chains derived from various disaccharides (lactose, cellobiose, maltose, gentibiose, and laminaribiose) have been synthesized (Table 3). Unfortunately, the synthesis of

TABLE 3
SOME COMPOUNDS SYNTHESIZED
BY REDUCTIVE AMINATION WITH CYANOBOROHYDRIDE

Laminaribiose -R_2 Maltose - R_1
Laminaribiose -R_3 Maltose - R_2
(Laminaribiose)$_2$ -R_2 Maltose - R_3
Laminaritriose -R_3 Maltotriose - R_3
Laminari(tetra-pentaose) -R_3 Maltotetrose - R_3
                                        Maltohexose - R_3
                                        Maltoheptaose - R_3
Cellobiose - R_3 (Maltose)$_2$ - R_2

Gentiobiose - R_1 Elicitor - R_3

R_1 = lysine
R_2 = lys-phe
R_3 = lysine methyl ester

the most interesting derivative, B-1,3-linked laminaribiose, is the slowest and least efficient reaction. Further studies of the linkage dependence of this reaction are in progress to improve efficiencies, along with synthesis of the entire series of oligosaccharide derivatives from laminaribiose to laminarin (dp 2->14). This procedure with modifications (adjustments of pH, buffer, solvent and salt amendments) has been successfully applied to natural elicitor and will be applied with mannan oligosaccharides purified from Pmg walls and glycoproteins to yield molecules with labels to facilitate binding and receptor identification (photoaffinity label).

A molecule with two laminaribiose residues conjugated to lysine methyl ester has been synthesized. This molecule was tested for elicitor activity in the soybean cotyledon assay (3). It was not active.

Radioactive photoaffinity (10) derivatives of lysine conjugated to various oligosaccharides have been constructed for use in identification of membrane receptors. These derivatives were made by reaction of the lysine derivatives with SNAP (cystamine (^{35}S)-3-[(2-nitro-4-azidophenyl)-2-aminoethyldithio]-N-succinimidyl propionate). These derivatives are now being used to identify putative elicitor receptors (11) on the plasma membranes of soybean roots.

Elicitor Derivatization.

Elicitor preparations from acid-hydrolyzed walls of Pmg have been successfully derivatized with lysine methyl ester using the reductive amination procedure. The derivatives retain their activity in the soybean cotyledon assay.

DISCUSSION

Characterization of the extracellular culture filtrates of three races of Pmg using antibodies and analytical techniques indicate that there are antigens common to different races, and also antigens specific to particular races. The antigens identified as race-specific are likely in most cases to represent random differences that happen to be associated with the particular pathogen isolates that were examined in this study. It will be important to test several isolates with the same avirulence genes, but from different geographical origin to provide adequate correlations between antigens and avirulence genes.

The observed immunodominance of minor proteins in a mixture, as observed in the case of the polyclonal rabbit antibodies to fungal glycoproteins is not unusual. This finding merely gives an indication of the results to be expected with mouse monoclonal antibodies.

The preliminary experiments on the use of monoclonal antibodies for the characterization of fungal glycoproteins indicate that the production of monoclonals and screening for various classes of antibodies is working. Monoclonal antibodies should be very useful tools in the identification of race-specific determinants and in the identification of membrane receptors. These reagents will also prove valuable in the purification of components, such as enzymes, elicitor and plasma membranes.

The oligosaccharide derivatives described here are examples of a class of very powerful reagents for investigating phenomena mediated by glycosyl moieties. The primary derivatives can be readily modified to yield multivalent, radiolabeled, or photoaffinity reagents. Identification of membrane receptors for various pathogen components should be greatly facilitated with these derivatives. Furthermore, the general applicability of the derivitization approach should make it useful in many fields of research.

ACKNOWLEDGMENTS

We wish to acknowledge the contributions of Dan Voytas and Mark Schaffer in the characterization of extracellular glycoproteins, Lisa Rayder and Doug Brugge for characterization of soybean membranes, and Lois Valenti for expert technical assistance in all phases of this project.

REFERENCES

1. Keen NT, Legrand M (1980). Surface glycoproteins. Evidence that they may function as the race specific phytoalexin elicitors of Phytophthora megasperma f.sp. glycinea. Physiol Pl Pathol 17:175.

2. Ziegler E, Pontzen R (1982). Specific inhibition of glucan-elicited glyceollin accumulation in soybeans by an extracellular mannan - glycoprotein of Phytophthora megasperma f. sp. glycinea. Physiol Pl Pathol 20:321.

3. Ayers AR, Ebel J, Finelli F, Berger N, Albersheim P (1976) Host-pathogen interactions IX. Quantitative assays of elicitor activity and characterization of the elicitor present in the extra-cellular medium of cultures of Phytophthora megasperma var. sojae. Pl Physiol 57:751.

4. Ayers AR, Ebel J, Valent B, Albersheim P (1976). Host-pathogen Interactions X. Fractionation and biological activity of an elicitor isolated from the mycelial walls of Phytophthora megasperma var. sojae. Pl Physiol 57:760.

5. Ayers AR, Valent B, Ebel J, Albersheim P (1976). Host-pathogen Interactions XI. Composition and structure of wall-released elicitor fractions. Pl Physiol 57:766.

6. Jeffrey AM, Zopf DA, Ginsberg V (1975). Affinity chromatography of carbohydrate-specific immunoglobulins: coupling of oligosaccharides to sepharose. Biochem. Biophys. Res. Comm. 62:608.

7. Borch RF, Bernstein MD, Durst HD (1971). The cyanohydriborate anion as a selective reducing agent. J Amer Chem Soc 93:2897.

8.   Gray GR (1974).   The direct coupling of
     oligosaccharides to proteins and derivatized gel.
     Arch Biochem Biophys 163:426.

9.   Jentoft N, Dearborn DG (1980).   Protein labeling by
     reductive methylation with sodium cyanoborohydride:
     Effect of cyanide and metal ions on the reaction.
     Anal Biochem 106:186.

10. Das M, Fox CF (1979).   Chemical cross-linking in
     biology. Ann Rev Biophys Bioeng 8:165.

11. Yoshikawa M, Keen NT, Wang M-C (1983). A receptor on
     soybean membranes for a fungal elicitor of phytoalexin
     accumulation. Pl Physiol 73: 497.

Cellular and Molecular Biology of Plant Stress, pages 459–473
© 1985 Alan R. Liss, Inc.

INDUCTION BY ELICITORS AND ETHYLENE OF PROTEINS
ASSOCIATED TO THE DEFENSE OF PLANTS

M.T. Esquerré-Tugayé, D. Mazau, B. Pélissier,
D. Roby, D. Rumeau, and A. Toppan

Centre de Physiologie Végétale - UA 241 CNRS
Université Paul Sabatier, 31062 Toulouse, France

ABSTRACT
Pathogen attack leads to high increases in proteins
involved in plant defense mechanisms. Among them,
hydroxyproline-rich glycoproteins (HRGP's), chitinases and proteolytic inhibitors are well characterized
in melons infected by *Colletotrichum lagenarium*. Cell
wall phosphoglycopeptides, called elicitors, isolated
from *Colletotrichum*, stimulate the synthesis of the
three proteins, and of ethylene. Elicitation seems to
be a general phenomenon : an elicitor of phytoalexins
from *Phytophtora* and an endogenous melon cell wall
elicitor have also the ability to elicit HRGP.
Ethylene plays a role in elicitation
of defense responses in melons. Elicitor-induced ethylene, HRGP, chitinase and proteolytic inhibitory activities, are lowered by AVG^2, a specific inhibitor
of ethylene synthesis ; the ethylene precursor ACC^2
partly restores them.
Prior to the elicitation of ethylene, fungal and
plant elicitors induce a rapid membrane depolarization
of plant cells. It is hypothesized that this electric
signal could be involved, together with ethylene, in
elicitation of defense responses in plants.

[1]This work was supported by contract "ATP CNRS Micro-
biologie" 1982 and by contract "ATP CNRS Biologie
Moléculaire Végétale" 1983.
[2]AVG = aminoethoxyvinylglycine ; ACC = 1-aminocyclo-
propane - 1-carboxylic acid.

INTRODUCTION

Pathogen attack is one of the major stresses to which plants are exposed and respond. Some of the responses involve proteins which are associated to defense mechanisms. They can be separated into :
- enzymes such as chitinase, β 1-3 glucanase, lysozyme, which are able to hydrolyze fungal and bacterial cell walls (1, 2, 3, 4);
- enzyme inhibitors, like inhibitors of microbial proteases and polygalacturonases (5, 6, 7) ;
-glycoproteins with agglutinating properties, such as lectins (8), and hydroxyproline-rich glycoproteins (9, 10, 11) ;
- proteins with antiviral properties (12), and possibly other proteins related to pathogenesis (13).

Hydroxyproline-rich glycoproteins, chitinases and proteolytic inhibitors have been closely associated to the defense of melon plants against *Colletotrichum lagenarium*. This paper reports on the characterization of the three proteins, and on the cellular and molecular events underlying their expression in plant-microorganism interaction.

PROTEINS INVOLVED IN THE DEFENSE
OF PLANTS

Hydroxyproline-Rich Glycoprotein

Several stresses have been shown to increase the hydroxyproline-rich glycoprotein (HRGP) content of the plant cell wall (Table 1).

TABLE 1
ENHANCEMENT OF HYP-RICH GLYCOPROTEINS IN PLANTS BY SEVERAL STRESSES

Nature of stress	Ref.
PHYSIOLOGICAL CONDITIONS	
Tissue culture (callus, cells)	(14)
Light (red)	(15)
CHEMICALS	
Ethylene, Ethrel	(16,17)
Boron deficiency, high Nitrate	(18,11)
INFECTIONS BY PATHOGENS	
Fungi, Bacteria	(19,20,21)
Viruses, Nematodes	(22,23)
MECHANICAL STRESS	
Wounding	(24)

Hydroxyproline is increased by abnormal growth conditions
(14), light (15), ethylene and ethrel (16, 17), boron deficiency (18), high nitrate (11), and mechanical wounding
(24). The stimulation of HRGP in infected plants is considered in this paper.
In 1973, we discovered that the cell wall hydroxyproline
content of melon seedlings is increased upon infection by
Colletotrichum lagenarium at about the time of symptom
appearance (Figure 1). Six to seven days after inoculation

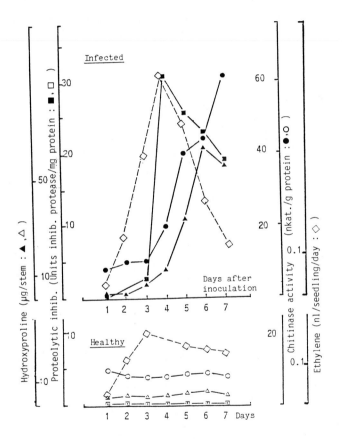

FIGURE 1. Time course measurement of Hydroxyproline
(HRGP), Chitinase, Proteolytic inhibitors, and
Ethylene, in melon seedlings after inoculation with
Colletotrichum lagenarium, and in healthy controls.

there is a ten time (or more) increase in hydroxyproline
which corresponds, in fact, to an increase in HRGP. This
increase occurs in the aerial part of the plant which is
affected by the disease, but it also occurs in the roots,
although roots, are not susceptible to the disease. A close
relationship is found between the amount of HRGP in melons
and their resistance to *Colletotrichum* : an increased HRGP
level correlates with on induced resistance, and a lowered
HRGP level correlates with an induced susceptibility (25).
HRGP is thought to be a structural protein (26) and was
recently shown to have agglutinating properties directed
towards negatively charged molecules (9, 10). These two
properties could account for the involvement of HRGP in
the defense of plants. This response to pathogen attack
is not restricted to melon plants ; it also occurs in other
systems, notably in other plants susceptible to the genus
Colletotrichum (21), in plants susceptible to different
pathogens either fungi or bacteria, and in resistant cultivars as well as in susceptible ones (11,27). So far, we
have not detected any significant changes in the cell wall
hydroxyproline content of infected Monocots.
Two different types of HRGP are present in melon callus :
type A and type B (Table 2).

TABLE 2
HYDROXYPROLINE-RICH GLYCOPROTEINS IN MELON TISSUES

Composition	A	B
Peptide moiety	6 % Hyp, Ser, Ala, Gly, Thr, account for 80 % of the total a.a. residues	36 % Hyp, Lys, Val, Ser, Pro, Tyr, account for 85 % of the total a.a. resi – dues
Sugar moiety	94 % Composed of Ara (66 %), Gal (34 %)	64 % Composed of Ara (77 %), Gal (22 %), Glc (1 %)

They differ by their peptide to sugar ratio, by their composition and by their glycosilation pattern. Both are hydroxyproline-rich ; 80 % of the peptide moiety of type A are composed of Hyp, Ser, Ala, Gly, and Thr, while 85 % of the
peptide moiety of type B are composed of Hyp, Lys, Val, Ser,
Pro, and Tyr. Actually, type A is thought to be an arabinogalactan protein. Type B has the same glycosilation pattern
of hydroxyproline as the glycoprotein which is increased in
the cell wall upon infection, in which most of the hydroxyproline residues are glycosilated by 1 to 4 arabinofuranose
residues (28, 29,30). Type A is probably present in the
cell wall or in the intercellular spaces, but in lower
amounts, than type B.
There is now evidence for the presence of more than one
HRGP in several plants (9,10, 31, 32). It was notably
reported recently by D.T.A. Lamport (33) that two different
glycoproteins can be eluted from tomato cells in culture.

Chitinases and Proteolytic Inhibitors

     These two kinds of proteins are believed to alter the
pathogenicity of plant pests which contain chitin (fungi,
insects), and use proteolytic enzymes to degrade their host
plant (fungi, bacteria,...). Chitin has not been found in
higher plants, and the proteolytic inhibitors produced upon
stress conditions are only effective against trypsin and the
trypsin-like proteases secreted by microorganisms. The level of chitinase, and of proteolytic inhibitors activity
is usually very low in plants, and is significantly increased in a number of stress situations : ethylene treatment
(1,3,4   and this paper) pathogen attack (2, 34, 35, 36, 37)
and mechanical wounding (36).
Chitinase and proteolytic inhibitory activities are sharply
increased in melon seedlings upon infection by *Colletotrichum
lagenarium* (Fig. 1), in a manner which parallels the accumulation of HRGP. The two activities are measured against
their putative targets *in vivo* : the chitin-containing cell
walls (3), and the proteolytic enzymes, of the fungus
(38, 6).
There is now evidence for the presence of several chitinases
and proteolytic inhibitors in plants (39,40). On the basis
of their chromatographic properties and of their affinity
to different substrats, infected melons contain 3 exo-, and
1 endo-chitinases (40). Preliminary data also suggest that,
as in the case of tomatoes, several proteolytic inhibitors
are present in melons.

It is interesting to note that the vacuole is the site of chitinases and proteolytic inhibitors accumulation (41,42). The vacuole content is readily released when cell breakage occurs upon infection by pathogens.

The fact that hydroxyproline-rich proteins, chitinases and proteolytic inhibitors are increased upon infection poses the question of their regulation at the cellular, and at the molecular, levels.

CELLULAR AND MOLECULAR EVENTS UNDERLYING DEFENSE RESPONSES

Molecular Aspects

In vivo labelling experiments have shown that the enrichment of the cell wall in HRGP is a protein synthesis dependent process (43). It has also been checked, by using cycloheximide, that the enhancement of exo- and endo-chitinase activities requires protein synthesis (40). The increased levels of HRGP, chitinases and proteolytic inhibitors, correlate with a simultaneous, transitory increase in poly(A)$^+$RNA in infected melons ; during that time, the amount of total RNA is lowered as a result of the disease. It thus seems that stimulation of defense reactions requires a modification of gene expression. The cellular events which lead to defense responses are now considered.

Cellular Events

Elicitation of defense responses. It has been shown by electron microscopy studies that, in many host-pathogen interactions, the host and the pathogen cell surfaces remain in close contact during the first few hours or days after inoculation. Several host responses start within that time. In melons infected by *C. lagenarium*, for example, the pathogen is restricted to the cell wall-plasmalemma area for 3 days (44). Figure 1 shows that the production of ethylene is highly increased during that time, prior to the stimulation of the defense responses. The hypothesis that cell surface interaction could elicit, through ethylene, HRGP's, chitinases and proteolytic inhibitors, originates from the fact that this hormone promotes several defense responses in plants (Table 1 and preceding paragraph).

We have demonstrated that elicitors, of fungal
and of plant origin have indeed this ability.

Phosphoglycopeptides elicitors have been isolated
from the cell walls of *Colletotrichum lagenarium* (45) ;
these elicitors induce the synthesis of ethylene and of
HRGP (21,46). Microgram amounts are sufficient to elicit
both responses in melon hypocotyls (Fig. 2 a). Elicitation

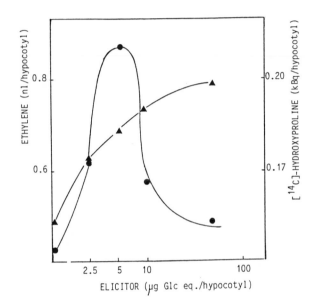

FIGURE 2a. Effect of *Colletotrichum lagenarium*
elicitors on the production of ethylene (▲) and
on the synthesis of HRGP (●). Melon petioles(2 g)
were incubated for 24 hours in the presence of
[^{14}C]-proline (37 kBq, precursor of hydroxyproline)
and of elicitors at increasing concentrations.

decreases at higher elicitor concentrations, as already
reported for phytoalexins by EBEL et al. (47). Similarly
an elicitor preparation from *Phytophtora megasperma* var.
glycinea (a gift of Pr. ALBERSHEIM'S laboratory) induces
ethylene and HRGP in soybeans (49), and the shape of
the curves is essentially the same as previously shown in
melons. It is interesting to note that the *Phytophtora*

elicitor induces phytoalexins in soybeans.
In addition, we have demonstrated that a melon cell wall
fraction, called endogenous elicitor, also elicits ethylene
and HRGP (46). Release of such an elicitor could account
for the increase of HRGP which occurs upon wounding.
The elicitors of *C. lagenarium* elicit chitinase and proteolytic inhibitory activities (6, 40) in melons as well as
HRGP (Fig. 2 b and c). Due to the leaf assay retained,

FIG. 2b   FIG. 2c

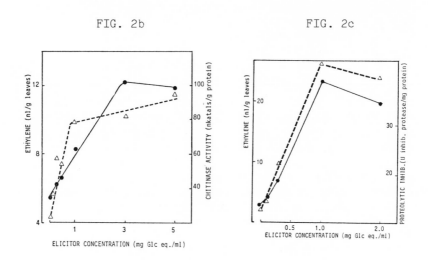

FIGURE 2b and c. Effect of *Colletotrichum lagenarium* elicitors on the production of ethylene
(△), and on chitinase (● , Fig. 2b) and proteolytic inhibitors measured against the pro-
tase of *C. lagenarium* (● , Fig. 2c). Melon
leaves were allowed to absorb elicitors for
24 hours.

higher elicitor concentrations were necessary in this case
(elicitor taken up through the cut surface of the petiole).
Elicitation of chitinase and proteolytic inhibitors have now
been reported in other laboratories, on other systems
(48, 49). The effect of elicitors on melons is a systemic
one : Figure 3 shows for example, that elicitors injected
into one cotyledon stimulate, for several days, the

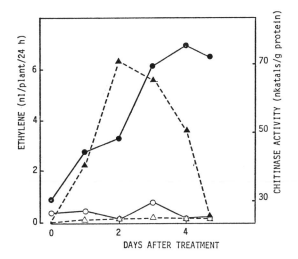

FIGURE 3. Time course measurement of ethylene production (▲) and chitinase elicitation (●) in
melon seedlings treated at day 0 with *C. lagenarium*
elicitors (220 μg Glc eq) through one of their cotyledons, and in untreated controls seedlings
(△ , O). The aerial part of the seedlings were
excised from the roots every day after the treatment,
and either enclosed in flasks for 24 hours in order
to measure ethylene production, or extracted for
chitinase measurement.

production of ethylene and the chitinolytic activity of
the aerial part of the seedling (40). This clearly indicates that either elicitors, or signals induced by elicitors, or both, are able to move within the plant, and to
trigger defense mechanisms.
    Possible signals. The possibility that ethylene is
one such singal has been investigated. In higher plants,
this hormone is synthesized from Methionine, through
S-Adenosyl Methionine and 1-Aminocyclopropane - 1-Carboxylic
acid : MET ⟶ SAM ⟶ ACC ⟶ C_2H_4 (50).
Elicitation of ethylene in melon petioles by *Colletotrichum*
elicitors occurs within 1 hour after the beginning of
the treatment (45). Elicitor-enchanced ethylene is markedly lowered by AVG, a specific inhibitor of ACC formation ;

at 10^{-4} M, there is a 60 % inhibition of ethylene, and a
simultaneous 40 % decrease of HRGP biosynthesis. Elicitor-
induced chitinase and proteolytic inhibitory activities
are also lowered, although to a lesser extent, by 10^{-4} M AVG.
The addition of ACC stimulates the production of ethylene
and partly restores the synthesis of HRGP (46) and of the
two other proteins.
The high levels of ethylene and of HRGP in infected melons,
are also markedly lowered by AVG(43).Alltogether, the data
suggest that ethylene is one of the signals which regula-
tes the synthesis of proteins associated to the defense of
plants. How does ethylene act ? Is it primarily, or indirectly, involved ? Is it part of a signal transduction mechanism ? Is it transported ? Although none of these questions
can be answered at the moment, recent evidence shows that
other possible signals are induced by elicitors. One such
possible signal is a rapid membrane depolarization. We have
found that the elicitor of *Colletotrichum* and the endogenous melon elicitor, induce a rapid, partial, and stable
depolarization in the roots of host and non host plants
(Figure 4).

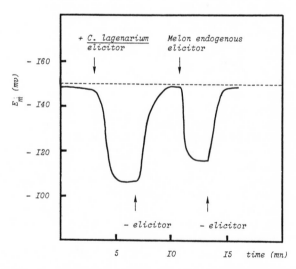

FIGURE 4. Depolarisation of transmembrane potential in melon roots treated with *C. lagenarium*
elicitor and melon endogenous elicitor ———— ;
untreated control ------- .

This effect is concerned with the electrogenic component
of the membrane potential, and is totally reversible once
the elicitor is withdrawn from the medium (51). Whether
or not elicitor-induced partial membrane depolarization
and ethylene production are linked to each other and participate to signal transduction is a question now under investigation.

CONCLUSION

Plants are able to overcome pathogen attack by displaying a variety of defense mechanisms. This paper reports
on three proteins associated to defense responses : hydroxyproline-rich glycoproteins, chitinases ans proteolytic inhibitors. Resistance is induced in melons against *Colletotrichum lagenarium* when the 3 proteins are enchanced prior
to inoculation with the fungus. This can be achieved by
treating melon seedlings with fungal elicitors, or ethylene (3, 21, 25). Their levels, and the production of ethylene, are also increased in control, susceptible melons
a few days after inoculation ; it is likely that these
increases occur too lately in order to be fully efficient.
This exemple shows that cellular and molecular studies
are necessary for a proper understanding and exploitation
of plant defense potentialities.

REFERENCES

1. Abeles FB, Bosshart RP, Forrence LE, Habig WH (1971).
   Preparation and purification of glucanase and chitinase
   from bean leaves. Plant Physiol 47: 129.
2. Pegg GF (1977). Glucanohydrolases of higher plants :
   a possible defense mechanism against parasitic fungi.
   In Solheim B, Raa J (eds) : "Cell Wall Biochemistry
   Related to Specificity in Host-Plant Pathogen Interactions", Oslo : Universitetsforlaget, p 305.
3. Toppan A, Roby D (1982). Activité chitinasique de plantes de melon infectées par *Colletotrichum lagenarium*
   ou traitées par l'éthylène. Agronomie 2: 829.
4. Boller T, Gehri A, Mauch F, Vogeli U (1983). Chitinase
   in bean leaves : induction by ethylene, purification,
   properties, and possible function. Planta 157: 22.

5. Ryan CA (1978). Proteinase inhibitors in plant leaves :
    a biochemical model for pest-induced natural plant pro-
    tection. Trends in Biochemical Sciences July:148.
6. Roby D, Toppan A, Esquerré-Tugayé MT (1981). A new bio-
    logical model for the study of proteinase inhibitors in
    plants. In Phytochemical Society of Europe Symposium
    (ed) "Seed Proteins", Abs. E$_{10}$, p 15.
7. Albersheim P, Anderson AJ (1971). Proteins from plant
    cell walls inhibit polygalacturonases secreted by plant
    pathogens. Proc Natl Acad Sci USA 68:1815.
8. Etzler ME (1981). Are lectins involved in plant fungus
    interactions ? Phytopath 71:744.
9. Leach JE, Cantrell MA, Sequeira L (1982). Hydroxyproline-
    rich bacterial agglutinin from potato : extraction,
    purification and characterization. Plant Physiol 70:1353.
10. Mellon JE, Helgeson JP (1982). Interaction of a hydro-
    xyproline-rich glycoprotein from tobacco callus with
    potential pathogens. Plant Physiol 70:401.
11. Esquerré-Tugayé MT (1983). Hydroxyproline-rich glyco-
    proteins and lectins in the defense of plants. In
    Goldstein IJ, Etzler ME (eds) : "Chemical Taxonomy,
    Molecular Biology, and Function of plant lectins",
    New York: Alan R. Liss, p 177.
12. Loebenstein G, Stein A (1984). Plant defense responses
    to viral infections. In Key JL, Kosuge T (eds): "Cel-
    lular and Molecular Biology of Plant Stress". New York:
    Alan R. Liss, this Symposium.
13. Van Loon LC (1977). Induction by 2-chloroethylphospho-
    nic acid of viral-like lesions, associated proteins
    and systemic resistance in tobacco. Virology 80:417.
14. Lamport DTA (1965). The protein component of primary
    cell walls. Advan bot Res 2:151.
15. Lang W (1976). Biosynthesis of extensin during normal
    and light-inhibited elongation of radish hypocotyls.
    Z. Pflanzenphysiol 78:220.
16. Ridge I., Osborne DJ (1970). Hydroxyproline and peroxy-
    dases in cell walls of *Pisum sativum* regulation by ethy-
    lene. J Exp Bot 21:843.
17. Sadava D, Chrispeels MJ (1973). Hydroxyproline-rich cell
    wall protein (extensin) : role in the cessation of
    elongation in excised pea epicotyls. Devel Biology
    30:49.
18. Shive JB Jr., Barnett NM (1973). Boron deficiency effects
    on peroxidase hydroxyproline and boron in cell walls and
    cytoplasm of *Helianthus annuus L.* hypocotyls. Plant and
    Cell Physiol 14:573.

19. Esquerré-Tugayé MT, Mazau D (1974). Effect of a fungal disease on extensin, the plant cell wall glycoprotein. J Exp Bot 25:509.
20. Esquerré-Tugayé MT, Lamport DTA (1979). Cell surfaces in plant-microorganism interactions. I. A structural investigation of cell wall hydroxyproline-rich glycoproteins which accumulate in fungus-infected plants. Plant Physiol. 64:314.
21. Esquerré-Tugayé MT, Mazau D, Toppan A, Roby D (1980). Elicitation, via l'éthylène, de la synthèse de glycoprotéines pariétales associées à la défense des plantes. Ann. Phytopathol. 12:403.
22. Brown RG. Kimmins WC (1973). Hypersensitive resistance. Isolation and characterization of glycoproteins from plants with localized infections. Can J Bot 51:1917.
23. Giebel J, Stobiecka M (1974). Role of amino acids in plant tissue response to *Heterodera rostochiensis*. I. Protein-proline and hydroxyproline content in roots of susceptible and resistant solanaceous plants. Nematologica 20:407.
24. Chrispeels MJ, Sadava D, Cho YE (1974). Enhancement of extensin biosynthesis in ageing disks of carrot storage tissue. J Exp Bot 25:1157.
25. Esquerré-Tugayé MT, Lafitte C, Mazau D, Toppan A, Touzé A (1979). Cell surfaces in plant-microorganism interactions. II. Evidence for the accumulation of hydroxyproline-rich glycoproteins in the cell wall of diseased plants as a defense mechanism. Plant Physiol 64:320.
26. Lamport DTA (1970). Cell wall metabolism. Annu Rev Plant Physiol 21:235.
27. Mazau D, Esquerré-Tugayé MT (1984). A survey of hydroxyproline-rich glycoprotein accumulation in the cell wall of infected plants. Plant Physiol (submitted).
28. Lamport DTA, Miller DH (1971). Hydroxyproline arabinosides in the plant kingdom. Plant Physiol 48:454.
29. Esquerré-Tugayé MT, Mazau D (1981). Les glycoproteines à hydroxyproline de la paroi végétale. Physiol Vég 19:415.
30. Akiyama Y, Mori M, Kato K (1980). ^{13}C-NMR analysis of hydroxyproline arabinosides from *Nicotiana tabacum*. Agric Biol Chem 44:2487.
31. Stuart DA, Varner JE (1980). Purification and characterization of a salt-extractable hydroxyproline-rich glycoprotein from aerated carrot discs. Plant Physiol 66:787.

32. Smith MA (1981). Characterization of carrot cell wallprotein I.Effect of α-α'dipyridyl on cell wall protein synthesis and secretion in incubated carrot discs.Plant Physiol 68:956.
33. Smith JJ, Muldoon EP, Lamport DTA (1984). Isolation of extensin precursors by direct elution of intact tomato suspension cell cultures. Phytochemistry 23:1233.
34. Wargo PM (1975). Lysis of the cell wall of *Armillaria mellea* by enzymes from forest trees. Physiol Plant Pathol 5:99.
35. Nichols IE, Beckman IM, Hadwiger LA (1980). Glycosidic enzyme activity in pea pod tissue and pea-*Fusarium solani* interactions. Plant Physiol 66:199.
36. Green TR, Ryan CA (1972). Wound-induced proteinase inhibitor in plant leaves : a possible defense mechanism against insects. Science 175:776.
37. Peng FH, Black LL (1976). Increased proteinase inhibitor activity in response to infection of resistant tomato plants by *Phytophtora infestans*. Phytopathology 66:958.
38. Pladys D, Esquerré-Tugayé MT, Touzé A (1981). Purification and partial characterization of proteolytic enzymes in melon seedlings infected by *Colletotrichum lagenarium*. Phytopath Z 102:266.
39. Nelson CE, Walker-Simmons M, Ryan CA (1981). Regulation of proteinase inhibitor synthesis in Tomato leaves. *In vitro* synthesis of inhibitors I and II with mRNA from excised leaves induced with PIIF. Plant Physiol 67:841.
40. Roby D, Toppan A, Esquerré-Tugayé MT (1984). Cell surfaces in plant-microorganism interactions. VI. Elicitors of ethylene from *Colletotrichum lagenarium* trigger chitinase activity in melon plants. Plant Physiol (submitted).
41. Shumway LK, Yang VV, Ryan CA (1976). Evidence for the presence of prteinase inhibitor I in vacuolar protein bodies of plant cells. Planta 129:161.
42. Boller T, Vogeli U (1984). Vacuole localization of ethylene-induced chitinase in bean leaves. Plant Physiol 74:442.
43. Toppan A, Roby D, Esquerré-Tugayé MT (1982). Cell surfaces in plant-microorganism interactions. III. *In vivo* effect of ethylene on hydroxyproline-rich glycoprotein accumulation in the cell wall of diseased plants. Plant Physiol 70:82.

44. Dargent R, Touzé A (1974). Etude cinétique, en miscros-
copie électronique,des interactions entre *Colletotrichum
lagenarium* et les cellules foliaires de *Cucumis melo*.
Can J Bot 52:1319.

45. Toppan A, Esquerré-Tugayé MT (1984). Cell surfaces in
plant-microorganism interactions. IV. Fungal glycopepti-
des which elicit the synthesis of ethylene in plants.
Plant Physiol (in press).

46. Roby D, Toppan A, Esquerré-Tugayé MT (1984). Cell surfaces in plant-microorganism interactions. V. Elicitors
of fungal and of plant origin trigger the synthesis
of ethylene and of cell wall hydroxyproline-rich glycoprotein in plants. Plant Physiol (submitted).

47. Ebel J, Ayers AR, Albersheim P (1976). Host-Pathogen
Interactions. XII. Response of suspension-cultured
soybean cells to the elicitor isolated from *Phytophtora
megasperma* var. *sojae*. a fungal pathogen of soybeans.
Plant Physiol 57:775.

48. Boller T (1984). Induction of hydrolases as a defense
reaction against pathogens. In Key JL, Kosuge TS (eds):
"Cellular and Molecular Biology of Plant Stress",
New York : Alan R. Liss, this Symposium.

49. Walker-Simmons M, Hadwiger L, Ryan CA (1983). Chitosans
and pectic polysaccharides both induce the accumulation
of the antifungal phytoalexin pisatin in pea pods and
antinutrient proteinase inhibitors in tomato leaves.
Biochem Biophys Res Comm. 110:194.

50. Adams DO, Yang SF (1981). Ethylene the gaseous plant
hormone ; mechanism and regulation of biosynthesis.
Trends Biochem Sci 6:161.

51. Pélissier B (1984). Etude d'effets membranaires des
éliciteurs fongiques et d'un fongicide systémique, le
Tris(-0-ethyl phosphonate) d'aluminium. Thèse de Spé-
cialité, Université Paul Sabatier, Toulouse.

Index

AAL-toxin, 369–371, 375, 377
  feedback loops, 376
  tricarballylic acid, 370
Aberrant ratio (AR) phenomenon,
    maize, 1–12
  A*/a stock, 5–10
  a locus, 2–4
  barley stripe mosaic virus infection,
      2–3, 12
  deficiency of colorless kernels, 4–5
  heterofertilization, genetic control, 7
  heterozygosity for two color factor
      loci, 4
  seed anthocyanin production, 2–3
  segregation for second
      complementary color factor,
      3–4
  self-pollination, 9–10
  cf. *shrunken-1* locus, 12
  zl linkage, 11
  zygotic lethal (*zl*), 10–11
Abscisic acid, 423–425
  and potato cold hardiness/freezing
      stress, 211–213
  suberin induction, 386
ACC, 424
  and chemical compartmentation, 360
  and ethylene biosynthesis, induction,
      264–265, 267, 270, 271
  and heat-shock induction of disease
      resistance, cucurbits, 297, 298
  induction of defense proteins, 467
ACC synthase, 255, 270, 271
Actinomycin, 101
Acylation and chemical
    compartmentation, 356, 357
Adaptation to stress and various levels
    of organization, 72–74; *see also*

Chemical conjugation and
    compartmentation, adaptive;
    specific types of stress
Adenine nucleotide ratios and anoxic
    stress, rice and lettuce, 228–232
  cf. nitrogen starvation, 229–230
  cf. temperature variation, 229
Age and disease resistance, 305–307
Alcohol dehydrogenase
  and anoxic stress, 243
  genes, anaerobic response, maize,
      217, 219; *see also under*
      Anaerobic response, maize
Alfalfa leaves, trypsin inhibitor, and
    wound induction of mRNA,
    326–329; *see also* Salinity effect
    on alfalfa cell growth and
    maintenance
Algal mutants and adaptation to light
    stress, 52
a locus, maize, 2–4
Alternaria alternata f. sp. *lycopersici*
    stem canker, 369
Amaranthus hybridus, and adaptation to
    light stress, 64
Anaerobic response, maize, 217–223
  Adh 1 and *Adh 2* genes, 217, 219
    cDNA clones, 221, 223
    localization, 221
    molecular analysis, 220–223
    single ancestral gene, 222
  anaerobic polypeptides, 218–220
  anaerobic-specific cDNA clones, 220
  flooding, 217
  fructose-1,6-diphosphate aldolase,
      219
  glucose phosphate isomerase, 219